Undergraduate Lecture Notes in Physics

For further volumes:
http://www.springer.com/series/8917

Undergraduate Lecture Notes in Physics (ULNP) publishes authoritative texts covering topics throughout pure and applied physics. Each title in the series is suitable as a basis for undergraduate instruction, typically containing practice problems, worked examples, chapter summaries, and suggestions for further reading.

ULNP titles must provide at least one of the following:

- An exceptionally clear and concise treatment of a standard undergraduate subject.
- A solid undergraduate-level introduction to a graduate, advanced, or non-standard subject.
- A novel perspective or an unusual approach to teaching a subject.

ULNP especially encourages new, original, and idiosyncratic approaches to physics teaching at the undergraduate level.

The purpose of ULNP is to provide intriguing, absorbing books that will continue to be the reader's preferred reference throughout their academic career.

Series Editors

Neil Ashby
Professor Emeritus, University of Colorado, Boulder, CO, USA

William Brantley
Professor, Furman University, Greenville, SC, USA

Michael Fowler
Professor, University of Virginia, Charlottesville, VA, USA

Michael Inglis
Professor, SUNY Suffolk County Community College, Selden, NY, USA

Elena Sassi
Professor, University of Naples Federico II, Naples, Italy

Helmy Sherif
Professor Emeritus, University of Alberta, Edmonton, AB, Canada

Eric L. Michelsen

Quirky Quantum Concepts

Physical, Conceptual, Geometric, and Pictorial Physics that Didn't Fit in Your Textbook

 Springer

Eric L Michelsen
Poway, California
USA

ISSN 2192-4791   ISSN 2192-4805 (electronic)
ISBN 978-1-4614-9304-4   ISBN 978-1-4614-9305-1 (eBook)
DOI 10.1007/978-1-4614-9305-1
Springer New York Heidelberg Dordrecht London

Library of Congress Control Number: 2013950791

Printed on acid-free paper

Springer is part of Springer Science+Business Media (www.springer.com)

To my wife, Laura, for all her patience and support, and to my children, Sarah and Ethan, for understanding of my absences while I was working on the book.

Frontispiece

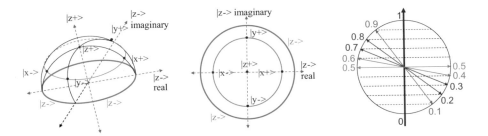

"Quantum Mechanics is a silly theory, perhaps the silliest theory to come out of the 20th century. The only reason it has any following at all is that it is completely supported by experiment." – *Unknown physicist*

"We are all agreed that your theory is crazy. The question that divides us is whether it is crazy enough to have a chance of being correct."—*Niels Bohr*

"Now in the further development of science, we want more than just a formula. First we have an observation, then we have numbers that we measure, then we have a law which summarizes all the numbers. But the real *glory* of science is that we can *find a way of thinking* such that the law is *evident*."—*Richard Feynman*

Preface

Why Quirky?

The purpose of the "Quirky" series is to help develop an accurate physical, conceptual, geometric, and pictorial understanding of important physics topics. We focus on areas that do not seem to be covered well in most texts. The Quirky series attempts to clarify those neglected concepts, and others that seem likely to be challenging and unexpected (quirky?). The Quirky books are intended for serious students of physics; they are not "popularizations" or oversimplifications.

Physics includes math, and we are not shy about it, but we also do not hide behind it.

> Without a conceptual understanding, math is gibberish.

We seek to be accurate, but not pedantic. When mathematical or physical words have precise meanings, we adhere to those meanings. Words are the tools of communication; it is impossible to make fine points with dull tools.

Who Is It For?

This work is one of the several aimed at graduate and advanced-undergraduate physics students, engineers, scientists, and anyone else who wants a serious understanding of Quantum Mechanics. The material ranges from fairly elementary (though often neglected) to graduate level. Go to http://physics.ucsd.edu/~emichels for the latest versions of the Quirky Series, and for contact information. We are looking for feedback, so please let us know what you think.

How to Use This Book

This book is an informal, topical review. We strive to be accurate, but not tedious.

This work is not a text book.

There are plenty of those, and they cover most of the topics quite well. This work is meant to be used *with* a standard text, to help emphasize those things that are most confusing for new students. When standard presentations do not make sense, come here. In short, our goal is to provide the foundation that enables future learning from other sources.

If you have some background in quantum mechanics, then most of the sections stand alone. The larger sections start by naming the prerequisites needed to understand that section. This work is deliberately somewhat redundant, to make sections more independent, and because learning requires repetition.

You should read all of Chap. 1, Basic Wave Mechanics Concepts, to familiarize yourself with the notation and contents. After the first two chapters, this book is meant to be read in any order that suits you. Each section stands largely alone, though the sections are ordered logically. You may read it from beginning to end, or skip around to whatever topic is most interesting. The "Desultory" chapter is a diverse set of short topics, each meant for quick reading.

We must necessarily sometimes include forward references to material which has not yet been covered in this book. If they are unfamiliar, most such references may be ignored without loss of continuity.

If you don't understand something, read it again *once*, then keep reading. *Don't get stuck on one thing.* Often, the following discussion will clarify things.

Scope
What This Text Covers

This text covers most of the unusual or challenging concepts in a first-year graduate course in nonrelativistic Quantum Mechanics (QM). Much of it is suitable for undergraduate QM, as well, because it provides a conceptual foundation for all of QM. We expect that you are taking or have taken such a QM course, and have a good text book. This text supplements those other sources.

What This Text Doesn't Cover

This text is neither a QM course in itself, nor a review of such a course. We do not cover all basic QM concepts; only those that are unusual or especially challenging (quirky?). There is almost no relativistic QM here.

What You Already Know

This text assumes you understand basic integral and differential calculus, partial differential equations, have seen complex numbers, and have some familiarity with probability. You must have a working knowledge of basic physics: mass, force, momentum, energy, etc. Further, it assumes you have a QM text for the bulk of your studies, and are using *Quirky Quantum Concepts* to supplement it. You must have been introduced to the idea of particles as waves, and photons as particles of light. Beyond that, different sections require different levels of preparation; some are much more advanced than others. Each section lists any particular prerequisites at the beginning. Some sections require some familiarity with classical Lagrangian and Hamiltonian mechanics, including canonical momentum.

Notation

Important points are highlighted in solid-border boxes.

Common misconceptions are sometimes written in dashed-line boxes.

References: As is common, we include references to published works in square brackets, where the abbreviations in the brackets are defined in the "References" section of this document. Where page numbers are given, they may be augmented by "t", "m", or "b", for "top", "middle", and "bottom" of the page.

Unit Vectors: We use $\mathbf{e}_x$, $\mathbf{e}_y$, etc. for unit spatial vectors. In other disciplines, these are more likely written as $\hat{\mathbf{x}}$, $\hat{\mathbf{y}}$, etc., but we reserve "hats" to indicate quantum operators.

Keywords are listed in **bold** near their definitions. All keywords also appear in the glossary.

We use the following symbols regularly:

Symbol	Name	Common scientific meanings
∇	Del	Gradient of a scalar field, divergence or curl of a vector field
$\equiv$	Identity	(1) is defined as; (2) identically (always) equal to
$\forall$	For all	for all
$\in$	Element	Is an element of (sometimes written as epsilon, ε)
$\approx$	Approximately equals	Approximately equals
$\sim$	Tilde	Scales like; is proportional to in some limit. For example, $1/(r+1) \sim 1/r$ for large r because $\lim_{r \to \infty} \dfrac{1}{r+1} = \dfrac{1}{r}$. Note that we want to preserve the scaling property, so we don't take such a limit all the way to $r \to \infty$ (limit of zero), which would hide any r dependence
!!	Double factorial	$n!! \equiv n(n-2)(n-4) \ldots (2 \text{ or } 1)$

Integrals: In many cases, we present a general integral, whose exact limits depend on what problem or coordinate system the integral is applied to. To provide a general formula, independent of such particulars, we give the limits of integration as a single "∞", meaning integrate over the entire domain appropriate to the problem:

$$\int_\infty f(x)\, dx \equiv \text{integrate over entire domain relevant to the given problem.}$$

Open and Closed Intervals: An **open interval** between c and d is written "(c, d)", and means the range of numbers from c to d *exclusive* of c and d. A **closed interval** between c and d is written "$[c, d]$", and means the range of numbers from c to d *including* c and d. A half-open interval "$[c, d)$" has the expected meaning of c to d including c but not d, and "$(c, d]$" means c to d excluding c but including d.

Operators: I write most operators with a "hat" over them, e.g. $\hat{x}$. Rarely, the hat notation is cumbersome, so I sometimes use the subscript $_{op}$ to denote quantum operators, as in [12]. Thus the symbol x is a real variable, $\hat{x}$ is the position operator, and $\left(p^2\right)_{op}$ is the operator for p^2.

Conjugates and Adjoints: We use "*" for complex conjugation, and "$\dagger$" for adjoint: $z* \equiv$ complex conjugate of the number 'z', $\hat{a}^\dagger \equiv$ adjoint operator of $\hat{a}$.

[Note that some math texts use a bar for conjugate: $\bar{a} \equiv$ complex conjugate of 'a', and a "*" for adjoint. This is confusing to physicists, but c'est la vie.]

[Interesting paragraphs that may be skipped are "asides," shown in square brackets, smaller font, and narrowed margins.]

[Short asides may be also be written in-line in square brackets.]

Vector Variables: In some cases, to emphasize that a variable is a vector, it is written in bold; e.g., $V(\mathbf{r})$ is a scalar function of the vector, $\mathbf{r}$. $\mathbf{E}(\mathbf{r})$ is a vector function of the vector, $\mathbf{r}$. We write a zero vector as $\mathbf{0}_v$ (this is different than the number zero).

Matrices: Matrices are in bold, **B**. A particular element of a single matrix may be specified with subscripts, e.g. B_{ij}. A particular element of a matrix expression uses brackets, e.g. $[\mathbf{AB}]_{ij} \equiv$ the ij^{th} element of the matrix product **AB**.

Tensor Products: Sometimes, we write a tensor product explicitly with the $\otimes$ symbol.

In-line derivatives sometimes use the notation d/dx and $\partial/\partial x$. There is not always a clear mathematical distinction between d/dx and $\partial/\partial x$. When the function arguments are independent, they are both the same thing. I use d/dx when a function is clearly a total derivative, and $\partial/\partial x$ when it is clearly a partial derivative. However, in some cases, it's not clear what arguments a function has, and it is not important. In that case, I tend to use $\partial/\partial x$ for generality, but do not worry about it.

Also, for the record, derivatives *are* fractions, despite what you might have been told in calculus. They are a special case of fraction: the limiting case of fractions of differentially small changes. But they are still fractions, with all the rights and privileges thereof. All of physics treats them like fractions, multiplies and divides them like fractions, etc., because they *are* fractions.

Greek Letters

The Greek alphabet is probably the next-best well known alphabet (after our Latin alphabet). But Greek letters are often a stumbling block for readers unfamiliar with them. So here are all the letters, their pronunciations, and some common meanings from all over physics. Note that every section defines its own meanings for letters, so look for those definitions.

The Greek alphabet has 24 letters, and each has both upper-case (capital) and lower-case forms. Not all can be used as scientific symbols, though, because some look identical to Latin letters. When both upper- and lower-case are useable, the lower-case form is listed first. Lower case Greek variables are italicized, but by convention, upper case Greek letters are not. Do not worry if you do not understand all the common meanings; we will define as we go everything you need to know for this book.

Letter	Name (pronunciation)	Common scientific meanings
α	Alpha (al'fuh)	Coefficient of linear thermal expansion. (Capital: A, not used.)
β	Beta (bae'tuh)	Velocity as a fraction of the speed of light ($\beta \equiv v/c$). (Capital: B, not used)
γ	Gamma (gam'uh)	The relativistic factor $(1-\beta^2)^{-1/2}$, aka time-dilation/length-contraction factor
Γ	Capital gamma	Christoffel symbols (General Relativity); generalized factorial function
δ	Delta (del'tuh)	The Dirac delta (impulse) function; the Kronecker delta; an inexact differential (calculus)

Letter	Name (pronunciation)	Common scientific meanings
∂	Old-style delta	Partial derivative (calculus)
Δ	Capital delta	A small change
ε	Epsilon (ep'si-lon)	A small value. (Capital: E, not used.)
ζ	Zeta (zae'tuh)	Damping ratio. (Capital: Z, not used.)
η	Eta (ae'tuh)	Efficiency; flat-space metric tensor. (Capital: H, not used.)
θ	Theta (thae'tuh)	Angle
Θ	Capital theta	Not commonly used. Sometimes angle
ι	Iota (ie-o'tuh)	Not commonly used. (Capital: I, not used.)
κ	Kappa (kap'uh)	Decay constant. (Capital: K, not used.)
λ	Lambda (lam'duh)	Wavelength
Λ	Capital lambda	Cosmological constant.
μ	Mu (mew)	Micro (10^{-6}); reduced mass. (Capital: M, not used.)
ν	Nu (noo)	Frequency. Not to be confused with an italic v: v vs. nu: ν. (Capital: N, not used.)
ξ	Xi (zie, sometimes ksee)	Dimensionless distance measure.
Ξ	Capital xi	Not commonly used
o	Omicron (oe'mi-kron)	Not used. (Capital: O, not used.)
π	Pi (pie)	Ratio of a circle's circumference to its diameter, $\approx 3.14159\ldots$
Π	Capital pi	Product (multiplication)
ρ	Rho (roe)	Mass density; charge density; correlation coefficient. (Capital: P, not used.)
σ	Sigma (sig'muh)	Standard deviation; surface charge density
Σ	Capital sigma	Sum (addition)
τ	Tau (rhyme: cow, or sometimes saw)	Time; torque. (Capital: T, not used.)
υ	Upsilon (oops'i-lon)	Not commonly used. (Capital: Y, not used.)
ϕ	Phi (fee or fie)	Angle
φ	Old-style phi	Angle
Φ	Capital phi	Electric potential; general potential.
χ	Chi (kie)	Degrees of freedom. (Capital: X, not used.)
ψ	Psi (sie)	Wave-function amplitude
Ψ	Capital psi	Not commonly used
ω	Omega (oe-mae'guh)	Angular velocity; angular frequency
Ω	Capital omega	Angle; solid angle; ohm (unit of electrical resistance)

Acknowledgements

Thank you, Herbert Shore, San Diego State University, Physics Department, for giving me a chance, and for being a good teacher. I am indebted to Daniel Arovas for his detailed comments. I also owe a big thank you to many professors at both SDSU and UCSD, for their generosity even before I was a real student: Peter Salamon, Arlette Baljon, Andrew Cooksy, George Fuller, Tom O'Neil, Terry Hwa, and many others. Thanks to Peter Betts, who first set the equations in MathType, and to Eric Williams, Yaniv Rosen, and Jason Leonard for their thorough comments, and to Kevin Duggento and Chad Kishimoto for their thorough discussions. Kim Griest gets a special mention for his unfailing moral support, which means so much. Also, all the students I have worked with over the years have helped, both knowingly and unknowingly, to make this work possible. As always, since I have collected all the input and put it into this form, any errors in the text are solely my own.

Please send any comments or corrections to me, at emichels@physics.ucsd.edu. Errata will be maintained at http://physics.ucsd.edu/~emichels/.

Contents

Chapter 1
Basic Wave Mechanics Concepts

The goal of this chapter is to convey a conceptual and pictorial understanding of the workings of quantum mechanics (QM). We introduce the experimental results that motivate QM, turning them into a reliable, quantitative model that will underpin all of our future developments. This chapter is conceptually the most difficult, but it is essential.

1.1 The Meaning of Science

Quantum theory is true. It is *not* a speculation, nor are the major parts of it in any doubt at all. In science, unlike ordinary conversation, a **theory** is the highest level of scientific achievement: a quantitative, predictive, and testable model that unifies and relates a body of facts. A theory becomes accepted science only after being supported by overwhelming evidence. A theory is not a speculation, e.g., Maxwell's electromagnetic theory. Note that every generally accepted theory was, at one time, *not* generally accepted. A **fact** is a small piece of information backed by solid evidence (in hard science, usually repeatable evidence). If someone disputes a fact, it is still a fact. ("If a thousand people say a foolish thing, it is still a foolish thing.") A **speculation** is a guess, perhaps motivated by facts, perhaps by intuition.

We must be careful when we talk of what is "really" happening in QM. Because QM is so far removed from everyday experience, there is no "really."

> All we can hope for are mathematical models which predict the outcomes of experiments. Any such model is valid (or "real").

Throughout this work, we adopt the simplest valid model we can. Other valid models exist. They may disagree on the wording or interpretation of the theory, but by definition, all valid models agree on known experimental results. When two models disagree on predictions that are not yet tested, then in principle, performing an experiment can falsify one model, and be consistent with the other. In general, it is not

E. L. Michelsen, *Quirky Quantum Concepts*, Undergraduate Lecture Notes in Physics, DOI 10.1007/978-1-4614-9305-1_1, © Springer Science+Business Media New York 2014

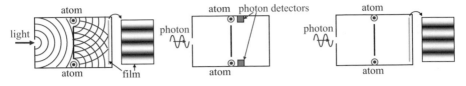

Fig. 1.1 (*Left*) A bright light fluoresces from both atoms and exposes an interference pattern. (*Middle*) A single photon at a time fluoresces off one atom or the other, but never both. (*Right*) A large number of individual photons still produce interference on the film

possible to prove that a theory is correct, only that it is consistent with all known facts (definitive experiments).

1.2   Not Your Grandfather's Physics: Quantum Weirdness

The following realistic (but here idealized) experiment demonstrates that QM is not your everyday physics.

Consider a box (Fig. 1.1, *left*), that admits light on the left, and has two identical atoms midway across, one each at the top and bottom. A black screen blocks light from passing in-between the atoms. On the right is photographic film which records the cumulative energy of light striking it at each point. The atoms fluoresce, i.e., they absorb photons, and quickly reradiate new photons of fixed energy. If we shine a bright light into the hole, the atoms fluoresce, and the film records a standard interference pattern (dark indicates higher exposure). Classical physics predicts this.

If we now place photon detectors in front of each atom (Fig. 1.1, *middle*), and send in a series of individual photons (one at a time), our detectors show that either one atom or the other fluoresces, but never both. Recall that all photons of a given frequency have the same energy. When an atom does fluoresce, it radiates a single photon with the full photon energy, just as when under a bright light. The energy is never split between the two atoms. This is quantum: an atom either absorbs and radiates or it does not. The event is quantized: there is no half-way. That is perhaps unexpected, but not yet truly weird.

Next, we remove the detectors and replace the film (Fig. 1.1, *right*). We again send in a series of individual photons (one at a time), and accumulate enough photons to expose the film. By extension of our previous single-photon experiment, we expect only one atom to radiate from each entering photon, and therefore we do not expect interference. However, the actual experimental result is an image of interference on the film.

It seems that the atoms behave differently when we are "looking at" them with detectors, than when we are not. Some physicists have called this "dual" behavior "quantum weirdness." This result says that QM is not just a modified set of laws from classical mechanics, but instead, the very nature of causality, time evolution, and observation are somehow different on the quantum scale. For example, this experiment goes far beyond saying "electrons are waves." Water travels in waves, too, but does not (macroscopically) exhibit this "weirdness."

QM is more than just a new "wavy model" of particles. QM also implies that the fundamental nature of causality, time evolution, and observation is different at the quantum scale than at the everyday macroscopic scale.

This experiment is essentially the famous double-slit experiment (each atom is a slit), but with the added twist that the atoms seems to change their behavior when the observational setup changes, even though that setup does not directly impinge on the process of photon absorption and reradiation: the observational setup only changes how we observe the results. A short time after a photon enters the box, we might think the two possible states of the system are "top atom is excited" or "bottom atom is excited." However, the recording of an interference pattern on the film indicates that the actual state is some kind of superposition of both.

This experiment is also directly analogous to the Schrödinger's cat experiment, wherein a radioactive element sits in a box with a cat. Over time, the element either decays or not, causing a mechanism to either kill the cat or not. After a time interval, the two states of the system are "the cat is alive" and "the cat is dead." In principle, we could add film to the box, and a "life-detector:" If the cat is alive, it radiates a photon from the top; if the cat is dead, it radiates a photon from the bottom. After killing a large number of cats, the idealized experiment would produce an interference pattern on the film: each cat has indeed been in a superposition of "alive" and "dead."

The reason we cannot successfully perform the cat experiment has nothing to do with life, death, or consciousness. It has to do with the cat being macroscopic. We describe all these aspects of QM in much more detail throughout this book.

Although the photon-atom-film experimental result demands a new mechanics, QM, I think it is not truly the essence of QM. QM is more about quantization of energy, angular momentum, and other properties of microscopic systems, which have a long history of experimental verification. In many ways, these effects are more important, though perhaps less exciting, than the mind-bending nature of causality and observation. Either way, it is the quantization effects that consume the vast majority of the study of QM, and to which we devote the vast majority of this book.

1.3   The Job of QM

What, then, does QM do? Like any physical theory, it predicts the results of experiments. All other physical theories, given enough information, predict the exact result of an experiment. The unique feature of QM theory is that predicting one single result is not generally possible. The universe itself is probabilistic. The *exact* same experiment can produce *different* results when repeated.

QM tells what results are possible, and with what probabilities.

In QM talk: QM defines the "spectrum" of results, and their probabilities of occurrence. Therefore, by combining this information, you can also predict the average value of a measurement A. We will have more on this later, but for now, recall some statistical facts about averages:

$$\langle A \rangle = \sum_{i=1}^{N} \Pr(a_i)\, a_i, \qquad \text{or} \qquad \langle A \rangle = \int_{-\infty}^{\infty} \mathrm{pdf}(a)\, a\, da,$$

where the a are possible values, $\Pr(a_j)$ is the probability of a discrete measurement, and $\mathrm{pdf}(a)$ is the probability distribution function of a continuous measurement.

1.3.1   The Premises of QM

We introduce here some of the underlying principles of QM, and then explore those principles more deeply in the "Axioms" section later.

QM says that, in general, systems exist in a linear superposition of oscillating states, i.e., they exist simultaneously in many different states.

These superpositions have two important characteristics:

First Each state in the superposition has its own mathematical weight. For example, a system might be 1/3 state A, 1/6 state B, and 1/2 state C. We says the system's "component states" are A, B, and C. Before a measurement, interactions occur with each component state acting independently, and as if it were the *whole* system. The resulting final state is a superposition of all the resulting component states.

For brevity, it is often said that taking an (ideal) measurement "collapses" the system to a state (possibly a superposition) consistent with the measurement. Often, the system then evolves into a broader superposition of states again, over time. We will see, however, that such a simple model cannot explain the full set of experimental observations, and leads to conceptual difficulties. A more complete model of measurements includes "decoherence," discussed later.

Second The states oscillate sinusoidally (in some way). Therefore, each state has not only weight (or "magnitude"), but **phase**. The phase of a state describes where in its periodic oscillation the state is, at a given moment. States oscillate sinusoidally, and since sin( ) has period 2π, the phase is often given as an angle, between 0 and 2π.

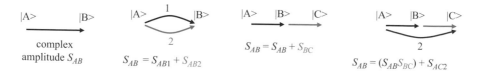

Fig. 1.2 The Laws of Quantum Mechanics for combining amplitudes

Note There is a big difference between a "superposition" state, and a "mixed" state. We will describe the difference in more detail later, but for now: superpositions include phase information in the set of component states, and describe a *single* system (e.g., particle). "Mixed states" have no phase information between the constituent states, and describe one of a set (ensemble) of multiple systems (e.g., particles), each in a different state. This is essentially a classical concept, but with very nonclassical (quantum) consequences. See "Density Matrices" later for more discussion.

1.3.2   The Success of QM

There are many fundamental successes of QM. The following rules of calculation follow from the axioms of linearity and superposition (described in more detail later):

1. Dynamic quantities are described by complex numbers, called **amplitudes** (or "complex amplitudes"), which have a magnitude and a phase. (This is different than some other applications where "amplitudes" are real numbers that quantify only the size of a sinusoidal oscillation.)
2. The probability of an observation is proportional to the squared-magnitude of the amplitude for that observation.
3. When there are two or more paths from an initial state to a final state, the complex amplitudes for each path add as complex numbers, called adding **coherently**.
4. When there is a path from initial state A, through an intermediate state B, to a final state C, the amplitude for that path is the complex product of the amplitudes for transitions from A→B and B→C.

These rules, shown diagrammatically in Fig. 1.2, are the basis for essentially all of QM. [Quantum Field Theory (QFT) uses them heavily, and Feynman diagrams build on the scheme depicted previously.]

1.3.3   The Failure of QM

The original QM was *nonrelativistic*, and it did not describe the creation and annihilation of particles. For relativistic quantum physics, which requires particle creation and annihilation, one must use QFT. QFT is the most successful physical theory of all time, in that it predicts the magnetic dipole moment (g-factor) of the electron to 13 digits. The reasons for the failures of QM are that it assumes that the number of particles is known and definite, and the Hamiltonian is nonrelativistic. Whether we use one-particle wave functions, or multiparticle wave functions, QM assumes a fixed set of particles. For nonrelativistic physics, a fixed set of massive particles is a good approximation, because the low energies have negligible probabilities of creating a particle/antiparticle pair, or other massive particles. However, at relativistic energies, the probability of massive particle creation can be significant. Particle creation (including photon creation) adds new "paths" for quantized interactions to take place, thus changing the probabilities of the outcomes. For example, the g-factor of the electron is measurably affected by these phenomena. QFT is the extension of QM to relativistic dynamics and particle creation.

1.4 Axioms to Grind: The Foundation of QM

One can develop most of QM from a small set of principles. Well-chosen axioms provide great power to work with QM, and learn new systems (unlike mathematical axioms, which are often cryptic and unenlightening). Each of our axioms is directly motivated by experimental results. Different references choose different axioms, and even different numbers of axioms. Their developments are generally not mathematically rigorous, and neither is ours. Instead, our development aims to illuminate the conceptual workings of QM. We reiterate that at the quantum level, we cannot talk of what is "really" happening; the best we can hope for is a simple, valid model. Here, we choose a model that is grounded in simple experiments, and therefore intuitive and instructive to the behavior of quantum systems.

This section assumes that you have seen some QM and wave-functions, and know the de Broglie relation between momentum and wavelength. It also refers to phasors, and their representation of oscillations, which are described in more detail later. Briefly, a phasor is a complex number that represents the amplitude and phase of a real sinusoid. The axioms later are written for a single-particle continuous system (wave-functions), but can be adapted easily to discrete systems (e.g., angular momentum). We also later extend them to multiple particles.

QM can be formulated from the following observations, taken as axioms:

1. Quantum systems oscillate sinusoidally *in space* with a frequency proportional to momentum: $\mathbf{k} = \mathbf{p}/\hbar$. In other words, quantum systems have a wavelength, given by de Broglie. $\mathbf{k}$ is the spatial frequency in (say) radians per meter, and points in the direction of momentum. This momentum–wavelength relation was known true for photons from simple interferometry (which means it is true relativistically, as well, since photons travel at light speed). This was demonstrated for electrons serendipitously by Davisson and Germer in 1927, after their equipment failed. No one knows *what* exactly is oscillating, as it cannot be directly

observed, however its interference patterns can be observed. We model the (unknown) thing that oscillates as the quantum state of the system. Initially, we consider quantum states which are wave-functions. Later, we will extend that to include discrete states.

2. Quantum systems oscillate sinusoidally *in time* with a frequency proportional to energy: $\omega = E/\hbar$. This was known true for photons from the photoelectric effect (which again means it is true relativistically). It is also experimentally confirmed for both electrostatic and gravitational potential energy [1 pp. 125–129]. Any temporal oscillation has a frequency, phase, and real amplitude. The real amplitude and phase can be described mathematically by a phasor: a complex number that represents a sinusoid. The frequency is not given by the phasor, and must come from other information, in this case, the energy. In QM, a phasor may be called a **complex amplitude**, or confusingly, often just "amplitude." Because quantum systems oscillate, and we represent those oscillations with complex-valued phasors, the wave-function is a complex-valued function of space (or a phasor-valued function of space). Since energy can be positive or negative, the frequency of oscillation can be positive or negative. We describe phasors and negative frequency later. [In a relativistic treatment, axioms (1) and (2) combine into a Lorentz covariant form that unifies the time and space pieces.]

3. Systems exist in a *linear* **superposition** of states. In other words, even a single, indivisible system behaves as if it is separated into pieces (components), each in a different state, and each of which behaves like the *whole* system. This is suggested by the observed interference of matter waves: electrons, atoms, molecules, etc. When measuring such a superposition, one will get a measurement consistent with *one* of the components of the system, but no one can predict exactly which component will be seen in the measurement. The "weights" of the pieces, though, determine the probability of measuring the system according to that component: larger weights have larger probabilities of being observed. For example, a spin-1/2 particle could be 1/3 spin up and 2/3 spin down. Furthermore, because systems oscillate *sinusoidally*, each piece has not only a magnitude, but a phase of its oscillation. This means we can represent each piece by a complex number (phasor), giving its weight and phase of oscillation. Again, the frequency of oscillation is given by the energy of each piece (Axiom 2). The concept of a superposition leads to the need for linear *operators*, which QM uses heavily.

4. QM is *consistent*, in the sense that if you measure something twice, you get the same answer (provided you measure quickly, before the system time evolves into a different state). This implies the collapse of the quantum state (aka, loosely, "collapse of the wave-function"). This was observed easily in many experiments, such as Stern–Gerlach filtering.

5. The weight of a given component of a superposition is proportional to the square of the amplitude of its oscillations, much like the intensity of an EM wave is proportional to the square of the E-field (or B-field, or A-field). A "point particle" actually exists in a superposition of locations. Thus each point in space has a phasor (complex number) representing the fractional *density* (per unit volume)

of the particle being at that point. This function of space is the particle's **wave-function**. The volume density is therefore given by the square of the wave function, which we normalize such that

$$\rho(\mathbf{r}) = |\psi(\mathbf{r})|^2 \quad \Rightarrow \quad \int_\infty |\psi(\mathbf{r})|^2 \, d^3\mathbf{r} = 1.$$

6. Note that the axioms of time and space oscillation, and superposition, imply the Schrödinger equation for classically allowed regions where the particle energy is above the potential energy. We could show this by considering particles of definite momentum, and noting that any particle can be written as a superposition of such momentum states. However, we must *postulate* that the Schrödinger equation is also valid in the classically forbidden regions, where the particle energy is *less than* the potential energy. In these regions, the quantum kinetic energy is negative, and the "momentum" is imaginary. This axiom is consistent with the empirically confirmed phenomenon of tunneling.

These axioms are the foundation of QM, and we rely on them implicitly and explicitly throughout our discussion. We also develop many of the implications of these axioms, which yield the incredibly large and diverse science of QM.

Note that the quantum formulas for energy use the classical nonrelativistic Hamiltonian, e.g., $E = p^2/2m + V(x)$, though x and p become operators. You might be surprised that QM is so weird, yet still uses a classical Hamiltonian. However, classical mechanics is the high-energy and/or large-system limit of QM, so perhaps it is to be expected that the classical Hamiltonian matches the underlying quantum Hamiltonian.

What is a wave function? A wave-function is some unknown "thing" that oscillates. That thing can be thought of as a "quantum field," similar to an electric or magnetic field, but with some important differences. (This is the basis of the more-complete theory of QFT.) Note that an EM field is a classical field on large scales, and also a quantum field on small scales, so the two types of fields are clearly related. In QM, though, a whole molecule (or other complex system) can be represented by a single "wave-function." This follows from the linear nature of superpositions. Such a representation is an approximation, but is quite good if the internal details of the system are not important to its observable behavior.

Collapse of the wave-function This is the most difficult concept of QM. It stems from the consistency postulate. There are several models of wave-function collapse, all of which predict the same results; therefore, all of them are valid. In this book, we choose the model we believe is simplest and most consistent over all sizes of systems, spanning a continuum from microscopic to macroscopic. We compare some other "collapse models" later. We introduce the ideas here, with more details to come.

Briefly, a wave-function is said to "collapse" when you make a measurement on it. If the quantum state was a superposition of multiple possibilities, your measurement changes the state to one consistent with your measurement. A more detailed analysis reveals that you can think of "collapse" as happening in two steps: first, the system "decoheres" by being measured (entangled) by *any* macroscopic measuring system (dead or alive). Note that there is no concern for sentience, just "bigness." Once a system has decohered, it follows classical probabilities, and there is no further possibility of quantum interference. In other words, decoherence eliminates any "quantum weirdness"(i.e., nonclassical behavior). However, in our model, time evolution is governed by the Schrödinger equation, and even a decohered system is in a superposition of states. Only when you observe a measurement do the other possibilities disappear (the quantum state collapses). This decoherence model is equivalent to one in which the quantum state *collapses upon decoherence*, but you simply do not know the result until you look at it. However, the decoherence model fully explains partial coherence with no additional assumptions, and is therefore preferable.

Note that since all sentient observers are macroscopic, they necessarily decohere a system before "seeing" the result of a measurement. Also, both the previous models have a collapse somewhere in them.

Before decoherence was widely recognized, scientists wondered about whether consciousness somehow affected quantum results. This was the puzzle of Schrödinger's cat: is it "sentient enough" to collapse a quantum state? One interpretation was that, regardless of whether a cat can collapse a wave function or not, an outside observer experiences the (idealized) "cat-in-the-box" as being in a superposition of states. This is still true even if the cat is replaced by an idealized physicist (whom we shall call "Alice"). However, realistic macroscopic systems (cats, physicists, or dial gauges) cannot exist in "coherent superpositions" for any noticeable time, because uncontrollable and unrepeatable interactions with the environment randomize the phase relationship between components of the superposition. This fact renders the riddle of Schrodinger's cat moot: decoherence leads to classical probabilities, and prevents any observable quantum effect. It is impossible to measure whether a cat collapses a wave-function, so the question is scientifically meaningless.

Uncertainty is not an axiom Note that there is no axiom about uncertainty, or wave-particle duality (complementarity). These concepts follow from the given axioms. The wave nature of the Schrödinger equation, combined with the collapse of the wave-function, allows us to summarize "duality" roughly as "particles propagate like waves, but measure like points." We examine the nature of measurements in some detail later.

Note on superpositions *Every* quantum state can be written as a superposition in some bases, and also as a (nonsuperposed) basis state in other bases. Thus, there is nothing fundamental about a state being either a "superposition" or a "basis" state. (We discuss bases in more detail later.) However, some bases are special, with special properties, such as energy, position, and momentum. Basis states in those bases have properties of particular interest, which we will discuss throughout this work.

1.5   Energy and Momentum Are Special

Energy and momentum are very special physical quantities: they are linearly related to a quantum particle's frequency and wave-number, by the constant $\hbar$ ("h-bar"):

$$E = \hbar\omega \quad \text{and} \quad p = \hbar k.$$

Energy and momentum relate directly to the wave nature of particles. This is not true of other physical quantities such as velocity, acceleration, etc. Therefore, energy and momentum get special treatment; we use them as much as possible, because they fit most easily into the wave equations. For example, we quantify kinetic energy as

$$T = \frac{p^2}{2m} \quad \text{rather than} \quad T = \frac{1}{2}mv^2,$$

because p (momentum) is directly related to wave functions and v (velocity) is not.

All other physical quantities have to be expressed in terms of energy and momentum. For example, to find the velocity of a particle (in the absence of a magnetic field), we find its momentum, which is fundamental, and divide by its mass:

$$v = \frac{p}{m} \quad \text{(absent magnetic fields).}$$

Not coincidentally, energy and momentum have a special place in relativity as well: they compose the energy–momentum four-vector, which is Lorentz invariant. In other words, relativity ties together energy and momentum in a special way, so they must *both* be special.

There is also a complication in that the quantum momentum p is *canonical* momentum, not kinetic momentum. As we will see, in systems with magnetic fields, this distinction becomes important.

1.6   Complex Numbers

1.6.1   *Fundamental Properties of Complex Numbers*

OK, so we admit complex numbers are not really all that quirky, but here is a summary of the characteristics of complex numbers that are most relevant to QM.

Why do we use complex numbers? We *could* write everything with only real numbers; after all, everything measurable uses real numbers to measure it. But using only real numbers makes the formulas very complicated.

We use complex numbers because they make mathematical formulas simpler.

Since we spend more time working with formulas than with the actual numbers, it is much more efficient to use complex numbers and simpler formulas, rather than simple (real) numbers and complicated formulas. We start by defining:

$$i \equiv \sqrt{-1}, \quad \text{an imaginary number,}$$

where i is a constant, whose square is -1. There is no *real* number with this property, so i is said to be **imaginary**. (Electrical engineers write "j" instead of "i," because "i" stands for electric current.) From the definition:

$$i^2 = (-i)^2 = -1 \quad \text{and} \quad -i^2 = 1.$$

Since all the rules of arithmetic, algebra, and calculus apply to imaginary numbers, you will commonly see these types of identities:

$$i = \frac{-1}{i}, \quad -i\hbar = \frac{\hbar}{i}, \quad \text{etc.}$$

where i can be multiplied by any real number, say y, to give iy. Any such real multiple of i is also imaginary, and its square is a negative number: $-y^2$. The sum of a real number and an imaginary number is a **complex number**: $x + iy$. The label "z" is often used for complex variables:

$$z = x + iy.$$

The graphical interpretation of complex numbers is critical. Recall that we can plot any real number as a point on a one-dimensional (1D) number line. Similarly, we can plot any complex number as a point, or a vector, on a two-dimensional (2D) plane (the **complex plane**); see Fig. 1.3.

The $z = x + iy$ form is called the **rectangular form** of a complex number. Adding (or subtracting) complex numbers is straightforward; the rectangular *components* add (or subtract), because complex numbers are vectors:

$$(a + ib) + (c + id) = (a + c) + i(b + d) \qquad\qquad (1 + 2i) + (3 + 4i) = 4 + 6i.$$

Besides rectangular form, complex numbers can also be expressed in **polar form**. Polar form is intimately tied to Euler's identity (pronounced "oilers"), which relates an angle to real and imaginary components (Fig. 1.4, *left*).

$$e^{i\theta} = \cos\theta + i\sin\theta, \quad \theta \text{ in radians.}$$

[Aside: Euler's identity can be derived from the Maclaurin expansion $e^x = \sum_{n=0}^{\infty} \frac{x^n}{n!}$, or from solving the second order differential equation, $\frac{d^2}{dx^2} f(x) = -f(x)$.]

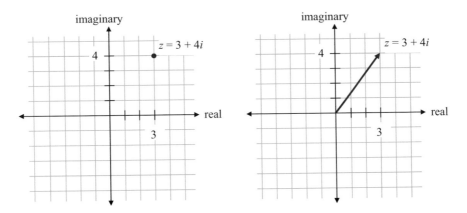

Fig. 1.3 Rectangular picture of z in the complex plane. (*Left*) As a point. (*Right*) As a vector.

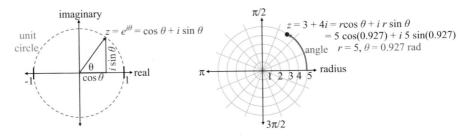

Fig. 1.4 (*Left*) Euler's identity in the complex plane. (*Right*) Polar picture of z in the complex plane

Multiplying $e^{i\theta}$ by a magnitude, r, allows us to write any complex number in **polar form**: $z = (r, \theta)$, by choosing r and θ appropriately (Fig. 1.4, *right*). Comparing the two graphs in Fig. 1.4, we see that we can relate the polar and rectangular forms by a mathematical expression:

$$z = (r,\theta) = re^{i\theta} = r(\cos\theta + i\sin\theta) = r\cos\theta + ir\sin\theta = x + iy .$$

$$x = r\cos\theta, \quad y = r\sin\theta, \quad \text{and} \quad r = \sqrt{x^2 + y^2}, \quad \theta = \tan^{-1}_{4q}(x,y),$$

where $\tan_{4q}^{-1}(x, y)$ is a full, 4-quadrant arctangent of (y/x). E.g., the usual 2-quadrant arctangent gives:

$$\tan^{-1}\left(\frac{-1}{-\sqrt{3}}\right) = \tan^{-1}\left(\frac{1}{\sqrt{3}}\right) = \frac{\pi}{6}, \quad \text{but} \quad \tan^{-1}_{4q}\left(-1,-\sqrt{3}\right) = \frac{-5\pi}{6}.$$

> Rectangular and polar forms are two different, but equivalent, ways of expressing complex numbers.

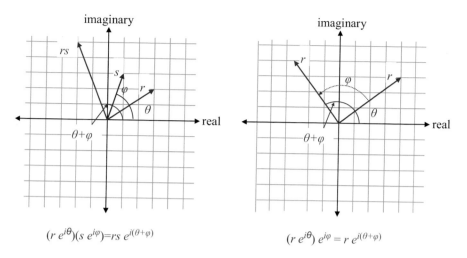

$$(r\,e^{i\theta})(s\,e^{i\varphi})=rs\,e^{i(\theta+\varphi)} \qquad\qquad (r\,e^{i\theta})\,e^{i\varphi} = r\,e^{i(\theta+\varphi)}$$

Fig. 1.5 (*Left*) Picture of complex multiply. (*Right*) Special case: multiplication by $e^{i\varphi}$

Note that θ may be positive or negative. Polar form is not unique, because:

$$e^{i\theta} = e^{i(\theta+2n\pi)}, \qquad n \text{ any integer.}$$

The angle of purely real numbers is either 0 or π, and the angle of purely imaginary numbers is $\pm \pi/2$.

All the rules of algebra and calculus work with complex numbers: commutativity, associativity, distributivity, exponentials, derivatives, and integrals:

$$e^z e^w = e^{z+w}, \qquad \frac{d}{dz}e^z = e^z, \qquad \int e^z\,dz = e^z + C, \text{ etc.}$$

Therefore, multiplying (or dividing) is easier in polar form, using Euler's identity (Fig. 1.5):

$$\left(re^{i\theta}\right)\left(se^{i\phi}\right) = rse^{i(\theta+\phi)}, \qquad \text{and} \qquad \frac{re^{i\theta}}{se^{i\phi}} = \frac{r}{s}e^{i(\theta-\phi)}.$$

When multiplying (or dividing) complex numbers, the radii multiply (or divide) and the angles add (or subtract). This fact suggests the concept of a "magnitude" for complex numbers. We define the **magnitude** (aka absolute value) of a complex number as its radius in polar form:

$$|z| \equiv r = \sqrt{x^2 + y^2} \qquad \text{(magnitude is a real number} \geq 0).$$

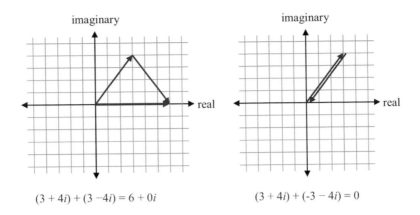

$(3 + 4i) + (3 - 4i) = 6 + 0i$ $(3 + 4i) + (-3 - 4i) = 0$

Fig. 1.6 A variety of results of adding complex numbers of equal magnitude

Thus when multiplying (or dividing) complex numbers, the magnitudes (radii) multiply (or divide):

$$|z_1 z_2| = |z_1| \, |z_2| \quad \text{and} \quad \left|\frac{z_1}{z_2}\right| = \frac{|z_1|}{|z_2|}.$$

Note that when z happens to be real, $|z|$ reduces to the familiar real-valued magnitude (aka absolute value).

An important special case When we multiply z by some unit-magnitude number $(e^{i\varphi})$, the magnitude of z does not change, but z gets rotated in the complex plane by the angle φ (Fig. 1.5, *right*). This is important for phasors, described shortly.

Using the polar form, we can easily raise a complex number to any real power:

$$z^a = \left(re^{i\theta}\right)^a = r^a e^{ia\theta} \quad \text{magnitude is raised to power } a; \text{ angle is multiplied by } a.$$

Adding complex numbers follows the "head-to-tail" rule for adding vectors. The two examples in Fig. 1.6 illustrate two different results possible from adding complex numbers of the same magnitude, but different angles:

A crucial fact is:

> When *adding* or *subtracting* complex numbers, there is no simple relationship between the magnitudes of the addends and the magnitude of the sum.

Their angles, or "phases," are important to the result. This is the essence of interference.

You can also define logarithms and exponents of complex numbers, but we do not need them at this level.

The angle of a complex number may be written as $\arg(z)$, and is sometimes called its **phase** (or its **argument**). As shown previously, when multiplying (or dividing) complex numbers, the angles add (or subtract):

$$\arg(z_1 z_2) = \arg(z_1) + \arg(z_2), \quad \text{and} \quad \arg\left(\frac{z_1}{z_2}\right) = \arg(z_1) - \arg(z_2).$$

The **complex conjugate** of a number z is written z^*, and is defined as follows:

$$\text{If } z = a + ib, \quad z^* \equiv a - ib, \quad \text{and in polar form,}$$

$$\text{if } z = re^{i\theta}, \text{ then } z^* = re^{-i\theta}.$$

Note that the conjugate of a *real* number is itself: $x^* = x$.
    An important identity is

$$|z|^2 = z^* z \quad \text{because } |z|^2 = r^2 = x^2 + y^2 = (x - iy)(x + iy) = z^* z,$$

$$\text{or in polar form: } |z|^2 = r^2 = \left(re^{i\theta}\right)\left(re^{-i\theta}\right) = z^* z.$$

Any complex function can be expressed by its real and imaginary parts, e.g., let $\psi(x)$ be a complex valued function of the *real* value, x:

$$\psi(x) = a(x) + ib(x).$$

Linearity then provides identities for derivatives and integrals, which are also complex valued functions:

$$\frac{d}{dx}\psi(x) = \frac{d}{dx}a(x) + i\frac{d}{dx}b(x), \qquad \int \psi(x)\, dx = \int a(x)\, dx + i\int b(x)\, dx.$$

The conjugate of a complex function $[\psi(x)]^*$ is written as $\psi^*(x)$. Important conjugation identities (derived simply from the previous rules):

$$(z + w)^* = z^* + w^* \qquad\qquad (zw)^* = z^* w^*$$

$$\frac{\partial}{\partial x}\psi^*(x) = \left[\frac{\partial}{\partial x}\psi(x)\right]^* \qquad\qquad \int \psi^*(x)\, dx = \left[\int \psi(x)\, dx\right]^*.$$

Let $v(z)$ be a complex valued function of the complex value, z. In general, $v(z^*) \neq v^*(z)$. A simple counter example is $v(z) = z + i$; then $v^*(z) = z^* - i \neq v(z^*)$. However, there are some special cases relating functions and conjugates. Most importantly:

For y real: $\quad \left(e^{iy}\right)^* = e^{-iy}$.

For $z = x + iy$: $\quad e^{(z^*)} = \left(e^z\right)^*$

because $\qquad e^{(z^*)} = e^{(x - iy)} = e^x \left(e^{+iy}\right)^* = \left(e^x e^{+iy}\right)^* = \left(e^{x+iy}\right)^* = \left(e^z\right)^*.$

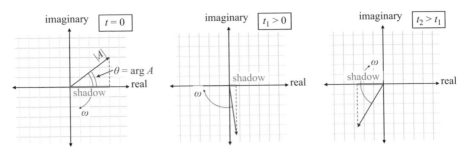

Fig. 1.7 For physicists, the stick usually rotates clockwise (as shown), per $e^{-i\omega t}$. Note that $\theta > 0$ in this example

The real part of a complex number, z, is written as Re(z). The imaginary part of "z" is written Im(z). Note that Im(z) is a *real* number: it is the coefficient of i in the complex number z. Thus,

$$z = \text{Re}(z) + i\,\text{Im}(z).$$

Derivatives and integrals of functions of *complex* values [say, $w(z)$] are uncommon in elementary QM, because dz has both a real and imaginary differential component. Though contour integrals of complex differentials do appear in more advanced studies, we will not need them here.

1.6.2 Phasors, Rotation in the Complex Plane, and Negative Frequency

Phasors are critical to nearly all aspects of oscillation in engineering and physics. The simple picture of a phasor starts with a clockwise rotating stick, and its projection (shadow) onto the horizontal axis (Fig. 1.7). The stick rotates at a constant angular velocity, ω. From the definition of cosine, the shadow traces out a sinusoid, of frequency ω, with amplitude equal to the length of the stick. We can add a phase to the sinusoid by having the stick start at some angle, θ, when $t=0$:

$$a(t) = (length)\cos(\omega t - \theta).$$

Rotation in the complex plane We have seen how multiplication by a unit-magnitude complex number simply rotates another complex number in the complex plane. Now imagine the unit-magnitude angle is not fixed, but changes linearly with time, i.e., rotates not by a fixed angle but by an increasing angle ωt. "ω" is the angular frequency, in rad/s. In physics, time-dependence usually goes as $e^{-i\omega t}$. If we multiply some complex number $r\,e^{i\theta}$ by $e^{-i\omega t}$, we get

$$\left(r\,e^{i\theta}\right)e^{-i\omega t} = r\,e^{i(\theta - \omega t)},$$

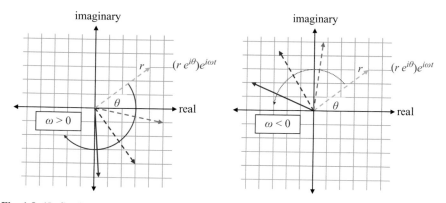

Fig. 1.8 (*Left*) Clockwise rotation in time at positive frequency $\omega > 0$. (*Right*) Counter-clockwise rotation in time at negative frequency $\omega < 0$

a complex function of time that rotates continuously in the complex plane. The magnitude of the result is fixed at $|r|$, because $|e^{-i\omega t}| = 1$ at all times. But the angle of the result decreases with time, at the rate ω (Fig. 1.8, *left*). (NB: Engineers, and some physics applications, use a different time dependence, $e^{+i\omega t}$, in which the complex vector rotates counter-clockwise for $\omega > 0$.)

We are now prepared to define a phasor: a **phasor** is a complex number which represents the amplitude and phase of a real-valued sinusoid. The phasor does not define the frequency of the sinusoid, which must come from other information. Specifically, the real-valued sinusoid $a(t)$ represented by a phasor A is:

$$a(t) = \mathrm{Re}\left\{Ae^{-i\omega t}\right\} = \cos\left(\omega t - \arg(A)\right)$$

The Re{} operator returns the "shadow" of the complex number on the real axis. Thus graphically, as before, $a(t)$ is the projection of a rotating vector onto the real axis; the vector's magnitude is $|A|$, and its starting angle ($t = 0$) is $\arg(A)$. A phasor is also called a "complex amplitude."

A key fact, proven in *Quirky Electromagnetic Concepts*, is that the sum of any two sinusoids of frequency ω is another sinusoid of frequency ω. Furthermore:

> The phasor for the sum of two sinusoids is the sum of the phasors of the two sinusoids.

Mathematically, for a given ω:

If $c(t) = a(t) + b(t)$,   then $C = A + B$

where $A, B, C \equiv$ phasors for $a(t), b(t), c(t)$.

Negative frequency In QM, system frequency is proportional to energy and energy can be negative. Therefore, instead of just oscillating, quantum systems can be thought of as *rotating* (in some unknown space). This requires both positive and negative frequencies (rotations in opposite directions). Again, no one knows *what* is rotating, but the interference consequences can be observed. Therefore, angular frequency is not constrained to be positive; it can just as well be negative. Rotation by a negative frequency rotates in the counter-clockwise direction, rather than clockwise (Fig. 1.8, *right*). Hence, both positive and negative frequencies occur in complex rotations, and in QM.

In QM, only frequency (and energy) *differences* are significant. Therefore we could, in principle, eliminate negative frequencies in all our calculations by adding a large, positive constant frequency to everything. That would be tedious, and problematic in future studies of QFT and antiparticles. Therefore, negative frequency gives us a convenient way to accommodate arbitrary frequency differences, while still allowing an arbitrary zero-point for frequencies. Since frequency is proportional to the system's energy ($E=\hbar\omega$), the zero point of frequency corresponds to the zero point of energy, and negative energies have negative frequencies. In other words, negative frequencies preserve our freedom to choose the zero-point of energy.

1.7   Probability, Density, and Amplitude

QM relies heavily on probability, probability density, and probability amplitude. Probability and probability density (henceforth often shortened to "density") are concepts common to many fields; "probability amplitude" is unique to QM. We briefly review here the concepts of probability, and precisely define our terms.

1.7.1   Probability of Discrete Events

Probability is a quantification of how likely a **random** event is to occur. A random event is one whose occurrence cannot be predicted exactly. The probability of an event is the fraction of the time that the event will occur in a large number of repetitions (**trials**) of an experiment. Each trial, of course, is set up exactly like all the others. It is in the nature of "random" events that even though all trials are set up identically, different trials may produce different events. So it is with QM measurements: most measurements of observables are random events. For experiments with **discrete** outcomes, such as rolling a die or measuring the energy of a particle in a superposition of discrete energy states, each possible outcome has a finite probability. For experiments with continuous outcomes, such as the position or momentum of a particle, each outcome has a finite probability density (more on this later).

Discrete example: measuring the spin of a spin-1/2 particle can produce only two possible values: parallel to the measuring equipment (call it "up"), or antiparallel to

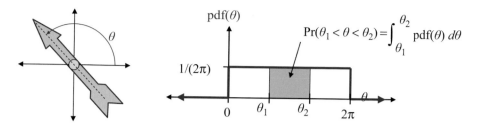

Fig. 1.9 (*Left*) A simple spinner. (*Right*) Example of pdf(θ), and Pr($\theta_1 < \theta < \theta_2$)

it ("down"). The two values, "up" and "down," are mutually exclusive. If in a given experiment the probability of "up" is 0.7, then the probability of "down" must be 0.3, because the sum of all possibilities must be 1 (i.e., the probability of measuring up *or* down is 1).

Note that an event is random if its outcome cannot be exactly predicted. All the possibilities need not be equally likely. Some references use the term "completely random," to mean all possibilities are equally likely. We find that term ambiguous. "Random" events are not exactly predictable, and the possible outcomes may or may not be equally likely.

A more complicated discrete example: the energy of a particle in a bound state has discrete values. However, though discrete, there may be an infinite set of possible values. For example, an electron bound to a proton (hydrogen) has an infinite set of possible energies, characterized by an integer, n, such that:

$$E_n = \text{Ry} / n^2 \quad where \quad \text{Ry} \equiv \text{Rydberg constant.}$$

Clearly, not all of them can be equally likely, since the infinite sum of their probabilities is 1.

The probability of an event, e, may be written as Pr(e), e.g., the probability of measuring E_n may be written "Pr(E_n)" or "Pr($E = E_n$)."

1.7.2   *Probability Density of Continuous Values*

We have seen that probability describes discrete random variables, such as rolling a die, or measuring (discrete) bound-state energy. But what about continuous random variables? Suppose I spin the pointer on a board game (Fig. 1.9, *left*), and consider the angle at which the pointer stops. That angle could be any of an infinite set of values, but a *continuous* infinite set, not a *discrete* infinite set. Angle, in this case, is a **continuous** random variable: between any two values are an infinite number of other possible values. There are many uncountable angle values that could occur. Probability density (aka "density") addresses such continuous random variables.

   In the case of a simple pointer that rotates in a plane, all angle values are equally likely; the pointer has no preference for one position over any other. The range of possible measurements is the half-open interval $[0, 2\pi)$, i.e., $0 \leq \theta < 2\pi$. The probability of measuring an angle <0 or $\geq 2\pi$ is zero:

$$\Pr(\theta < 0) = 0, \quad \text{and} \quad \Pr(\theta \geq 2\pi) = 0.$$

However, the probability of measuring the pointer between a range of allowed angles is finite. Suppose we divide the angles into two halves: upper $[0, \pi)$, and lower $[\pi, 2\pi)$. The probability of measuring in the upper half is 0.5 and the probability of measuring in the lower half is also 0.5. We could divide the circle into quadrants, and the probability of measuring in any quadrant is 0.25. In fact, for any finite angular range, there is a finite probability of measuring an angle in that range. In the simple pointer example, the probability of measuring in a range is proportional to the size of the range, i.e., all angles are equally likely. Noting that $\Pr(0 \leq \theta < 2\pi) = 1$, and that all angles are equally likely, it must be that (given $0 \leq \theta_1 \leq \theta_2 < 2\pi$):

$$\Pr(\theta_1 < \theta < \theta_2) = \frac{\theta_2 - \theta_1}{2\pi}.$$

We can find and graph a function of angle, pdf(θ), such that (Fig. 1.9, *right*):

$$\Pr(\theta_1 < \theta < \theta_2) = \int_{\theta_1}^{\theta_2} \text{pdf}(\theta) \, d\theta \qquad \text{for all } \theta, \, -\infty < \theta < \infty$$

Since the angle must be somewhere in $[0, 2\pi)$,

$$\int_0^{2\pi} \text{pdf}(\theta) \, d\theta = 1 \qquad \Rightarrow \qquad \text{pdf}(\theta) = \frac{1}{2\pi}.$$

Note that the probability of measuring any given, specific value (with no interval around it) is vanishingly small, i.e., zero. Therefore, $\Pr(\theta \text{ in } [a, b]) = \Pr(\theta \text{ in } (a, b))$.

   As a slightly more complicated example, let us measure the 1D position of a hypothetical particle. Position is a continuous random variable, and with many real-world measurements, the probability of finding the particle is concentrated near a point, though finite probabilities extend to infinity in both directions.

   Within a differentially small region dx, around a value x, the probability of finding the particle is proportional to dx.

This is the relationship between probability and probability density:

$$\Pr(\text{random value being in the region} \left[x, x + dx \right]) = \text{pdf}(x) \, dx.$$

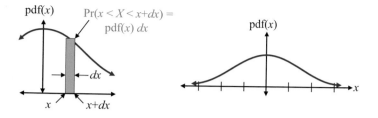

Fig. 1.10 Typical probability distribution functions (PDFs)

pdf(x) varies for different values of x and is called the **probability distribution function**, or **PDF**:

$$\text{pdf}(x) \equiv \lim_{dx \to 0} \frac{\Pr(x < X < x + dx)}{dx}, \quad \text{where } X \text{ is a random variable described by pdf}(x).$$

Thus we can graph the PDF (proportionality factor) as a function of x (Fig. 1.10).

The PDF is like a smoothed-out histogram of samples of the random variable. If you made billions of trials, and a fine-grained histogram of the results, you would get (essentially) the PDF.

What about the probability of the random value being in a finite interval $[a, b]$? From:

$$\Pr(\text{random value being in the region } [x, x + dx]) = \text{pdf}(x)\, dx,$$

it follows that:

$$\Pr(\text{random value being in the region } [a, b]) = \int_a^b \text{pdf}(x)\, dx$$

Furthermore, since the particle must be somewhere in $(-\infty, +\infty)$,

$$\int_{-\infty}^{\infty} \text{pdf}(x)\, dx = 1$$

When there is more than one PDF under consideration, we use a subscript to distinguish them. For example, the PDF for a random variable X might be given as $\text{pdf}_X(x)$.

(Sometimes, one encounters a function whose value is *proportional* to the probability density, i.e., the function is some constant times the PDF. In that case, $\int_{-\infty}^{\infty} f(x)\, dx \neq 1$. Such a function is called an **unnormalized** PDF.) In this book, we always use normalized PDFs (and wave-functions).

1.7.3   Complex (or "Probability") Amplitude

In QM, probabilities are given by a wave-function, such as $\psi(x)$. $\psi(x)$ is a complex-valued function of the real position x: for every real number x, $\psi(x)$ gives a complex number. $\psi(x)$ is related to pdf(x), i.e., the probability density function for measuring the particle at the point x:

$$|\psi(x)|^2 = \text{pdf}(x), \qquad \text{and recall} \qquad |\psi(x)|^2 = \psi^*(x)\psi(x).$$

Because PDFs must integrate to 1, we say $\psi(x)$ is "normalized" if:

$$\int_{-\infty}^{\infty} dx\; \psi^*(x)\; \psi(x) = 1.$$

$\psi(x)$ is the complex-valued **probability amplitude** [1, p8t], or just **amplitude**, at each point, x.

Why do we need a complex valued function to define a simple real-valued PDF? We need it because of the way probability amplitudes combine when a wave-function is a superposition of two or more possible states.

The two states of a superposition combine as if their amplitudes were oscillating sinusoids at every point, with both an amplitude and a phase. We can think of $\psi(x)$ as a phasor-valued function of space. This means the aggregate probability is *not* the sum of the (real) probabilities of the component states. Instead, the aggregate (complex) probability amplitude is the sum of the (complex) amplitudes of the component states. An example illustrates this:

Suppose there exists a state, ψ, which at some point a is: $\psi(a) = (1+i)$. Then $|\psi(a)|^2 = 2 = \text{pdf}(a)$. Suppose another state exists, φ, which at the point a is the same as ψ: $\varphi(a) = (1+i)$. Then $|\varphi(a)|^2 = 2 = \text{pdf}(a)$. Now suppose a particle is in an equally weighted superposition of states ψ and φ (call this new state χ). What is the PDF of $\chi(a)$? Let us start by noting that, if $\psi(x)$ is orthogonal to $\varphi(x)$:

$$\chi(x) = \frac{\psi(x) + \varphi(x)}{\sqrt{2}} \qquad \text{(we insert the } \sqrt{2} \text{ to keep } \chi \text{ normalized).}$$

Then $\chi(a) = \dfrac{(1+i)+(1+i)}{\sqrt{2}} = \sqrt{2} + \sqrt{2}i$, and $\text{pdf}(a) = |\chi(a)|^2 = 4$, which is twice the PDF of either ψ or φ. This is because ψ and φ reinforce each other at that point.

Now let us suppose that ψ and φ are different at the point a. We will leave $\psi(a) = (1+i)$, but let $\varphi(a) = (1-i)$. Then $|\varphi(a)|^2 = 2 = \text{pdf}(a)$, just as before. But what happens now if a particle is in an equally weighted superposition of states ψ and φ?

Then, the state χ is given by (again assuming orthogonal ψ and φ):

$$\chi(a) = \frac{\psi(a) + \varphi(a)}{\sqrt{2}} = \frac{(1+i) + (1-i)}{\sqrt{2}} = \sqrt{2}$$

Then $(a) = |\chi(a)|^2 = 2$. In this case, even though the magnitudes of both ψ and φ are the same as before, the magnitude of their sum is *different* than before. This is because they only partially reinforce each other; the *phase* of the complex values of ψ and φ are different, so their magnitudes do not simply add.

Finally, suppose $\varphi(a) = (-1-i)$. Then $|\varphi(a)|^2 = 2 = \text{pdf}(a)$, as before, but $\chi(a) = 0$ and $\text{pdf}(a) = 0$. In this case, the magnitudes of ψ and φ are still the same as before, but they are of opposite phase, so they cancel completely in an equally weighted superposition of states.

> It is this adding of complex valued probability amplitudes that accounts for most quantum weirdness.

In other words, QM is weird because of interfering sinusoids.

Note that this kind of interference(that of adding complex amplitudes) is identical to phasor computations in fluids, electromagnetics, or any other kind of wave interference. In other words:

> The algebra of adding sinusoids of a fixed frequency, but arbitrary amplitude and phase, is the same as the algebra of complex numbers.

1.7.4 Averages vs Expectations

It can be shown that the average value of a random variable (averaged over many trials) is:

$$\langle x \rangle = \int_{-\infty}^{\infty} dx\, \text{pdf}(x)\, x, \qquad \text{or} \qquad \langle x \rangle = \sum_{i=1}^{N} \Pr(x_i)\, x_i$$

It is clear that the average value of many trials may be an impossible value for any single trial; e.g., the average value of the roll of a die is 3.5, but no single role can produce that value. Often, the term "expectation value" or "expected value" is used to mean "average value." This can be confusing, because the "expected value" of a die roll is 3.5, yet you would *never* expect the value of a die roll to be 3.5. Since the term "average" is clear and precise, we do not use the term "expectation value."

1.8   The Origin of Schrödinger's Equation

The Schrödinger equation is the equation of motion (EOM) for a wave-function, as well as for other quantum systems, including spin and multiparticle systems. Like all EOMs, it predicts the future state from an initial state. The Schrödinger equation is completely deterministic; it is the measurement of quantum systems that is probabilistic. Where does Schrödinger's equation come from? Many books note that it cannot be *derived* from anything; it is a *discovery* of QM. Nonetheless, it must be *motivated* by something; Erwin Schrödinger did not just pull it out of thin air.

1.8.1   The Wave Equation

Wave Mechanics began when experimentalists noticed that particles had wave properties; most notably, they exhibited wavelike diffraction and interference. (Davisson and Germer discovered electron waves by accident after their equipment malfunctioned.) Moreover, the wavelength of particles has the same relationship to their momentum as photons, namely:

$$p = \frac{h}{\lambda} = \hbar k \qquad where \qquad k \equiv \text{spatial frequency (aka wave-number), in rad/m.} \quad (1.1)$$

Since one wavelength, λ, is 2π radians, $k = 2\pi/\lambda$.

Since photons are traveling waves, and particles had similar wavelike properties, physicists wondered if particles could be represented by similar traveling waves. A real-valued wave, traveling in the x-direction, can be represented by:

$$a(t, x) = B\cos(kx - \omega t) \qquad where \qquad \omega \equiv \text{frequency in time, measured in rad/s}$$
$$B \equiv \text{real amplitude of the wave.}$$

Thus at a fixed point in space, the amplitude varies sinusoidally over time; at a fixed point in time, the amplitude varies sinusoidally over space. Viewed over space and time, the wave travels smoothly in the positive x direction (Fig. 1.11).

> Particles were found experimentally to interfere like sinusoids. The mathematics of interfering sinusoids is, in fact, the mathematics of complex arithmetic.

(See phasors section earlier). This means that particles' waves are well-described by complex-valued traveling waves. Complex-valued wave-functions add as complex numbers and then square to produce probability densities. The complex wave equation then has both real and imaginary parts as traveling waves:

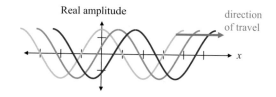

Real amplitude

direction of travel

x

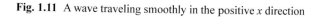

Fig. 1.11 A wave traveling smoothly in the positive x direction

$$\psi(t,x) = Ae^{i(kx-\omega t)} \qquad where \qquad A \equiv \text{complex amplitude of the wave.}$$

In other words, A is a phasor. The instantaneous value of a wave at a given point in space at a given point in time is sometimes called the **instantaneous amplitude**. Thus, $\psi(t, x)$ is the instantaneous (complex) amplitude of the wave-function at the point (t, x).

1.8.2   Energy, the Hamiltonian, and the Wave Equation

Further work found that for a fixed-energy quantum particle or system, the phase of its representing sinusoid shifted in time at a constant rate: $\phi = \omega t$. This is equivalent to the complex wave function rotating in the complex plane over time as $e^{-i\omega t}$ (i.e., the complex value, at each point in space, had its phase decrease continuously with time). Here again, the analogy to photons was useful. The frequency of rotation(in the complex plane) of the complex value of a wave-function is proportional to its energy, just like a photon:

$$E = \hbar\omega \qquad where \qquad E \text{ is the } total \text{ energy of the particle } (\text{or photon}). \qquad (1.2)$$

Total energy is kinetic energy plus potential energy: $E = T + V$. For a particle, kinetic energy is $T = p^2/2m$. The potential energy of a particle is a given function of its position in space; call it $V(x)$. (For simplicity, we have assumed V is independent of time, but all of this works even if $V = V(t, x)$). Therefore:

$$E = T + V = \frac{p^2}{2m} + V(x) \quad (\text{just as in classical mechanics}).$$

We will see that Schrödinger's equation is just this equation, written as quantum operators acting on the wave-function:

$$\hat{E} = \hat{T} + \hat{V} \quad \rightarrow \quad \underbrace{i\hbar\frac{\partial}{\partial t}\psi}_{\hat{E}} = \left[\underbrace{-\frac{\hbar^2}{2m}\frac{\partial^2}{\partial x^2}}_{\hat{T}} + \underbrace{V(x)}_{\hat{V}}\right]\psi \qquad (\text{Schrödinger's equation}).$$

If a particle of fixed energy is represented by a traveling wave-function, $\psi(t, x)$ = $A\,e^{i(kx-\omega t)}$, then its derivative with respect to time is:

$$\frac{\partial}{\partial t}\psi(t,x) = -i\omega A e^{i(kx-\omega t)} = -i\omega\psi(t,x) = -i\left(\frac{E}{\hbar}\right)\psi(t,x),$$

and rearranging:

$$i\hbar\frac{\partial}{\partial t}\psi(t,x) = E\psi(t,x). \tag{1.3}$$

The derivative with respect to time multiplies the wave function by a value proportional to the particle's total energy.

The derivative with respect to x of a simple traveling wave function [which has definite momentum, Eq. (1.1)] is:

$$\frac{\partial}{\partial x}\psi(t,x) = ikAe^{i(kx-\omega t)} = \left(\frac{ip}{\hbar}\right)\psi(t,x) \quad\text{and}\quad \frac{\partial^2}{\partial x^2}\psi(t,x) = -\left(\frac{p}{\hbar}\right)^2\psi(t,x).$$

Rearranging the latter equation, to get kinetic energy:

$$-\hbar^2\frac{\partial^2}{\partial x^2}\psi(t,x) = p^2\psi(t,x), \quad\text{and dividing by } 2m\ (m \equiv \text{particle mass})$$

$$\frac{-\hbar^2}{2m}\frac{\partial^2}{\partial x^2}\psi(t,x) = \frac{p^2}{2m}\psi(t,x) = T\psi(t,x).$$

The derivative with respect to x multiplies the wave function ψ by a value proportional to the particle's momentum, and the second derivative multiplies ψ by a value proportional to p^2, and thus proportional to the particle's kinetic energy. Now simply plug $E = T + V(x)$ into the previous time derivative, Eq. (1.3):

$$i\hbar\frac{\partial}{\partial t}\psi(t,x) = [T + V(x)]\,\psi(t,x) = \frac{-\hbar^2}{2m}\frac{\partial^2}{\partial x^2}\psi(t,x) + V(x)\psi(t,x).$$

Voila! Schrödinger's equation! It is nothing more than a Hamiltonian operator $(T+V)$ applied to the wave function, using the experimental relationship between total energy and temporal frequency (ω), and momentum and spatial frequency (k). The Hamiltonian produces a particle's total energy, kinetic plus potential [even if $V(t, x)$ varies in time]. On the left side of Schrödinger's equation is the time derivative, producing total energy times ψ: $E\psi$. On the right side is the spatial second derivative, producing $T\psi$, plus the potential energy, $V(x)\psi$. (Magnetic fields complicate this simple picture somewhat.)

The final piece of the puzzle is that a particle usually cannot be in a state of a single traveling sinusoidal wave function (because then it would be equally likely to be found anywhere in infinite space). It must be in a superposition of many (even infinitely many) sinusoidal wave functions. But each component traveling wave function individually satisfies Schrödinger's equation. And Schrödinger's equation is a *linear* differential equation. So given a set of functions which satisfy the equation, any superposition (linear combination) of those functions also satisfies Schrödinger's equation. Thus:

> In nonrelativistic QM, Schrödinger's equation applies to all massive particles and all wave-functions.

1.9   The Meaning of the Wave-Function

1.9.1   *Where Are We, and Where Do We Go From Here?*

The most evident feature of the wave function is that it tells the probability density of finding the particle at any point in space; in 1D the unit of probability density is m^{-1}; in 2D the unit of probability density is m^{-2}; and in 3D the unit of probability density is m^{-3}. But the wave function is much more [15, p59m]:

> A particle's wave-function defines *all* the spatial properties of the particle, not just its position. It defines it momentum, orbital angular momentum, kinetic energy, etc.

To compute things involving external fields, such as potential energy, we must know (in addition to the wave-function) the potential field $V(\mathbf{r})$. Note that even the probability density of a particle is more than just the probability of measuring it to be somewhere: for many quantum purposes, a single particle behaves as if it were *actually distributed* throughout space, as defined by its wave function. (In fact, in QFT, they often call it "particle density" or "number density" instead of "probability density." Particle density for antiparticles can be taken as negative.)

For example, if the particle has charge q, we compute the potential produced by it from the charge distribution defined by $\rho_{\text{charge}}(x) = q|\psi(x)|^2$ [2, p434b]. (However, the distribution of charge from a *single* particle does not push on itself.) If the particle has mass m, we distribute its mass according to $\rho_{\text{mass}}(x) = m|\psi(x)|^2$, as well. If a particle has spin, it is spin angular momentum, and therefore also magnetic dipole moment, is actually a spin (and dipole moment) *density*, spread out according to $\rho_{\text{spin}}(x) = \dfrac{\hbar}{2}|\psi(x)|^2$ [1, 14–92 p326m].

However, there are limits to the model of a physically distributed particle. When things like mass appear in a denominator, as with kinetic energy $=p^2/2m$, we cannot consider an infinitesimal point as having an infinitesimal fraction of the mass. If it did, and the momentum was macroscopic, the kinetic energy would blow up everywhere.

You can localize a particle to a very small region of space with a good measurement, but until you do so, it may be spread out. Also, immediately after you measure it, it will start spreading out again, and become distributed over time. (The concept of actually *being* distributed in space gets trickier with entangled multiparticle states, which we address later.) (In QM, a particle can be localized to an arbitrarily small region, but in reality, as described by QFT, even a so-called "point particle" has a limit to how tightly you can localize it, but that is beyond our scope.)

The wave function does not define the particle's spin (if any), nor any angular momentum or energy associated with the spin (such as magnetic-moment/B-field interactions). Spin-1/2 and related values are defined by "spinors," described later.

Because the wave-function defines a particle's momentum, it tells us not only where it is, but where it is going. The question of where it is going is tricky in QM, though, because any real particle is in a superposition of momentum states, i.e., it is moving in multiple ways. For example, consider bound states: a particle is tied to some region by an inescapable potential. For a **stationary** state (where the particle properties do not change with time), the average momentum must be zero, otherwise the particle would be moving, and escape the potential.

Nonstationary states and unbound states can be subtle; we discuss these in the text as the need arises.

1.9.2   Gross Anatomy of a Wave Function

We now describe some important qualitative properties of wave-functions, whose understanding helps make QM more sensible and intuitive. We first introduce a particle at a high energy above the potential, which behaves fairly classically, then discuss a more "quantum-like" energy, to reveal some of the nonclassical consequences of QM.

High energy wave-function   Consider a particle in a box, with a potential step at the bottom (Fig. 1.12). What does the wave function look like?

The stationary state $\psi(x)$ is real, and is everywhere a superposition of positive and negative momenta. The real-valued sinusoids oscillate according to $|k(x)| = \left[2m(E-V)\right]^{1/2}/\hbar \equiv p_{cl}/\hbar$. (We will see that since p^2 can be thought of as an average, $p_{cl} \equiv (p^2)^{1/2}$ is not the same as p.) On the left, the kinetic energy$(T = E - V = p^2/2m)$ is low, so p^2 is low, and $k = p_{cl}/\hbar$ (rad/m) is low (low spatial frequency). On the right, T is higher, p_{cl} is higher, and k is higher (higher spatial frequency). Not only is $\psi(x)$ a superposition of momenta, but it is important that we can associate *different* momenta to different *locations* on the wave-function. p_{cl} is lower on the left, and higher on the right. This introduces the concept of "local momentum," and "local properties" in general.

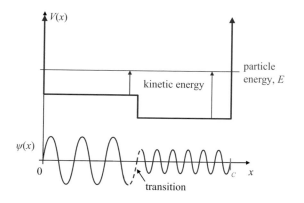

Fig. 1.12 Particle of energy E, in a box with a step potential bottom. From the correspondence principle, we expect the wave-function to be larger on the left, where the particle is moving "slower."

Concerning the amplitude of $\psi(x)$: on the left of the step-bottom box, the kinetic energy $(E-V)$ is lower, so a classical particle moves "slowly." On the right, the kinetic energy is higher, so a classical particle moves "quickly." Classically, then, the particle is more likely to be measured on the left half than the right. The correspondence principle says that at high energy, these classical results must be born out quantum mechanically. The particle "spends more time" on the left side, where it is moving slowly, so the amplitude of ψ is larger (more likely to find the particle there). Conversely, the particle "spends less time" on the right, where it is moving quickly, so the amplitude of ψ is smaller (less likely to find the particle there).

As always, the allowed energies of this potential are quantized by the boundary conditions on ψ, in this case, that $\psi(x)=0$ at the edges.

If the potential box is wide (many cycles of ψ in each half), then the potential step has only a small transient effect on ψ, indicated as a transition from the sine wave on the left to the sine wave on the right. The perturbation on ψ from the discontinuity decays rapidly away from the transition.

Low energy wave-function We compute the potential produced by oscillator [8, 7.18 p144]:

$$\psi(x) = \left(\frac{m\omega}{\hbar\pi}\right)^{1/4} \exp\left(-\frac{1}{2}(x/x_0)^2\right) \qquad where \quad x_0 \equiv \sqrt{\hbar/m\omega},$$

and also consider the first excited state (both shown in Fig. 1.13).

There are several important points to notice, which illustrate general quantum mechanical principles. We discuss the following points:

1. All bound states can have $\psi(x)$ real.
2. At a peak, the local momentum $p=0$, but $p^2 \neq 0$.

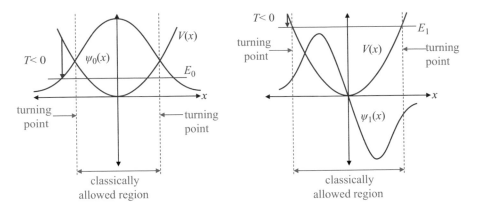

Fig. 1.13 Ground state (*left*), and first excited state (*right*) of a harmonic oscillator. $E_1 = 3E_0$

3. In the classically allowed region, $\psi(x)$ turns toward zero: down when it is positive (negative curvature), and up when it is negative (positive curvature).
4. Beyond the classical turning points, $p^2 < 0$, and p is imaginary.

It is less obvious than with the previous wave-function, but each point on this wave-function also has a value of momentum associated with it. We call that the "local momentum," and define it more precisely later, when we examine operators.

Bound states can have $\psi(x)$ real Simple Hamiltonians, $\hat{H} = \hat{p}^2 / 2m + V(x)$, are symmetric with respect to p: any solution for p also has $-p$ as a solution, with the same energy. Since Schrödinger's equation is linear, then the superposition wave function for $p + (-p)$ is also a solution. Their sum is real. For example, ignoring normalization (and dropping $\hbar$ for simplicity), we have:

$$\psi_p(x) = e^{ipx}, \qquad \psi_{-p}(x) = e^{-ipx} \qquad \text{are wavefunctions for } p \text{ and } (-p).$$

$$\psi(x) = e^{ipx} + e^{-ipx} = 2\cos(px), \quad \text{is also a solution, and is real .}$$

[With magnetism, the Hamiltonian is:

$$\hat{H} = \frac{1}{2m}\left(\hat{p} - \frac{q}{c}\mathbf{A}(x)\right)^2 + V(x) \qquad where \qquad \mathbf{A}(x) \text{ is the magnetic vector potential.}$$

The Hamiltonian no longer commutes with $\hat{p}$, and the situation is more complicated.]

Ultimately, all Hamiltonians are real, so the time-independent Schrodinger eigenvalue equation is real and the existence theorem of differential equations says it has a real solution. Therefore, all bound states can be real.

At a peak of $\psi(x)$, $\hat{p}\psi(x) = \dfrac{\hbar}{i}\dfrac{\partial \psi(x)}{\partial x} = 0$. We will see that the momentum operator at a point x multiplies the wave-function by the local value of the momentum at that point. When $\partial\psi/\partial x=0$, there are equal contributions at x from each of $+p$ and $-p$, so they cancel. However, $(p^2)_{local}$ is *not* generally zero at such a point.

In the classically allowed region, $\psi(x)$ turns toward zero: down when it is positive and up when it is negative. In other words, when the spatial frequency k (aka wave number) is real, ψ oscillates (i.e., we have a *wave*, given enough room). In the ground state of Fig. 1.13, there is not enough room in the classically allowed region for ψ to actually cross zero, which is why the original claim is about ψ "turning toward 0," instead of "oscillating." This means every zero of $\psi(x)$ is also an inflection point. From the local momentum, we will find that:

$$\frac{\partial^2\psi(x)}{\partial x^2} = -\frac{\left(p^2\right)_{local}}{\hbar^2}\,\psi(x).$$

In the classically allowed region of space, kinetic energy $T>0$, so $(p^2)_{local}>0$. Then, when $\psi(x)>0$, $\partial^2\psi/\partial x^2<0$, and ψ curves downward. When $\psi(x)<0$, $\partial^2\psi/\partial x^2>0$, and ψ curves upward.

Negative kinetic energy and imaginary momentum Related to wave-function curvature, we see that *beyond the classical turning point*, $(p^2)_{local}<0$! This surprising result is consistent with $\psi(x)$ being an energy eigenstate: the energy operator, $\hat{H}=\hat{T}+\hat{V}(x)$, must evaluate to the same local energy E at every point of $\psi(x)$. But in the classically disallowed region (beyond the classical turning point), $V(x)>E$, so kinetic energy $T=p^2/2m$ must be less than zero. This is crucially important to orbital angular momentum, which we discuss later. Furthermore, the classically disallowed region adds a negative contribution to the overall averages of T and of p^2.

A negative kinetic energy implies that formally, in the disallowed region, p itself is imaginary. For an energy eigenstate, the local momentum is either real or purely imaginary. For superpositions of energy eigenstates, the local momentum may be complex. We will see later, in the discussion of probability current, that the *real* part of the local momentum has a direct physical meaning. However, negative kinetic energy and imaginary momentum have no classical or simple interpretation. Nonetheless, we accepted their validity as an axiom earlier (£6, p. 9), and their consequences are fully verified by experiment.

Also, at the classical turning points, $\psi(x)$ has inflection points. (Recall that a change in curvature is defined as an inflection.) In the classically allowed region, ψ curves toward the x-axis, as noted previously. In the classically forbidden region, ψ is asymptotic to the x-axis, and therefore curves away from the axis.

1.10   Operators

Operators are an essential part of QM.

> The existence of superpositions of states leads to the need for operators to compute the properties of such superpositions, because an operator can associate a different number with each component of the superposition.

In this section, we focus on operators on functions. Later, we consider operators on discrete-state systems. Before discussing the physical meaning of operators, we must first give some mathematical description.

Operators turn a function of space into another function of space (or more generally, a vector into another vector). In QM, there are three main uses for operators:

1. To extract observable (measurable) attributes of a particle from a wave function. These are *Hermitian* operators. The functions resulting from such operators are *not* quantum states.
2. To compute new states from old states, such as the state of a particle or system after some time, or after a rotation. These are *unitary* operators, and their results *are* quantum states.
3. To perform mathematical operations which yield valuable information about QM, and which are much more difficult to discover with nonoperator methods. These *algebraic* operators are usually neither Hermitian nor unitary, but some of them might be either. Their results are not generally quantum states.

We describe each of these uses in more detail shortly. Examples of the three uses of operators:

Observable operators	State transformation operators	Algebraic operators
$\hat{x}, \hat{p}, \hat{H}, \hat{L}^2, \hat{L}_z, \hat{s}^2, s_z...$	$\hat{T}(a), \hat{R}(\theta,\phi), \hat{U}(t)$	$\hat{a}, \hat{a}^\dagger, \hat{J}_+, \hat{J}_-$

> An operator acting on a spatial function $\psi(x)$ produces another spatial function.

Operators are part of the algebra of QM; operators are written in both wave-function notation and Dirac notation (described later). Operators exist for both spatial wave-functions and discrete states, such as particle spin. In QM, all operators (except time-reversal) are linear operators.

> A linear operator acting on a superposition, say $\psi(x)$, produces a superposition of *results* based on the superposition of *functions* which compose $\psi(x)$.

We distinguish between a **quantum state**, which is wave-function that describes all the spatial properties of a particle, and more general functions of space which

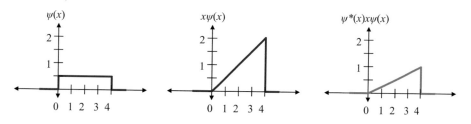

Fig. 1.14 (*Left*) A normalized wave-function, $\psi(x)$. (*Middle*) $x\psi(x)$. (*Right*) $\psi^*(x)x\psi(x)$

have other uses but are *not* quantum states. The result of an operator may be either a quantum state, or some other function of space that is not a quantum state. For example,

$$\hat{p}\psi(x) = \frac{\hbar}{i}\frac{d}{dx}\psi(x) \quad where \quad \hat{p} \text{ is a hermitian operator,}$$

is a function of space (function of x), but is not a quantum "state," because it is not a wave-function. It is not normalized, and has the wrong units for a wave-function. In contrast, given a 2D wave-function, $\psi(x,y)$, which *is* a state, we can rotate it in the x-y plane:

$$\hat{R}(\theta)\psi_1(x, y) \equiv \psi_2(x, y) = \psi_1\left(x\cos\theta - y\sin\theta, x\sin\theta + y\cos\theta\right)$$

where $\hat{R}(\theta)$ is a unitary operator.

This *is* a new quantum state, computed from the old one.

Some references do not properly distinguish between a "spatial function" and a quantum "state." Some even define an "operator" as a thing which acts on a "state" to produce another "state". This is incorrect. An operator acts on a function to produce another function. Either one or both functions may or may not be quantum "states."

For example, imagine a particle state $\psi(x) = 1/2$ between [0, 4] and 0 elsewhere (Fig. 1.14, *left*). Consider also the position operator acting on it, $\hat{x}\psi(x) = x\psi(x) = x/2$ between [0, 4] and 0 elsewhere (Fig. 1.14, *middle*).

Note that, though $\psi(x)$ *is* a quantum state, $\hat{x}\psi(x)$ is not a state, because it is not a wave-function: it does not define the properties of a particle or system, it is not even normalized, and it has the wrong units. However, it can be used to calculate things related to the particle's position, as shown later.

Operators may also act on a given function which is *not* a quantum state, e.g., the result of some other operator on a state. Such a given function may still be a superposition of results. A second operator acting on such a function which is *not* a state produces a new superposition of new results based on the superposition of old results in the given function. For example, in two dimensions, we have

wave-functions $\psi(x, y)$. The $\hat{L}_z$ operator (angular momentum about the z-axis) is a composition of $\hat{x}, \hat{p}_y, \hat{y}$ and $\hat{p}_x : \hat{L}_z = \hat{x}\hat{p}_y - \hat{y}\hat{p}_x$. $\hat{L}_z$ typically acts on a state, which means that in the first term, $\hat{p}_y = \dfrac{\hbar}{i}\dfrac{\partial}{\partial y}$ acts first on the given state to produce a function of space which is *not* a state. Then $\hat{x}$ acts on *that* spatial function to produce a new spatial function, which is also *not* a state. Similarly for the second term, $-\hat{y}\hat{p}_x$. The result of $\hat{L}_z$ acting on a state is a spatial function that is not a state.

1.10.1 Linear Operators

In QM, all operators (except time-reversal) are linear operators. A linear operator produces a superposition of *results* based on the superposition of *components* in the given function.

> A **linear operator** *distributes* over addition, and *commutes* with scalar multiplication:

$$\hat{L}[a\psi(x)+b\phi(x)] = a\hat{L}\psi(x)+b\hat{L}\phi(x).$$

Examples: Is multiplication by x a linear operator? Let us see:

$$x[a\psi(x)+b\phi(x)] = ax\psi(x)+bx\phi(x). \quad \text{It is linear!}$$

How about multiplication by x^2? $x^2[a\psi(x)+b\phi(x)] = ax^2\psi(x)+bx^2\phi(x)$. It is also linear.

How about multiplication by $\cos(x)$? Or multiplication by any arbitrary function $f(x)$?

$$f(x)[a\psi(x)+b\phi(x)] = af(x)\psi(x)+bf(x)\phi(xs). \quad \text{It is linear!}$$

Is squaring a spatial function a linear operation?

$$[a\psi(x)+b\phi(x)]^2 = a^2\psi^2(x)+2ab\psi(x)\phi(x)+b^2\phi^2(x)$$
$$\neq a\psi^2(x)+b\phi^2(x). \quad \text{It is } not \text{ linear.}$$

Note that complex conjugation, the Re{}, and Im{} operators are all *nonlinear*.

Composition of operators: We frequently operate on a function with one operator, and then operate on that result with another operator. For example, we may first operate with the operator "x," which means multiply the function by "x" everywhere: $\psi(x) \to x\psi(x)$. We may then operate on that result by differentiating w.r.t. x: $\frac{\partial}{\partial x} x \psi(x)$. Acting with one operator and then another, is called **composition** of operators. The composition of $\hat{A}$ on $\hat{B}$ is $\hat{A}\hat{B}$. Because operators act to the right, this means act with $\hat{B}$ *first*, then act with $\hat{A}$. Many references call the composition of two operators the "product" of the two, and sometimes even "multiplying" the two. In such a case, there is generally no multiplication involved (unless one of the operators is a "multiply by" operator).

1.10.2   Operator Algebra

Operators have their own algebra, or rules, for performing mathematical operations and for being manipulated. In wave mechanics, operators are linear operators on continuous functions. (Matrix mechanics is discussed later.) Some operators are simple and straightforward:

$\frac{\partial}{\partial x}$ This is a linear operator: it takes a derivative with respect to "x".

However, some other operators are more involved. For example, "$\left(\frac{\partial}{\partial x}\right) x$" can be a compound operator. It means "multiply by x, and then take the derivative with respect to x." As an operator, it is *not* an arithmetic expression (where $\left(\frac{\partial}{\partial x}\right) x = 1$). Instead,

$$\left(\frac{\partial}{\partial x}\right) x = 1 + x\frac{\partial}{\partial x} \quad \text{(an operator equation)}.$$

How can this be? What happened to basic calculus? The key here is to distinguish the *operator* from the *operand*. The **operator** is the action to be taken. The **operand** is the object on which the action is taken. For example,

$\frac{\partial}{\partial x}$,   "$\frac{\partial}{\partial x}$" is an operator. There is no operand or result here.

$\left(\frac{\partial}{\partial x}\right) x = 1$ The operator is "$\frac{\partial}{\partial x}$", the operand is "$x$", and the result is "1".

But in the case of the operator equation:

$\left(\frac{\partial}{\partial x}\right) x = 1 + x\frac{\partial}{\partial x}$,   "$\left(\frac{\partial}{\partial x}\right) x$ is an operator, and "$1 + x\frac{\partial}{\partial x}$" is an *equivalent* operator.

In other words, this is an **operator identity**. The previous one is not an *arithmetic* equation; it is a statement of equivalence of two operators. There is no operand in the previous equation, and no result (just like writing "∂/∂x" by itself). Let us establish the equivalence in the previous **operator equation** (aka **operator identity**):

$$\left(\frac{\partial}{\partial x}\right)x\,\psi(x)$$ We insert an arbitrary operand, $\psi(x)$, so we can operateon it.

$$=\left(\frac{\partial}{\partial x}\right)(x\,\psi(x))$$ The operator says to first multiply by x, then differentiate w.r.t. x.

$$=\left[\left(\frac{\partial}{\partial x}\right)(x)\right]\psi(x)+x\frac{\partial}{\partial x}\psi(x)$$ Using the product rule for derivatives.

$$=\psi(x)+x\frac{\partial}{\partial x}\psi(x)$$ because $\left(\frac{\partial}{\partial x}\right)x=1$, x is an *operand* here.

$$=\left(1+x\frac{\partial}{\partial x}\right)\psi$$ "factoring out" ψ, because linear operators distribute over addition.

Thus the operator "$\left(\frac{\partial}{\partial x}\right)x$" is exactly equivalent to the operator "$1+x\frac{\partial}{\partial x}$."

When you see something like "$\left(\frac{\partial}{\partial x}\right)x$," how do you know if this is an arithmetic expression (with an operator "$\frac{\partial}{\partial x}$" and operand "$x$"), or just an operator? You can only tell from context. In other cases, though, operators may have a "hat" over them, to explicitly indicate they are operators.

Let us now derive an example of operator equivalence without actually inserting an arbitrary operand. We keep the operand silently in our minds, and write our steps as we operate on this hypothetical operand:

$$\left(\frac{\partial}{\partial x}\right)x^2$$ What is an equivalent operator to this?

$$=\left[\frac{\partial}{\partial x}x^2\right]+x^2\frac{\partial}{\partial x}$$ Using product rule for derivatives; in the 1st term,

x^2 is an *operand*.

$$=2x+x^2\frac{\partial}{\partial x}$$ Because $\frac{\partial}{\partial x}x^2=2x$, where x^2 is an *operand* here.

Notice how there are fewer steps compared to the previous derivation, because we did not actually insert an operand and then remove it at the end. However, this shortcut method may take some getting used to.

Beware that many pairs of operators do not commute, i.e., $\hat{A}\hat{B} \neq \hat{B}\hat{A}$, so keep operators in order. In ordinary algebra, we often write:

$$(a+b)^2 = a^2 + 2ab + b^2.$$

When "a" and "b" are numbers (or functions), this is fine. But if instead of numbers, we had operators $\hat{A}$ and $\hat{B}$, this would be wrong, because we changed the order of "a" and "b" in the second term. Instead, we should write:

$$\left(\hat{A}+\hat{B}\right)^2 = \hat{A}^2 + \hat{A}\hat{B} + \hat{B}\hat{A} + \hat{B}^2.$$

For example, the Hamiltonian of a charged particle in a magnetic field includes:

$$\left(\hat{\mathbf{p}} - \frac{q}{c}\mathbf{A}(x)\right)^2 = \hat{\mathbf{p}}^2 - \frac{q}{c}\left(\hat{\mathbf{p}}\mathbf{A}(x) + \mathbf{A}(x)\hat{\mathbf{p}}\right) + \frac{q^2}{c^2}\mathbf{A}^2(x)$$

where $\hat{\mathbf{p}}$ and $\mathbf{A}(x)$ do not commute.

This preservation of ordering is also required for matrix operators in discrete-state systems. (See the later section on commutators for more on operator identities.)

1.10.3 Operators: What is the Meaning of This?

We continue our focus here on wave-mechanics, i.e., operating on wave-functions such as $\psi(x)$. Later, we will discuss operators in Dirac notation, and in discrete state QM.

1.10.3.1 Think Globally, Act Locally: Local Values, Local Operators

Just what do operators do? What is the meaning of some arbitrary operator, $\hat{o}\psi(x)$? Let us start simply, with a spatial wave function $\psi(x)$, in the position basis. At each point in space, the wave function defines the particle density (or probability density of finding the particle). Also, for a given $\psi(x)$, a given *operator* has a **local value** at each point in space (i.e., for each x), which generally depends on $\psi(x)$. In other words, at each point in space, the wave function and the operator *together* define the local value of the operator. We could write this local value as a function of position, say $o_\psi(x)$, where the subscript ψ indicates that the local value of the operator at each point x was computed from a given $\psi(x)$. The result of an operator acting on a spatial function $\psi(x)$ is to simply multiply $\psi(x)$ at each point in space by the local value of the operator at that point in space, $o_\psi(x)$:

$$\hat{o}\psi(x) = o_\psi(x)\psi(x). \tag{1.4}$$

The operator $\hat{o}$ is a mathematical operation that turns $\psi(x)$ into $o_\psi(x)\psi(x)$.

For example, in the position basis, the simplest operator is the position operator, $\hat{x}$:

$$\hat{x}\psi(x) = x_{local}(x)\psi(x).$$

But what is the local position associated with the position "x?" It is simply "x." Therefore,

$$x_{local}(x) = x \quad \Rightarrow \quad \hat{x}\psi(x) = x\psi(x).$$

We will show shortly that we can use this to compute the *overall* average value of "x" (the particle's position), from its wave-function $\psi(x)$:

$$\langle \hat{x} \rangle = \int_\infty \psi^*(x)\, x\, \psi(x)\, dx$$

A more complicated example is "local energy," which is an important quantity. In computational quantum chemistry, one can find numerical solutions to the time-independent Schrödinger equation by starting with a trial wave-function (a trial solution), and adjusting it according to its local energy at each point, to construct the next iteration of trial wave-function. But how can we compute $E_{local}(x)$? We use the definition of an operator:

$$\hat{H}\psi(x) = E_{local}(x)\psi(x) \quad \Rightarrow \quad E_{local}(x) = \frac{\hat{H}\psi(x)}{\psi(x)}.$$

For the nonmagnetic Hamiltonian, we have:

$$\hat{H} = \frac{\hat{p}^2}{2m} + V(x) = -\frac{\hbar^2}{2m}\frac{\partial^2}{\partial x^2} + V(x)$$

$$\Rightarrow E_{local}(x) = \frac{\hat{H}\psi(x)}{\psi(x)} = \frac{\left[-(\hbar^2/2m)\dfrac{\partial^2}{\partial x^2} + V(x)\right]\psi(x)}{\psi(x)} = \frac{-(\hbar^2/2m)\dfrac{\partial^2}{\partial x^2}\psi(x)}{\psi(x)} + V(x).$$

Given some trial wave-function $\psi(x)$, one can evaluate the local energy from this formula.

In general, then, we *define* the local value of any operator as:

$$o_{local}(x) \equiv \frac{\hat{o}\psi(x)}{\psi(x)}, \qquad \text{provided } \psi(x) \neq 0$$

There is a subtlety if $\psi(x)$ is zero somewhere, because then the local value is undefined. This is usually not a problem, because the "local density" *is* well defined, as we will see later. However, for well-behaved wave-functions, a zero of $\psi(x)$ is a "removable singularity" in the local value, and can be "filled" by its limiting value:

$$\text{If} \quad \psi(x_0) = 0, \quad \text{then} \quad o_{local}(x_0) \equiv \lim_{x \to x_0} o_{local}(x).$$

A **local operator** is an operator determined only by ψ at x, or ψ in an infinitesimal neighborhood of x. Thus $\partial/\partial x$ is a local operator. Some operators depend on more than one point of ψ, or even on *all* of $\psi(x)$; they are **nonlocal operators**.

> All common observables are local operators.

In general, an operator can act on any function of space; it need not be a quantum state. If a spatial function is not a state, then what is it? It is usually the result of some other operator acting on a state, but it could instead be a given potential (e.g., $V(x)$ or $\mathbf{A}(x)$), or one of various other functions of space. Such a spatial function encodes information. To get at that information, you usually have to take a dot product with another relevant spatial function.

Summary *An operator acting on a given spatial function $\psi(x)$ produces another spatial function.* The given function may or may not be a quantum *state* and the resulting function may or may not be a quantum *state*, i.e., it may or may not represent a quantum state that a particle or system could be in.

Here is a summary of some common operators:

Operator	Position basis representation	Comments
Position, $\hat{x}$	x	Multiplies $\psi(x)$ by position x. The value of x does *not* depend on ψ.
Potential energy, $\hat{V}$	$V(x)$	Multiplies $\psi(x)$ by potential $V(x)$. The value of $V(x)$ does *not* depend on ψ.
Momentum, $\hat{p}$	$-i\hbar\, \partial/\partial x$	Multiplies $\psi(x)$ by momentum $p_{local}(x)$. The value of $p_{local}(x)$ *does* depend on ψ.
Kinetic energy, $\hat{T}$	$\dfrac{\hat{p}^2}{2m} = \dfrac{-\hbar^2}{2m}\dfrac{\partial^2}{\partial x^2}$	Multiplies $\psi(x)$ by kinetic energy $T_{local}(x)$. The value of $T_{local}(x)$ *does* depend on ψ.
Hamiltonian, $\hat{H}$	$\dfrac{\hat{p}^2}{2m} + \hat{V}(x) = \dfrac{-\hbar^2}{2m}\dfrac{\partial^2}{\partial x^2} + V(x)$	Just a sum of kinetic and potential energy.
Total energy, $\hat{E}$	$i\hbar\, \partial/\partial t$	Multiplies $\psi(t, x)$ by energy $E_{local}(t, x)$. The value of $E_{local}(t, x)$ *does* depend on $\psi(t, x)$. Note that $\partial/\partial t$ works in the momentum basis, and other bases, as well.

1.10.3.2   Operators as Eigenfunction/Eigenvalue Weighting

Another way to think of a linear operation on a given function is this: the operator mentally decomposes the given function into a superposition (weighted sum) of the operator's own eigenfunctions, then multiplies each eigenfunction component by its eigenvalue, and then sums up the result, i.e.,

$$\text{Given:}\quad \phi_n(x) \equiv \text{eigenfunctions of } \hat{o};\quad o_n \equiv \text{eigenvalue for } \phi_n(x),$$

$$\text{and}\quad \psi(x) = \sum_{n=1}^{\infty} c_n \phi_n(x), \quad \text{then} \quad \hat{o}\psi(x) = \sum_{n=1}^{\infty} c_n o_n \phi_n(x).$$

In other words, the operator "factors in" its eigenvalues as additional weights to the eigenfunction components of $\psi(x)$. We will examine more details of operator behavior in discrete bases in the chapter on matrix mechanics.

1.10.3.3   Computing Measurable Results, Local Density

Example: position   As an example of extracting a measurable attribute of a particle from a wave function, let us consider the position operator, $\hat{x}$. From our earlier discussion, we know that:

$$\hat{x}\psi(x) = x_{local}(x)\psi(x) = x\psi(x).$$

We use this to compute the *overall* average value of "x", the particle's position, from its wave function $\psi(x)$:

$$\langle \hat{x} \rangle = \int \psi^*(x)\, x\, \psi(x)\, dx$$

For example, in Fig. 1.14 (*left*), we had $\psi(x) = 1/2$ between $[0, 4]$ and 0 elsewhere. The average value of the particle's position is:

$$\langle \hat{x} \rangle = \int_{-\infty}^{\infty} \psi^*(x)\, x\, \psi(x)\, dx = \int_{0}^{4} \frac{1}{2} x \frac{1}{2}\, dx = \frac{1}{8} x^2 \Big|_{0}^{4} = 2$$

which we can also see by inspection.

You can think of the "average value" equation, $\langle \hat{o} \rangle = \int \psi^*(x)\hat{o}\,\psi(x)\,dx$, as a direct statistical computation of an average value from a PDF, where $\text{pdf}(x) = |\psi(x)|^2 = \psi^*(x)\psi(x)$. Then statisticians would write:

$$\langle \hat{o} \rangle = \int o_{local}(x)\, \text{pdf}(x)\, dx = \int o_{local}(x)\, \psi^*(x)\psi(x)\, dx$$

So why do we write "$\hat{o}$" in between ψ^* and ψ? Because the operator notation of QM is that an operator acts on the function to its right (or sometimes in a different way, it acts to the left). Therefore,

$$\int \psi^*(x)\left(\frac{\hbar}{i}\frac{\partial}{\partial x}\right)\psi(x)\,dx \text{ is quite different from } \int\left(\frac{\hbar}{i}\frac{\partial}{\partial x}\right)(\psi^*(x)\psi(x))\,dx.$$

In QM, we must write the operator between ψ^* and ψ, so that it acts on ψ:

$$\langle\hat{o}\rangle = \int o_{local}(x)\,\text{pdf}(x)\,dx = \int o_{local}(x)\,\psi^*(x)\psi(x)\,dx$$

$$= \int \psi^*(x)\,\underbrace{o_{local}(x)\,\psi(x)}_{\hat{o}\psi(x)}\,dx = \int \psi^*(x)\,\hat{o}\psi(x)\,dx. \tag{1.5}$$

Example: momentum in position basis An operator such as momentum includes a derivative: $\hat{p} = (\hbar/i)\partial/\partial x$. [Therefore, the value of the momentum at any point in space depends not only on the value of $\psi(x)$ at x, but on $\psi(x)$ in an infinitesimal neighborhood *near x*.] How can we understand this operator for momentum? Recall that one of the starting points of wave mechanics was that particles of definite momentum have a definite spatial frequency (wave-number), $k = p/\hbar$. This implies the eigenfunctions of momentum are (ignoring normalization):

$$\phi_p(x) = e^{ipx/\hbar} \quad \text{for a definite momentum, } p.$$

This particle has the same momentum, p, everywhere, so $p_{local}(x) = p$ (a constant). Then by the definition of operators, the momentum operator must take $\phi_p(x)$ to $p\phi_p(x)$. That is:

$$\hat{p}e^{ipx/\hbar} = p_{local}(x)\psi(x) = pe^{ipx/\hbar}.$$

What linear operator brings down the p from the exponent, as a multiplier in front? The derivative w.r.t. x:

$$\frac{\partial}{\partial x}e^{ipx/\hbar} = \frac{i}{\hbar}pe^{ipx/\hbar}.$$

This is almost what we need. We fix the prefactor by simply multiplying by $\hbar/i$

$$\frac{\hbar}{i}\frac{\partial}{\partial x}e^{ipx/\hbar} = pe^{ipx/\hbar} \quad \Rightarrow \quad \hat{p} = \frac{\hbar}{i}\frac{\partial}{\partial x}.$$

We derived the momentum operator from the definition of operators, and the empirical fact that particles of definite momentum have definite wave-number k.

Then for a general state $\psi(x)$, at each point x there is a local value of momentum:
$$p_{local}(x) = \frac{\hbar}{i}\frac{\psi'(x)}{\psi(x)}, \text{ provided } \psi(x)\neq 0. \text{ We can then compute the average value of } p$$

for the particle in the standard way:

$$\langle \hat{p} \rangle = \int_{-\infty}^{\infty} p_{local}(x)\, \mathrm{pdf}(x)\, dx = \int_{-\infty}^{\infty} \psi^*(x) p_{local}(x)\psi(x)\, dx$$

$$= \int_{-\infty}^{\infty} \psi^*(x)\left(\frac{\hbar}{i}\frac{d}{dx}\right)\psi(x)\, dx = \int_{-\infty}^{\infty} \psi^*(x)\,\hat{p}\psi(x)\, dx.$$

It is instructive to see how this works when $\psi(x)$ is a superposition of two momenta, p_1 and p_2:

$$\psi(x) = A\exp(ip_1 x/\hbar) + B\exp(ip_2 x/\hbar), \quad A,B \text{ complex}, \ |A|^2 + |B|^2 = 1 \ \Rightarrow$$

$$p_{local}(x) = \frac{\hbar}{i}\frac{\psi'}{\psi} = \frac{A\exp(ip_1 x/\hbar)p_1 + B\exp(ip_2 x/\hbar)p_2}{A\exp(ip_1 x/\hbar) + B\exp(ip_2 x/\hbar)}.$$

This is a weighted average of p_1 and p_2, where the weights depend on both the coefficients A and B, and on the values of the momentum eigenfunctions at x. The weights are complex, and given explicitly by:

$$p_{local}(x) = \frac{w_1 p_1 + w_2 p_2}{w_1 + w_2} \quad where \quad w_1 = A\exp(ip_1 x/\hbar), \quad and$$

$$w_2 = B\exp(ip_2 x/\hbar).$$

Momentum squared We noted earlier that at an extremum of a *real*-valued $\psi(x)$, $\partial\psi/\partial x = 0$ and the local momentum $p_{local}(x) = 0$. This can be thought of as a weighted average of two different momenta: one at $+p$ and one at $-p$. But when considering momentum-squared at that same extremum, $\hat{p}^2\psi(x) = -\hbar^2\dfrac{\partial^2\psi(x)}{\partial x^2}$, the contributions from $(+p)^2$ and $(-p)^2$ add; they do not cancel. In other words:

$$\left(p_{local}(x)\right)^2 \neq p^2{}_{local}(x) \quad where \quad p^2{}_{local}(x) \equiv \frac{\hat{p}^2\psi(x)}{\psi(x)} = \frac{-\hbar^2\psi''(x)}{\psi(x)}$$

This implies that globally, $\langle p\rangle^2 \neq \langle p^2\rangle$, as is well-known. A similar situation exists with angular momentum $\left(\hat{L}^2\right)$, which we will examine more closely later.

Local density We have noted that the overall average of an operator is an integral over all space of the local value, weighted by the particle density (PDF) at each point:

$$\langle \hat{o} \rangle = \int_{\infty} o_{local}(x)\,\psi^*(x)\psi(x)\, dx = \int_{\infty} \psi^*(x)\,\hat{o}\psi(x)\, dx$$

This suggests we define a **local density** of $\hat{o}$ as:

$$\rho_o(x) = o_{local}(x)\,\psi^*(x)\psi(x) = \psi^*(x)\,\hat{o}\psi(x), \quad e.g., \quad \rho_p = \psi^*(x)\hat{p}\psi(x).$$

A local density is analogous to physical densities such as mass density. Then the global average is just the integral over all space of the local density:

$$\langle \hat{o} \rangle = \int_{\infty} \rho_o(x)\, dx = \int_{\infty} \psi^*(x)\, \hat{o}\psi(x)\, dx, \quad \text{e.g.,} \quad \langle \hat{p} \rangle = \int_{\infty} \rho_p(x)\, dx = \int_{\infty} \psi^*(x)\, \hat{p}\psi(x)\, dx$$

Conceptually, the local density can be thought of, in most single-particle cases, as describing the actual distribution in space of a physical quantity, such as momentum density. Also:

> Local density has an advantage over the local value, $o_{local}(x)$, because it is well defined, even when $\psi(x)=0$.

When $\psi(x)=0$, the local density is zero, regardless of the local value at that point, since (roughly) no part of the particle exists at that point.

Even nonwave properties, such as a particle's intrinsic spin, have local densities given by the wave-function. For example, for a spin-up electron, its spin angular momentum density is [from 1, 14–92 p362m]:

$$\rho_{spin}(x) = \frac{\hbar}{2}\psi^*(x)\psi(x).$$

1.10.3.4 New States From Old States: Time Evolution and More

An example of an operator producing a new state from an old state is the time-evolution operator. This section requires an understanding of composing an arbitrary state from energy eigenstates.

The time evolution operator takes a wave-function at some time t_0, and produces what the wave-function will be at a later time t_1. Therefore, the time-evolution operator is really a family of operators, parameterized by both t_0 and t_1. In the simple case where the Hamiltonian is not explicitly time-dependent, the time evolution operator depends only on the time difference, $\Delta t \equiv t_1 - t_0$. Then, for arbitrary t:

$$\psi(t+\Delta t, x) = \hat{U}(\Delta t)\psi(t,x) \quad where \quad \hat{U}(\Delta t) \equiv \text{time evolution operator,}$$

$$\hat{H} \text{ independent of time.}$$

For this simple case of time-independent Hamiltonian, we can derive the time-evolution operator from the fact that energy eigenstates $u_n(t, x)$, with energy E_n, evolve in time with a particularly simple form. The spatial form of energy eigenstates does *not* change with time. By definition, energy eigenstates satisfy the time-independent Schrödinger equation $\hat{H}u_n(t,x) = E_n u_n(t,x)$. Then for the full Schrödinger equation, we have:

$$i\hbar\frac{\partial}{\partial t}u_n(t,x) = \hat{H}u_n(t,x) = E_n u_n(t,x) \quad \Rightarrow \quad \frac{\partial}{\partial t}u_n(t,x) = -\frac{iE_n}{\hbar}u_n(t,x).$$

This is of the form $y'(t) = ky(t)$, with solution (by inspection) $y = y(0)e^{kt}$. Also, by a simple time-shift, $y(t + \Delta t) = y(t)e^{k\Delta t}$. Applying this form to our equation for $u_n(t, x)$, we get the time evolution for an energy eigenstate:

$$u_n(t + \Delta t, x) = e^{-iE_n \Delta t / \hbar} u_n(t, x) \quad \text{time evolution of an energy eigenstate} \quad (1.6)$$

Time evolution simply multiplies the wave-function by a complex phase, $\exp(-iE_n t/\hbar)$. Since phase does not affect any quantum property, all the properties of such a system are independent of time; such a state is called **stationary**. Note that stationary does not imply "static:" a stationary state can be moving (such as an electron orbiting a nucleus).

Decomposing an arbitrary state $\psi(t, x)$ into energy eigenstates yields the general time evolution. At an arbitrary time t:

$$\psi(t, x) = \sum_n c_n u_n(t, x) \quad \Rightarrow \quad \psi(t + \Delta t, x) = \hat{U}(\Delta t)\psi(t, x)$$

$$= \sum_n e^{-iE_n \Delta t / \hbar} c_n u_n(t, x). \quad (1.7)$$

This gives the explicit form of time evolution of a state from its energy eigenstate components. We now show how to write this as an operator, independent of components. Recall that the exponential of a linear operator is another linear operator, and is defined by the power series expansion of the exponential. For example:

$$e^{\hat{a}} = \hat{1} + \hat{a} + \frac{\hat{a}^2}{2!} + \frac{\hat{a}^3}{3!} + \ldots \quad \Rightarrow$$

$$e^{-i\hat{H}\Delta t/\hbar} = \hat{1} + \frac{-i\Delta t \hat{H}}{\hbar} + \frac{1}{2!}\left(\frac{-i\Delta t \hat{H}}{\hbar}\right)^2 + \frac{1}{3!}\left(\frac{-i\Delta t \hat{H}}{\hbar}\right)^3 + \ldots.$$

Thus, the exponential of the Hamiltonian operator is a power series of Hamiltonian operators. When acting on an energy eigenstate $u_n(t, x)$, each appearance of $\hat{H}$ gets replaced by the energy E_n. Therefore,

$$e^{-i\hat{H}\Delta t/\hbar} u_n(t, x) = \left[\hat{1} + \frac{-i\Delta t E_n}{\hbar} + \frac{1}{2!}\left(\frac{-i\Delta t E_n}{\hbar}\right)^2 + \frac{1}{3!}\left(\frac{-i\Delta t E_n}{\hbar}\right)^3 + \ldots\right] u_n(t, x)$$

$$= e^{-iE_n \Delta t / \hbar} u_n(t, x)$$

Thus, we see that for an energy eigenstate, the exponential of the Hamiltonian (times some factors) gives the simple form for time evolution of an energy eigenstate shown in Eq. (1.6). But again, any state can be written as a sum of energy eigenstates, and $e^{-i\hat{H}\Delta t/\hbar}$ is linear, so $e^{-i\hat{H}\Delta t/\hbar}$ gives the time evolution of an *arbitrary* state shown in Eq. (1.7):

$$e^{-i\hat{H}\Delta t/\hbar}\psi(t,x) = \sum_n e^{-i\hat{H}\Delta t/\hbar} c_n u_n(t,x) = \sum_n e^{-iE_n\Delta t/\hbar} c_n u_n(t,x) = \psi(t+\Delta t, x).$$

$$= \hat{U}(\Delta t)\psi(t,x).$$

Therefore, the general time evolution operator is:

$$\hat{U}(\Delta t) = e^{-i\hat{H}\Delta t/\hbar} \quad \text{time evolution operator for time-independent hamiltonian}$$

This is a pure operator equation, and is independent of any representation of the wave-function.

We note in passing that, in general, the future value of ψ at some point x_0, $\psi(t, x_0)$, depends on *all* the values of $\psi(0, x)$ for all x. Therefore:

> The time-evolution operator, $\hat{U}(\Delta t)$, is a nonlocal operator.

More on nonlocal operators later.

In the general case, where the Hamiltonian explicitly depends on time, we must keep t_0 and t_1:

$$\psi(t_1, x) = \hat{U}(t_1, t_0)\psi(t_0, x)$$

where $\hat{U}(t_1, t_0) \equiv$ time evolution operator, $\hat{H}$ depends on time.

Time evolution for a time-dependent Hamiltonian is much more complicated, and we do not address that here.

A second example of "new states from old states" is rotation operators, e.g., in 3D, $\psi_{rotated}(\mathbf{r}) = \hat{R}(\theta, \phi)\psi(\mathbf{r})$. Here again, the rotation "operator" is really a family of operators, parameterized by the rotation angles, θ and ϕ. We return to rotations later when considering generators.

1.10.3.5   Simplifying Calculations

A third use for operators is to aid and simplify calculations. For example, raising and lowering operators($\hat{a}^\dagger$, $\hat{a}$, $\hat{J}_+$, $\hat{J}_-$) do not represent observables, and have nothing to do with physically changing a state, or adding/removing energy (or angular momentum). They are used for computing matrix elements, for analysis, proving relationships and theorems, perturbation theory, etc.

The harmonic oscillator raising and lower operators, $\hat{a}$ and $\hat{a}^\dagger$ (often written without hats), are usually defined as acting on entire states:

$$\hat{a}|n\rangle = \sqrt{n}|n-1\rangle \qquad \hat{a}^\dagger|n\rangle = \sqrt{n+1}|n+1\rangle.$$

From this definition, you might think that they are nonlocal operators (since they seem to depend on the entire state). However, it turns out that these definitions are achieved with local operators. We can see this from the formulas for $\hat{a}$ and $\hat{a}^\dagger$ in terms of $\hat{x}$ and $\hat{p}$, which are both local operators:

$$\hat{a} = \sqrt{\frac{m\omega}{2\hbar}}\hat{x} + \frac{i}{\sqrt{2m\hbar\omega}}\hat{p}, \quad \hat{a}^\dagger = \sqrt{\frac{m\omega}{2\hbar}}\hat{x} - \frac{i}{\sqrt{2m\hbar\omega}}\hat{p} \Rightarrow \hat{x} = \sqrt{\frac{\hbar}{2m\omega}}\left(\hat{a}+\hat{a}^\dagger\right).$$

These definitions and formulas allow us to evaluate common inner products (aka "matrix elements") without any integration:

$$\langle 2|\hat{x}^2|0\rangle = \langle 2|\left[\sqrt{\frac{\hbar}{2m\omega}}(a+a^\dagger)\right]^2|0\rangle = \frac{\hbar}{2m\omega}\langle 2|\left(a^\dagger\right)^2|0\rangle = \frac{\hbar\sqrt{2}}{2m\omega}\langle 2|2\rangle = \frac{1}{\sqrt{2}}\frac{\hbar}{m\omega}.$$

This would be substantially more work to evaluate by integrating products of Hermite polynomials and exponentials.

1.10.3.6 Nonlocal Operators

The time evolution operator, $\hat{U}(t_1, t_0)$ (described previously), takes a quantum state (a function of space) into another quantum state (another function of space). Recall that for time-independent Hamiltonians:

$$\hat{U}(\Delta t) = e^{-i\hat{H}\Delta t/\hbar} = \hat{1} + \frac{-i\Delta t\hat{H}}{\hbar} + \frac{1}{2!}\left(\frac{-i\Delta t\hat{H}}{\hbar}\right)^2$$

$$+ \frac{1}{3!}\left(\frac{-i\Delta t\hat{H}}{\hbar}\right)^3 + \dots \qquad \text{(time-independent } \hat{H}).$$

For a finite time interval, the new value of the wave-function at a given point x depends not only on the infinitesimal neighborhood of x, but on values of ψ far away from x. Therefore, the time evolution operator for a finite time interval is a *non*local operator. It may be surprising that the time evolution operator is nonlocal, since it is written as a Taylor series sum of (local) Hamiltonian operators. This is a subtle issue that requires careful examination, but briefly, the Hamiltonian requires an infinitesimal *neighborhood* around the point x to evaluate the momentum. Then, integrating over an infinite number of such infinitesimal neighborhoods makes the time evolved state at any point x depend on distant points of the original state. (The space translation operator has a similar characteristic.) [This nonlocality is evident in the more advanced concept of "propagators." Propagators are the time evolution of a localized particle (delta-function) in space, and for all finite times, they extend to infinity.] We do not consider this nonlocality of operators further.

1.10.4 *Commutators*

The commutator of two operators is also an operator. Commutators are compound operators built from other operators. A commutator is written with square brackets, and defined as the difference between the operators acting in both orders:

$$\left[\hat{A}, \hat{B}\right] = \hat{A}\hat{B} - \hat{B}\hat{A}.$$

We usually think of operators as acting to the right, so the first term in the previous equation has $\hat{B}$ acting first.

Commutators are often a convenient way to specify operator identities:

$$\left[\hat{x}, \hat{p}\right] = i\hbar \quad \leftrightarrow \quad \hat{x}\hat{p} - \hat{p}\hat{x} = i\hbar \quad \leftrightarrow \quad \hat{x}\hat{p} = i\hbar + \hat{p}\hat{x}.$$

As with all operators, their algebraic forms are generally *dependent* on the representation basis (bases are described in some detail later). The only commutators that are *independent* of representation are constants, e.g., $\left[\hat{x}, \hat{p}\right] = i\hbar$. It is mildly interesting to evaluate this explicitly in both the x (position) and p (momentum) bases:

$$x \text{ basis: } \left[\hat{x}, \hat{p}\right] = x\frac{\hbar}{i}\frac{\partial}{\partial x} - \frac{\hbar}{i}\frac{\partial}{\partial x}x = \frac{\hbar}{i}\left(x\frac{\partial}{\partial x} - \mathbf{1}_{op} - x\frac{\partial}{\partial x}\right) = i\hbar,$$

where $\quad \mathbf{1}_{op} \equiv$ the identity operator .

$$p \text{ basis: } \left[\hat{x}, \hat{p}\right] = i\hbar\frac{\partial}{\partial p}p - p\left(i\hbar\frac{\partial}{\partial p}\right) = i\hbar\left(\mathbf{1}_{op} + p\frac{\partial}{\partial p} - p\frac{\partial}{\partial p}\right) = i\hbar.$$

We have noted that a commutator is an operator, so is there a physical meaning to the function $\left[\hat{x}, \hat{p}\right]\psi(x)$? Let us see: the first term of the commutator is $\hat{x}\hat{p}$. Recall that $\hat{p}$ multiplies a function by the (local) momentum at each point, and produces a function of space which is *not* a quantum state. Similarly, $\hat{x}$ multiplies a function by the position "x" everywhere, and also produces a function of space which is *not* a state. What does it mean then for $\hat{x}$ to act on $\hat{p}\psi(x)$, since $\hat{p}\psi(x)$ is not a state? In other words, what is the physical meaning of $\hat{x}\hat{p}$? Answer: nothing. This kind of commutator, taken as an operator, has no direct physical interpretation. It is simply an algebraic relationship that is used to derive extremely important results (such as the uncertainty principle).

Many commutators, taken as operators, have no direct physical interpretation; they are simply algebraic relationships that are used to help derive physical results.

A second use for commutators relates to simultaneous eigenstates. The uncertainty principle says that some pairs of dynamic quantities cannot have both definite values. It is important to know which physical properties a quantum system can have with simultaneously definite values. A set of quantum numbers, one for each possible definite value, then fully specifies the state of the system, and therefore *all* its properties, definite or not. When two operators **commute**, it means they produce the same result acting in either order:

$$\hat{a}\hat{b} = \hat{b}\hat{a} \quad \Rightarrow \quad \left[\hat{a},\hat{b}\right] = \mathbf{0}_{op} \quad \text{(the zero operator)}.$$

Note that the "zero" on the right-hand side is an operator, not exactly a number. It means to multiply the function on which it acts by the number 0, which always returns the zero-valued function of space. Similarly, in a general vector space (say, spin states), it means to multiply the ket(i.e., vector) on which it acts by the number 0, which always returns the zero-vector, $\mathbf{0}_v$.

When two operators commute, they have a common set of eigenvectors, and this set is complete: it forms a basis for constructing any vector in the vector space.

Units of commutators The units of a composition of operators is the product of the units of the constituent operators. Therefore, the units of each term of a commutator are the same: the product of the commutator's constituent operators. For example, the units of $\left[\hat{x},\hat{p}\right]$ are:

$$\left[\hat{x},\hat{p}\right] = \left[\hat{x}\hat{p}\right] = \text{m kg - m/s} = \text{kg - m}^2/\text{s}.$$

Commutators are linear Commutators are linear in both arguments (where the arguments are themselves operators). Recall that linearity implies commuting with scalar multiplication and distributivity over addition:

$$\left[k\hat{a},\hat{b}\right] \equiv k\hat{a}\hat{b} - \hat{b}k\hat{a} = k\left[\hat{a},\hat{b}\right], \quad \left[\hat{a},j\hat{b}\right] \equiv \hat{a}j\hat{b} - j\hat{b}\hat{a} = j\left[\hat{a},\hat{b}\right] \text{ (scalar multiplication)},$$

$$\left[\hat{a}+\hat{c},\hat{b}\right] \equiv (\hat{a}+\hat{c})\hat{b} - \hat{b}(\hat{a}+\hat{c}) = \hat{a}\hat{b} - \hat{b}\hat{a} + \hat{c}\hat{b} - \hat{b}\hat{c}$$

$$= \left[\hat{a},\hat{b}\right] + \left[\hat{c},\hat{b}\right]$$

$$\left[\hat{a},\hat{b}+\hat{d}\right] \equiv \hat{a}(\hat{b}+\hat{d}) - (\hat{b}+\hat{d})\hat{a} = \hat{a}\hat{b} - \hat{b}\hat{a} + \hat{a}\hat{d} - \hat{d}\hat{a} \qquad \text{(distributes). (1.8)}$$

$$= \left[\hat{a},\hat{b}\right] + \left[\hat{a},\hat{d}\right]$$

There is also a composition rule (loosely, a "product rule") for commutators:

$$\left[\hat{a}\hat{c},\hat{b}\right] = \hat{a}\hat{c}\hat{b} - \hat{b}\hat{a}\hat{c} = \hat{a}\hat{c}\hat{b} \underbrace{- \hat{a}\hat{b}\hat{c} + \hat{a}\hat{b}\hat{c}}_{=0} - \hat{b}\hat{a}\hat{c} = \hat{a}\left[\hat{c},\hat{b}\right] + \left[\hat{a},\hat{b}\right]\hat{c}, \quad \text{and similarly,}$$

$$\left[\hat{a},\hat{b}\hat{d}\right] = \hat{b}\left[\hat{a},\hat{d}\right] + \left[\hat{a},\hat{b}\right]\hat{d}.$$

You can remember the commutator composition rule as the sum of two terms, which you create as follows: "pull the left factor out to the left, and pull the right factor out to the right" (Fig. 1.15).

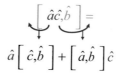

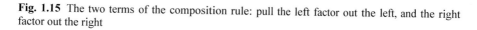

Fig. 1.15 The two terms of the composition rule: pull the left factor out the left, and the right factor out the right

1.10.5   External Fields

A quantum system may include "external fields," e.g., an externally applied magnetic fields whose value at all points is a given, and *not* part of the state vector $|\psi\rangle$. Such fields are good approximations to the more precise quantized EM field. In such a case, the value of the field is embedded in the operators we define for observables of the system. For example, the energy operator of an electron in a given B-field would be:

$$\hat{H}_{magnetic} = -\boldsymbol{\mu}\cdot\mathbf{B}(\mathbf{r}) = -\left(g_e\mu_B\frac{\hat{\mathbf{s}}}{\hbar}\right)\cdot\mathbf{B}(\mathbf{r})$$

$$= -\left(-2\mu_B\frac{1}{2}\boldsymbol{\sigma}\right)\cdot\mathbf{B}(\mathbf{r}) = +\mu_B\boldsymbol{\sigma}\cdot\mathbf{B}(\mathbf{r}), \quad \text{using} \quad g_e \approx -2.$$

Similarly, the (canonical) momentum operator would include the given potential function $\mathbf{A}(\mathbf{r})$.

Including a "given" field in the operators of the system is in contrast to descriptions of a system where the field itself is quantized, and its value is part of the system state. Quantizing such fields involves QFT. Then, a function such as $\mathbf{B}(\mathbf{r})$ is replaced by the field operator, $\hat{\mathbf{B}}$, which supplies the value of the field from the quantized field state vector.

1.11   From Schrödinger to Newton: Deriving $F=ma$

We now show the quantum mechanical statement of Newton's second law, $dp=F\,dt$ (which is still true relativistically), and which is (nonrelativistically) equivalent to $F=ma$. Consider a 1D particle with definite momentum $p(t_0)=p_0$, at some time t_0. Its wave-function is $\psi(t_0, x)=\exp[(i/\hbar)p_0x]$ (dropping the normalization factor, for simplicity) (Fig. 1.16).

Now apply a force $F=-dV(x)/dx$ to it for a time dt. From Newton's second law, we expect that its new momentum and wave-function are:

$$p(t_0 + dt) = p_0 + F\,dt \quad \Rightarrow \quad \psi(t_0 + dt, x) = \exp\left[(i/\hbar)(p_0 + F\,dt)x\right]. \quad (1.9)$$

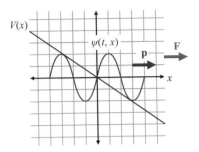

Fig. 1.16 $\psi(t, x)$ is a 1D wave-function moving right. The force is also to the right

We now derive that result from quantum dynamics. Recall that the Schrödinger equation implies the time evolution operator, which tells how a wave-function evolves in time:

$$\psi(t_0 + dt) = \exp\left[-(i/\hbar)\hat{H}\,dt\right]\psi(t_0)$$

$$= \exp\left[-(i/\hbar)\left(\frac{p_0^{\,2}}{2m} + V(x)\right)dt\right]\exp\left[(i/\hbar)p_0 x\right]$$

where $p_0 \equiv$ initial momentum.

The force is constant everywhere along the wave-function, so $V(x) = -Fx$. Rearranging the exponentials, and replacing $V(x)$ with $-Fx$, we get (to first order in dt):

$$\psi(t_0 + dt) = \exp\left[-(i/\hbar)\frac{p_0^{\,2}}{2m}dt\right]\exp\left[+(i/\hbar)Fx\,dt\right]\exp\left[(i/\hbar)p_0 x\right]$$

$$= \exp\left[-(i/\hbar)\frac{p_0^{\,2}}{2m}dt\right]\exp\left[(i/\hbar)(p_0 x + Fx\,dt)\right]$$

$$= \exp\left[-(i/\hbar)\frac{p_0^{\,2}}{2m}dt\right]\exp\left[(i/\hbar)(p_0 + F\,dt)x\right].$$

This is of the form expected in Eq. (1.9). The first factor is the usual complex rotation in time according to the energy, and does not affect the observable properties of the particle. The second factor is the wave function throughout space of momentum $(p_0 + F\,dt)$. Thus, QM has proven the impulse law, $dp = F\,dt$. Then nonrelativistically, as in classical mechanics:

$$F = \frac{dp}{dt} = m\frac{dv}{dt} = ma.$$

Thus, Schrödinger's equation, applied to a momentum eigenstate, reproduces Newton's second law. Isn't it amazing that Newton could find this 300 years before QM was discovered?

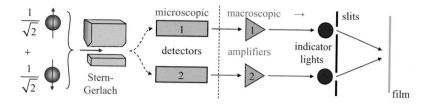

Fig. 1.17 A typical measurement process: is the spin up or down?

1.12 Measurement and Loss of Coherence

We now discuss the important phenomena of measurement and observation. In theoretical QM, we usually focus on perfect systems, and pure states. We frequently say that a measurement "collapses" the quantum state vector to one agreeing with the measurement, and this is often a useful simplification of the measurement process. However, in practice, the measurement process is more complicated than that because most measuring equipments and all observers are macroscopic. The "decohered" state is the norm; you must work hard to achieve even an approximately pure entangled state. We show here that elementary QM can explain some of the features of real measurements; however, the full explanation of decoherence is beyond our scope. (The term "decoherence" has a specific meaning: the process of a system becoming entangled with its environment in irreversible ways, resulting in the loss of a consistent phase relationship between components of the system state. We therefore use the more general term "loss of coherence" for both decoherence and other processes).

Most *macroscopic* measurements do not show quantum interference. Why not? One reason is that macroscopic bodies suffer unknowable and unrepeatable energy interactions, i.e., they gain or lose an unknowable amount of energy due to uncontrollable interactions with their environments. In other words, they are subject to simple "noise." This results in the loss of a consistent phase relationship between components of a superposition state. We discuss below how such a loss of consistent phase leads to classical probabilities (e.g., [1], pp. 26–27).

Walk-Through of a Real Measurement: Let us walk through a plausible measurement, and consider the elementary quantum mechanics involved. The system of Fig. 1.17 is a macroscopic version of Fig. 1.1, the quantum experiment that demonstrated the need for new concepts of measurement in quantum mechanics.

Suppose we start with a particle which can be in either of two states, |s1> or |s2>, such as polarization (horizontal or vertical), or spin (up or down). A general particle state is then:

$$|\psi\rangle = a|s_1\rangle + b|s_2\rangle \quad where \quad a, b \equiv \text{complex coefficients}, |s_1\rangle, |s_2\rangle \equiv \text{basis states}.$$

This is called a coherent superposition, because a and b have definite phases. (This is in contrast to a mixed state or incoherent mixture, where a and b have unknown phases.) All that is required for loss of coherence is for the relative phases of a and b to become unknown. For simplicity, we take $|s_1\rangle$ and $|s_2\rangle$ to be energy eigenstates, and the particle is spread throughout our measurement system.

According to the Schrödinger equation, every state time-evolves with a complex phase, $\alpha(t)$, determined by its energy:

$$\alpha(t) = \alpha(0) + \frac{E}{\hbar}t \quad \text{(phase accumulation of a system according to its energy).}$$

Then our 2-state system time evolves according to:

$$|\psi(t)\rangle = \exp(-iE_1 t/\hbar)|s_1\rangle + \exp(-iE_2 t/\hbar)|s_2\rangle.$$

Since the energies E_1 and E_2 are quantized, the complex phases multiplying $|s_1\rangle$ and $|s_2\rangle$ maintain a precise (aka coherent) relationship, though the relative phase varies with time.

When we measure the particle state, the state of the measuring device becomes entangled with the measured particle. Let $|M_1\rangle$ and $|M_2\rangle$ be states of the whole measuring system in which either detector 1 or detector 2 detected the particle. If we look directly at the indicator lights, we will observe only state 1 or state 2, but never both. This means that $|M_1\rangle$ and $|M_2\rangle$ are orthogonal. As the measuring system first detects the particle, the combined state of the particle/measuring-device starts out as a coherent superposition:

$$|\Psi\rangle = c|M_1\rangle|s_1\rangle + d|M_2\rangle|s_2\rangle \quad where \quad c,d \equiv \text{complex coefficients.}$$

The combined system time evolves according to its new energies:

$$|\Psi(t)\rangle = \exp[-i(E_1 + E_{M1})t/\hbar]\,|M_1\rangle|s_1\rangle + \exp[-i(E_2 + E_{M2})t/\hbar]|M_2\rangle|s_2\rangle.$$

If the energies of the two measuring device states fluctuate even a tiny bit, the two components of the superposition will rapidly wander into an unknown phase relation. They will lose coherence.

> *Every* macroscopic system suffers unrepeatable and unknowable energy fluctuations due to its environment.

We estimate a typical coherence loss rate shortly.

Let us examine the effects of various kinds of energy transfers between a system and its environment. In our two-path experiment, the interference pattern is built up over many trials, by recording detections on film. Now, suppose one path suffers an energy transfer to/from its environment before recombining and interfering. There are four possibilities:

1. The energy transfer is knowable and repeatable. Then one can predict and see an interference pattern in the usual way.
2. The energy transfer is unknowable, but repeatable. Then we can record an interference pattern, and from that, determine the relative phases of the two paths (mod 2π), and therefore the relative energies (mod $2\pi\hbar/t$) from $\Delta\alpha = \omega t = (\Delta E / \hbar)t$.
3. The energy transfer is knowable for each trial, but not repeatable. Essentially, each trial has its own position for the interference pattern. One can then divide the detection region into intervals of probability calculated for each trial, and then show consistency with QM predictions, but contrary to classical probability
4. The energy transfer is unknowable and unrepeatable. Then there will be no interference pattern, and repeated trials do not allow us to measure any quantum effects, since the phase is unknown on each trial. Therefore, the measurements are equivalent to classical probabilities: it is as if a single path was chosen randomly, and we simply do not know which path it was.

This fourth condition, of unknowable and unrepeatable energy transfer, causes loss of coherence and the randomization of phase of components of a superposition. Loss of coherence makes measurements look like the system behaves according to classical probabilities, with no "wave" effects. Loss of coherence destroys the interference pattern when we try to measure through "which slit" a particle passes.

Full loss of coherence leads to classical probabilities.

Our example process leading to loss of coherence follows directly from the Schrödinger equation and unknown energy transfers. There is no need to invoke any "spooky" quantum effects.

Note that although accounting for loss of coherence, quantum theory *still* requires the axiom of collapse of the wave-function upon observation. When a particle's wave splits, then passes through both detector 1 and detector 2, and then loses coherence because of entanglement with a macroscopic measuring device, the system is still left in a superposition of both slits:

$$\left|\Psi\left(t_{after}\right)\right\rangle = f\left|M_1\right\rangle\left|s_1\right\rangle + g\left|M_2\right\rangle\left|s_2\right\rangle;$$

we just do not know f or g. We cannot generate an interference pattern from multiple trials, because each trial has a different phase relation between f and g, putting the peaks and valleys of any hoped-for interference pattern in a random place on each trial. These shifts average over many trials to a uniform distribution. Nonetheless, each trial evolves in time by the Schrödinger equation, which still leaves the system in a superposition. Once we "see" the result, however, the unobserved component of the wave-function disappears, i.e., the wave-function **collapses**.

Collapse of the wave-function is outside the scope of the Schrödinger equation, but *within* the scope of QM, because collapse is a *part* of QM theory. It is one of our axioms. Some references confuse this issue: they try to avoid assuming such a collapse as an axiom, but cannot derive it from other axioms. From this, they conclude that QM is "incomplete." In fact, what they have shown is that the axiom of collapse completes QM.

Note that once the measuring system *fully* loses coherence, we could just as well say that the wave-function has then collapsed, because from then on the system follows classical probabilities (equivalent to a collapsed, but unknown, wave-function). However, we now show that a binary model of "collapse or not" cannot explain partial coherence.

Partial Coherence: What if we start with a microscopic system, such as that in Fig. 1.1, but replace our microscopic atoms with *mesoscopic* things—bigger than microscopic, but smaller than macroscopic? Mesoscopic things might be a few hundred atoms. These are big enough to lose coherence much faster than single atoms, but still slowly enough that *some amount* of interference is observed. However, the interference pattern is weaker: the troughs are not as low, and the peaks are not as high. A superposition leading to a *weak* interference pattern is said to be **partially coherent**. We describe partial coherence in more detail in Sect. 8.4. The simple model that the wave-function either collapsed or did not collapse, cannot describe the phenomenon of partial coherence.

The larger the mesoscopic system, the more uncontrollable interactions it has with its environment, the faster it loses coherence, and the less visible is any resulting interference pattern. We can estimate the time-scale of coherence loss from our example energy fluctuations as follows: a single $10\ \mu m$ infrared photon is often radiated at room temperature. It has an energy of $\sim 0.1\ eV = 1.6 \times 10^{-20}$ J. This corresponds to $\omega = E/\hbar \sim 2 \times 10^{14}$ rad/s. When the phase of the resulting system has shifted by an unknowable amount $> \sim 2\pi$, we can say that the system has completely lost coherence. At this ω, it takes $\sim 4 \times 10^{-14}$ s. In other words, thermal radiation of a single IR photon causes a complete loss of coherence in about 40 fs. In practice, other effects cause macroscopic systems to lose coherence in dramatically shorter times.

Summary: A measurement entangles a measuring device with the measured system. The entangled state of device and system time evolves according to the SE. Macroscopic devices lose coherence due to interactions with the environment. Lack of coherence prevents any interference pattern within the system. Therefore, measurement by a macroscopic device produces subsequent results that are classical, as if the system collapsed into a definite state upon measurement, but observers only "see" which state it is when they look at the measuring device. Any observation by a person is necessarily macroscopic, because people are big. Such an observation collapses the (incoherent) device/system/world state to that observed. Quantum interference can only be seen if it occurs *before* any entanglement with a macroscopic system (and therefore before any loss of coherence in the system).

The model of "collapse of the wave-function" is a binary concept: either the wave-function collapses or it does not. Such a model cannot account for the phenomenon of partial coherence. Loss of coherence is a continuous process, taking a fully coherent state through less and less partially coherent states and eventually to incoherent (aka "mixed") states. Continuous loss of coherence fully explains partial coherence and the varying visibility of interference patterns.

Some quantum effects such as the spectrum of atoms do not rely on interference, and are therefore macroscopically observable. In fact, measurement of such effects led to the development of QM.

1.13   Why Do Measurements Produce Eigenvalues?

This section is preparation for the next, which discusses wave-function collapse. QM is consistent in this way:

> If you make a measurement, and then quickly repeat the same measurement, you will get the same result (quickly, so the system does not "move" in between).

(Similarly, if you make a measurement, then wait and account for the system time evolution, and then make another measurement, your second result will be consistent with the first.) If you believe the postulates of QM, specifically that the wave function defines the spectrum (the set) of possible results, and their probabilities, then you must conclude that the wave function after a measurement must allow only a *single* result for a remeasurement (the result already measured).

Since the wave function before a measurement may be a superposition of possibilities, it must be that the act of measuring "collapses" the wave-function to a state that can produce only one result. Recall that a spatial wave-function (i.e., a wave-function of position in space) associates a local value for an observable with each point of the wave-function. Assume a wave function, $\psi(x)$, and an observable operator, $\hat{o}$. "$\hat{o}$" has some local value associated with each position in space; call it $o_{local}(x)$. After a measurement, the particle has a definite value of the observable, so it is reasonable that all points of the wave-function should contribute the same local value: that which was already measured. This means $o_{local}(x)$ is constant throughout space; call it o_{obs}. Then from the definition of local values, Eq. (1.4), we have:

$$o_{local}(x)\psi(x) = o_{obs}\,\psi(x) = \hat{o}\psi(x)$$

The last equality requires that the wave function after a precise measurement be an eigenfunction of the observed operator, and the value measured must be its eigenvalue.

This phenomenon of a measurement causing a "collapse of the wave function" into an eigenstate is directly related to the uncertainty principle. The collapse is an inevitable part of taking a measurement; it derives from the consistency of measuring the same observable twice. It has nothing to do with clumsy photons, or other "probing" particles. Since noncommuting observables rarely have simultaneous eigenstates, a single wave function usually cannot have definite values of

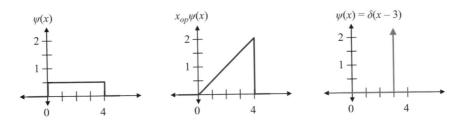

Fig. 1.18 (*Left*) A normalized wave-function, $\psi(x)$. (*Middle*) $\hat{x}\psi(x)$. (*Right*) $\psi(x)$ after measuring the particle to be at position $x=3$

noncommuting observables. The lack of a common eigenfunction means that a definite value of one observable forces some uncertainty in the other. (See later section on Generalized Uncertainty for more details.)

Observable Operators Do *Not* Produce States Resulting From Measurements

> The mathematical result of an observable operator acting on a state is very different from the state resulting from actually measuring that observable.

> Many people confuse the result of an observable operator on a state with the act of measuring that observable. These are very different things!

The action of an observable operator on a state produces a mathematical result, a function of space (a "ket" in Dirac notation), which is one step in calculating the statistics of possible measurements of that observable on a system in that state. Such a result is *not* a quantum state. As we saw earlier, $\hat{x}\psi(x)$ yields a function of space (*not* a state) which is part of calculating possible values of the position of a particle in state $\psi(x)$ and their probabilities. This is quite different from the resulting *state* after an actual measurement of the particle's position. Recall the example $\psi(x)=1/2$ between [0, 4] and 0 elsewhere (Fig. 1.18, *left*).

Figure 1.18, middle, is the function $\hat{x}\psi(x)$, which is a mathematical calculation. If we actually measure such a particle to be at position $x=3$, the resulting state of the particle is (essentially) a delta function at $x=3$ (Fig. 1.18, right):

$$\psi(x)=\delta(x-3).$$

(This is *not* square integrable, and so uses "delta-function normalization." More on this later.)

Note that the act of measurement is a *nonlinear* operation on the wave function; it cannot be represented by a linear operator acting on the wave function. Recall that the whole point of a linear operator is to produce a superposition of *results* based on the superposition that composes the given function (or ket). In contrast, the consequence of a measurement is to *choose one specific state* out of a superposition of eigenstates.

> A measurement eliminates a superposition, in favor of a more definite state. Therefore, a measurement is *not* a linear operation on the state; it is inherently nonlinear.

We must be careful in interpreting this statement, because the term "superposition" is relative: it depends on the basis in which we write our function. A state of definite energy is a superposition of position states. A state of definite position is a superposition of energy states. *Any* spatial function is always a superposition in some bases, and a single component in other bases. However, when we take a measurement, one basis is special: the eigenbasis of the operator whose value we are measuring. If we measure energy, then the energy basis is special. If we measure position, then the position basis is special. A better statement of the effect of a measurement is then:

A measurement of a property eliminates a superposition of eigenstates of that property. The resulting ("collapsed") state is then an eigenstate of the operator corresponding to the measurement.

We now describe some ways of thinking about this effect.

1.14   The Collapse of QM

Or, "Shut Up and Calculate." (Thanks to Andrew Cooksey for that sympathetic advice. This section is inspired by Sidney Coleman, Heinz R. Pagels, and all the other dedicated physicists trying to make QM seem sensible.)

What is *really* going on inside the collapse of the wave-function? We must be careful with the word "really:" science is the ability to predict future observations based on past ones. Science uses **theories** (detailed quantitative models) to make these predictions. Theories usually include a *conceptual* model to help us remember the theory, but theories *always* include a mathematical model to compute predictions.

> Any theory which computes predictions that agree with experiment is a "valid" theory.

Sometimes, more than one theory is valid. Because we are simple-minded, between two competing valid theories, we choose the simpler one. (This criterion is known as Occam's Razor.) We now compare two models, both of which agree with the following facts. These facts are verified countless times by experiment, and are undisputed by serious physicists:

1. If Alice takes a measurement, and now wants to predict future measurements, she *must* use, as her starting point, the quantum state which is consistent with her first measurement. No other quantum state will work for her. Calculations

starting from any state other than that demanded by her first measurement will be *wrong*. Absolutely, completely, no "ifs, ands, or buts" wrong.

2. If Alice measures an observable for which the system is known *not* to be in an eigenstate, then her answer is determined by probability, and Alice cannot predict for certain the result of any such single measurement.

3. Alice prepares a quantum system in state A. She allows it to time-evolve into a superposition of two states, B or C [perhaps, (B)=decayed and (C)=not-decayed], but she does not measure the system to determine which state it is in. She waits further, and finally measures it in state D. To predict the probability of starting in A and ultimately measuring D, Alice *must* use as the intermediate state a coherent superposition of B and C, i.e., she must use the complex amplitudes of states B and C, to predict the measurement D. She *cannot use* classical probabilities.

Fact #1 says that QM is self-consistent: no measurement will contradict another. It is why measurements always produce eigenvalues as their result (see "Why Do Measurements Produce Eigenvalues?" on page 42).

Fact #2 says that *from the point of view of an observer*, QM is inescapably probabilistic.

Fact #3 says that probability for an observer *only comes into play when that observer actually observes something*. Until then, there is deterministic time evolution of the quantum state, including fully complex-valued amplitudes, and therefore (critically) *the possibility of quantum interference*. (However, as noted earlier, large systems inevitably suffer from decoherence: uncontrollable interactions with their environments.)

What is the "explanation" of these facts? There is no single, definitive explanation. There are many competing, valid theories. Here are two:

Dr. Xavier E. Rox believes in predictability. There can be no collapse, no random events. Physics is deterministic, just like the old classical physicist, Dr. Diehard, says. Dr. Rox observes, "Diehard's only problem is that his math is wrong. Physics follows the Schrödinger equation." Of course, to be consistent with experiment, Dr. Rox must assume that at every instant, the quantum state of the entire universe, including himself(!), splits into a new superposition of all possible results. "Better complexity and confusion than uncertainty," he declares, much to the dismay of Werner Heisenberg. On Sunday, Dr. Rox goes to the Church of Duplicity, and worships a rapidly growing list of very similar gods.

Dr. Ophelia C. Cam retorts, "Stuff and nonsense! I can only ever perceive *one* world, so it is unscientific to talk about others. They are, by definition, outside the possibility of observation, and therefore, also by definition, outside the realm of science." She believes that each observer, with each observation, collapses his or her own wave-function of the universe. That is, to be consistent with experiment, she must assume that each observer has his or her own wave-function for the universe, which collapses only when its owner makes an observation. This means the quantum state of the universe is *different for different observers*. How can a wave-function collapse? How can a wave-function be subjective, and not absolute?

"I don't know, and I don't care," says Dr. Cam. "Like it or not, it is what it is. The *measured* results provide a single reality for all observers, so there is no physical consequence of personalized wave-functions." On Wednesdays, Dr. Cam goes to the Church of One Mind, where she prays to a very lonely God.

Who is right, then, Dr. Rox or Dr. Cam? This is not a scientific question, since both professors make the same experimental predictions. Whom you believe depends on which church you attend.

(There are many other conceptual models ("churches") of QM which agree with the facts, but since they have no observable consequences, we maintain that they have no scientific substance.)

Chapter 2
Riding the Wave: More on Wave Mechanics

2.1 Units of Measure

Many quantum mechanics (QM) references ignore the units of measure of the components of a calculation. For example, the most common element of QM calculations is the one-dimensional (1D) x-representation wave-function, $\psi(x)$. What are its units? Answer: $m^{-1/2}$, or per-square-root-meters. Surprised? So were we.

In the following, we use square brackets to mean "the units of." For example, $[x]$ means "the units of x."

Let us start with the basics: in the macroscopic universe there are exactly four fundamental quantities: distance, mass, time, and charge. (One can reasonably argue for a fifth: angle.) In the MKSA system, the corresponding units are meters (m), kilograms (kg), seconds (s), and coulombs (C). We stick mostly with MKSA in this text. As is common, we use the terms "units" and "dimensions" interchangeably in this context.

For the units of $\psi(x)$, recall that the dot product of a normalized wave-function with itself is a dimensionless 1:

$$\int_{-\infty}^{\infty} \psi^*(x)\psi(x)\,dx = 1 \quad \text{(dimensionless)}$$

Since dx is in meters (m) and the units of ψ^* are the same as ψ, then $\psi^*\psi$ must be in m^{-1}, and thus ψ is in $m^{-1/2}$.

Equivalently, if x is in meters and we compute the average of x:

$$\langle x \rangle = \int_{-\infty}^{\infty} \psi^*(x)\, x\, \psi(x)\,dx,$$

then the units of ψ must be $m^{-1/2}$.

What about the momentum representation, $a(p)$? The same normalization process starts with

$$\int_{-\infty}^{\infty} a^*(p)a(p)\,dp = 1 \quad \text{(dimensionless)} \quad \text{where} \quad p \text{ is in } \frac{kg \cdot m}{s}.$$

E. L. Michelsen, *Quirky Quantum Concepts*, Undergraduate Lecture Notes in Physics, DOI 10.1007/978-1-4614-9305-1_2, © Springer Science+Business Media New York 2014

Then $a(p)$ must be in $[\text{momentum}]^{-1/2} = \left[\dfrac{s}{kg \cdot m}\right]^{1/2}$, or $s^{1/2}\,kg^{-1/2}\,m^{-1/2}$, or "inverse square-root momentum."

Recall that mathematically, exponentials and logarithms are dimensionless and their arguments must be dimensionless. Also, the unit "radian" is equivalent to dimensionless, because it is defined as arc-length/radius$=$m/m$=$dimensionless.

What about three-dimensional (3D) wave-functions? Given $\psi(x, y, z)$, its units are m$^{-3/2}$. Why? We refer again to the normalization integral, which says that the particle must be somewhere in the universe, i.e.,

$$\Pr(\text{particle is somewhere in the universe}) = 1$$

$$= \iiint_{universe} \psi^* \psi \, dx \, dy \, dz \quad (\text{dimensionless}).$$

The units of $dx\,dy\,dz$ are m^3, so ψ must be in m$^{-3/2}$. Often, for spherically symmetric potentials, ψ is a function of r, ϕ, and θ: $\psi\,(r,\,\theta,\,\phi)$. Then it must have units of m$^{-3/2}$ rad^{-1}:

$$\iiint_{universe} \psi^* \psi\, r^2 \sin\theta\, dr\, d\phi\, d\theta = 1 \quad \text{and} \quad r^2\, dr\, d\phi\, d\theta \text{ is in m}^3\text{-rad}^2.$$

However, since rad is dimensionless, this is the same as before: m$^{-3/2}$. Thus, as expected, the units of ψ are independent of the units of its arguments.

The unit of two-dimensional (2D) $\psi(x, y)$ is left as an exercise for the reader.

2.1.1 Dimensions of Operators

Operators also have dimensions.

Let us consider the momentum operator. $\dfrac{d}{dx}$ is like dividing by x, so it has units of $(1/m)$, or m^{-1}. Planck's constant h, or $\hbar \equiv h/2\pi$, is a quantum of action, (energy) (time), or of angular momentum (distance) (momentum); the units are thus joule-seconds (J-s), or in purely fundamental terms, kg-m^2/s. Then:

$$\hat{p} = \frac{\hbar}{i}\frac{d}{dx} \quad \Rightarrow \quad \text{units of} \left(\frac{kg \cdot m^2}{s}\right)\left(\frac{1}{m}\right) = \frac{kg \cdot m}{s}, \quad \text{consistent with } p = mv.$$

The momentum operator has units of momentum.

In fact, all observable operators have the units of the observable. We can see this from the average value formula:

$$\langle \hat{o} \rangle = \int_\infty \psi^* \hat{o} \psi \, dx, \quad \text{and} \quad [\psi^* \psi\, dx] = \text{dimensionless} \quad \Rightarrow \quad [\hat{o}] = [o].$$

When composing operators, their units multiply. Thus, we see that $\dfrac{d^2}{dx^2}$ has units of m^{-2}, etc.

Commutators are compositions of other operators, so the units of commutators are the composition of the units of the constituent operators. (More on commutators elsewhere.) Perhaps the most famous quantum commutator is:

$$[\hat{x}, \hat{p}] = \hat{x}\hat{p} - \hat{p}\hat{x} = i\hbar.$$

The units of $\hat{x}$ are m. (Note that the units of $\hat{x}\psi(x) = $ m(m$^{-1/2}$)=m$^{1/2}$, *not* m.) The units of $\hat{p}$ are kg-m/s. The units of $\hat{x}\hat{p}$ are simply the product of the units of $\hat{x}$ and $\hat{p}$: (m)(kg-m/s) $= \dfrac{kg \cdot m^2}{s}$. This must be, because the commutator in this case works out to a constant, $i\hbar$, with those units.

Note that the units of operators do not change with the representation basis. For example, $\hat{x}$ in the momentum representation is still meters:

$$\hat{x} = i\hbar \frac{d}{dp} \qquad \Rightarrow \qquad \text{units of} \left[\frac{\dfrac{kg\ m^2}{s}}{\dfrac{kg\ m}{s}} \right] = m.$$

2.2 The Dirac Delta Function

The Dirac delta function is used heavily all over physics, engineering, and mathematics. Thoroughly understanding it is essential for anyone in those fields. Read more about the δ-function in *Funky Mathematical Physics Concepts* (http://physics.ucsd.edu/~emichels/FunkyMathPhysics.pdf). The δ-function is also called an "impulse" or "impulse function."

The Dirac delta function is really a pseudofunction: it implies taking a limit, but without the bother of writing "$\lim_{\Delta x \to 0}$" all the time. The Dirac delta function is often formally defined as the limit of a Gaussian curve of (a) infinitesimal width, (b) unit integral $\left(\int_{-\infty}^{\infty} \delta(x)\,dx = 1 \right)$, and thus (c) infinite height. This is somewhat overkill for our purposes, and it may be simpler to think of the delta function as a *rectangular* pulse of (a) infinitesimal width, (b) unit area, and thus (c) infinite height, located at zero (Fig. 2.1, *left*):

Mathematically, we could write this *simplified* (asymmetric) delta function as

$$\text{simplified } \delta(x) = \lim_{\Delta x \to 0} f(x) \qquad where \qquad f(x) = 0, \qquad x < 0$$

$$f(x) = 1/\Delta x, \quad 0 \le x \le \Delta x$$
$$f(x) = 0, \qquad x > \Delta x.$$

Though the previous works for any well-behaved (i.e., continuous) function, the delta function is usually considered an even function (symmetric about 0), so it is sometimes better to write (Fig. 2.1, *right*):

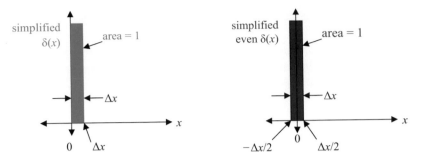

Fig. 2.1 (*Left*) The δ-function can be written as one-sided in most cases. However, (*right*) it is usually considered even. In any case, we take the limit as $\Delta x \to 0$

$$\text{simplified } \delta(x) = \lim_{\Delta x \to 0} f(x) \qquad where \qquad f(x) = 0, \qquad x < -\Delta x / 2$$

$$f(x) = 1 / \Delta x, \quad -\Delta x / 2 \leq x \leq \Delta x / 2$$
$$f(x) = 0, \qquad x > \Delta x / 2.$$

However, in spherical polar coordinates, the radial delta function at zero requires the *asymmetric* form, and cannot use the symmetric form (see *Funky Mathematical Physics Concepts*).

Both of the previous simplified versions of the delta function require special handling for more advanced applications where we need to take derivatives of $\delta(x)$; we will not use such derivatives in this book.

2.2.1   Units of the Delta Function

Another surprise: the δ-function is *not* dimensionless.

The Dirac delta function has units!

Usually, such mathematically abstract functions are dimensionless, but the key property of the delta function is that its *area* is 1 and dimensionless. This means:

$$\int_{-\infty}^{\infty} \delta(x) \, dx = 1 \quad \text{(dimensionless)}.$$

So if x (and thus dx) is in m, $\delta(x)$ must be in m^{-1}. But we use the delta function for all sorts of measures, not just meters: radians, momentum, etc. So by definition, the delta function assumes units of the inverse of its argument. Given a radian, the units of $\delta(\theta)$ are inverse radians (rad^{-1}, equivalent to dimensionless); given a momentum, the units of $\delta(p)$ are $\dfrac{s}{kg \cdot m}$; and so on. Also, $\delta^{(3)}(\mathbf{r})$ has units of the inverse cube of the units of $\mathbf{r}$ ($\mathbf{r}$ in m $\Rightarrow \delta^{(3)}(\mathbf{r})$ in m^{-3}), and $\delta^{(4)}(x^\mu)$ has units of the inverse fourth power of the units of x^μ.

An important consequence of the definition of $\delta(x)$ is that, because $\delta(x)=0$ except near $x=0$,

$$\int_{-\varepsilon}^{\varepsilon} \delta(x) \, dx = 1, \quad \forall \varepsilon > 0.$$

Note that $\delta(x)$ is *not* square integrable, because

$$\int_{-\infty}^{\infty} \delta^2(x) \, dx = \lim_{\Delta x \to 0} \int_0^{\Delta x} \left(\frac{1}{\Delta x}\right)^2 dx = \lim_{\Delta x \to 0} \left[\frac{x}{(\Delta x)^2}\right]_0^{\Delta x} = \lim_{\Delta x \to 0} \frac{1}{\Delta x} \to \infty.$$

Interestingly, though the delta function is often given as a Gaussian curve, the precise form does not matter, so long as it is analytic (i.e., infinitely differentiable or has a Taylor series), unit integral, and infinitely narrow [15, p. 479b]. Other valid forms are:

$$\delta(x) = \lim_{\lambda \to 0} \frac{1}{\pi} \frac{\lambda}{\lambda^2 + x^2}, \qquad\qquad \delta(x) = \frac{1}{2\pi} \int_{-\infty}^{\infty} e^{ikx} \, dk.$$

This latter form is extremely important in quantum field theory, and QM in the momentum representation.

2.2.2   Integrals of δ-Functions of Functions

When changing variables, we sometimes need to know what is $\int dx \, \delta(f(x))$?

Let $u = f(x), \quad du = f'(x) \, dx$ and define x_0 s.t. $f(x_0) = 0$.

Then $\displaystyle\int_{-\varepsilon}^{\varepsilon} dx \, \delta(f(x)) = \int_{-\varepsilon'}^{\varepsilon'} du \, \frac{1}{|f'(x_0)|} \delta(u) = \frac{1}{|f'(x)|}.$

We must take the magnitude of the derivative, because $\delta(x)$ is always positive, and always has a positive integral. The magnitude of the derivative scales the area under the delta function.

If the interval of integration covers multiple zeros of $f(x)$, then each zero contributes to the integral:

Let x_i = zeros of f, i.e. $f(x_i) = 0, \quad i = 1,\dots n.$

Then $\displaystyle\int_{-\varepsilon}^{\varepsilon} dx \, \delta(f(x)) = \sum_{i=1}^{n} \frac{1}{|f'(x_i)|}.$

2.2.3   3D δ-function in Various Coordinates

See *Funky Mathematical Physics Concepts* for a more complete description, but
note that $\delta^3(\mathbf{r})$ has a simple form *only* in rectangular coordinates:

$$\delta^3(x, y, z) = \delta(x)\delta(y)\delta(z),$$

but:

$$\delta^3(r, \theta, \phi) \neq \delta(r)\delta(\theta)\delta(\phi). \quad \text{(It is more complicated than this.)}$$

2.3   Dirac Notation

Dirac notation is a way to write the **algebra** of QM bras, kets, and operators. It is
widely used, and essential to all current QM. It applies to both wave-mechanics and
discrete-state mechanics (discussed in a later chapter).

You are familiar with the ordinary algebra of arithmetic. You may be familiar
with Boolean algebra. There are also algebras of modular arithmetic, finite fields,
matrix algebra, vector spaces, and many others. All algebras are similar to arithmetic algebra in some ways, but each is also unique in some ways. In general,
an **algebra** is a set of rules for manipulating symbols, to facilitate some kind of
calculations. We here describe Dirac notation and its associated Dirac algebra. Included in Dirac algebra is the algebra of operators (covered in a later section). Dirac
algebra also brings us closer to the concept of kets and bras as vectors in a vector
space (see p. 81).

2.3.1   Kets and Bras

For wave mechanics, **kets** and **bras** are complex-valued functions of space (spatial
functions), such as quantum states, and the results of operators on states. In Dirac
notation, kets are written as $|name\rangle$, where "name" identifies the ket. The ket is a
shorthand for the spatial wave-function, say $\psi(\mathbf{r})$. The "name" is arbitrary, much
like the choice of letters for variables in equations. However, there are some common conventions for choosing ket names, again similar to the conventions for using
letters in equations. In this section, we discuss only the spatial kets.

As a ket example, suppose we have a 1D spatial wave-function, $\psi(x)$. Since any
wave-function can be written as a ket, we might write the ket for $\psi(x)$ as $|\psi\rangle$ (as-
suming some notational license for now):

$$|\psi\rangle \equiv \psi(x) = \text{complex-valued function of } x.$$

Note that the ket $|\psi\rangle$ stands for the *whole* wave-function; it does *not* represent the
value of the wave-function at any particular point. One of the key benefits of Dirac
notation is that *kets, bras, and operators are independent of any representation basis*.

Since they always represent the *entire* spatial function, there's no question of "what is the basis for a ket?" More on representations (decomposition in different bases) later.

Some might object to equating a ket to a function, as we did previously: $|\psi\rangle = \psi(x)$. More specifically, $\psi(x)$ is a particular representation of the quantum state $|\psi\rangle$, so it would perhaps be more explicit to say "$|\psi\rangle$ can be represented as $\psi(x)$," but that seems pedantic. We all agree that "5=4+1," yet the symbol "5" is different than the symbol "4+1." They are two representations of the same mathematical quantity, 5. Furthermore, since any function of position, say $\psi_x(x)$, can be written as a function of momentum, $\psi_x(x)$, our flexible notation would say that $\psi_x(x) = \psi_p(p)$, which is OK with us. This simply means that $\psi_x(x)$ and $\psi_p(p)$ both represent the same mathematical entity. I am therefore content to say:

$$|\psi\rangle = \psi_x(x) = \psi_p(p) = \text{any other representation of the ket } |\psi\rangle.$$

Dual to kets are **bras**. Bras are written as $\langle name|$, where "*name*" identifies the bra. Bras are also a shorthand for complex-valued functions of space. The same function of space can be expressed as either a ket or a bra. The difference is the ket is shorthand for the spatial function itself; the bra is shorthand for the complex conjugate of the function. Thus (continuing our flexible notation),

$$\langle\psi| = \psi^*(x) \qquad \text{(complex conjugate)}.$$

For example, suppose we have two wave-functions over all space, $\psi(x)$ (in one dimension) and $\varphi(x)$. (The generalization to higher dimensions is straightforward.) It is frequently useful to determine the **dot product** of two wave-functions, which is a single complex number, defined as:

$$\psi \cdot \varphi = \int_{-\infty}^{\infty} \psi^*(x)\varphi(x)\, dx \qquad \text{(a complex number)}.$$

Notice that the first wave-function, ψ, is conjugated. Now the bra representation of ψ^* is just $\langle\psi|$ and the ket representation of φ is $|\varphi\rangle$. When written next to each other, bra–ket combinations are defined as the dot product integral, i.e.,

$$\langle\psi|\varphi\rangle \equiv \psi \cdot \varphi = \int_{-\infty}^{\infty} \psi^*(x)\varphi(x)\, dx.$$

(A bra–ket combination is a bra–c–ket, < >. Get it?)

When writing a bra–ket combination, use only one vertical bar between them: $\langle\psi|\varphi\rangle$, *not* $\langle\psi||\varphi\rangle$.

As a related example, using our new Dirac shorthand, we can write the "squared-magnitude" (sometime called "squared-length") of ψ as the dot product of ψ with itself:

$$magnitude^2 = \int_{-\infty}^{\infty} \psi^*(x)\psi(x)\, dx = \langle\psi|\psi\rangle.$$

2.3.1.1   Summary of Kets and Bras

> The ket shorthand for $\psi(x)$ is $|\psi\rangle$. The bra shorthand for $\psi*(x)$ is $\langle\psi|$. Combining a bra with a ket, $\langle\psi|\varphi\rangle$, invokes the dot product operation.

Quite simply, a **ket** is a function of space; a **bra** is the complex conjugate of such a function. A bra–ket is the dot product of the bra and ket, yielding a complex number. Recall that the QM dot product is *not* commutative (discussed elsewhere):

$$\langle\psi|\varphi\rangle \equiv \int_{-\infty}^{\infty} \psi*(x)\varphi(x)\,dx = \langle\varphi|\psi\rangle*$$

(reversing the operands conjugates the dot product).

We have seen that kets and bras can be wave-functions which are quantum states, but as noted earlier, kets and bras are more general than that. A ket or a bra can be either a quantum state, or the result of *operations* on a state. In other words, a ket or bra can be most any function of space. (Recall that a **quantum state** defines everything there is to know about a particle, including probabilities of finding it anywhere in space. A particle spatial quantum state (i.e., excluding its spin part), can be expressed as a complex-valued function of position, say $\psi(x,y,z)$.) Therefore,

> All states are kets, but not all kets are states.

For example, a particle can be in a *state* $|\psi\rangle$, but no particle state can be given by the *ket* $\hat{p}|\psi\rangle$.

A note about spin: Wave-functions alone may not fully define a quantum state, because they do not define the **spin** of a particle, i.e., its intrinsic angular momentum. Therefore, a full quantum state, for a particle with spin, is a combination of the wave-function (spatial state) and its **spin-state**. More on this later.

2.3.2   Operators in Dirac Notation

This section repeats much of the information in the previous "Operators" section, but in Dirac notation.

An operator acting on a ket $|\psi\rangle$ *produces another ket.* The given ket $|\psi\rangle$ may or may not be a *state*, and the resulting ket may or may not be a *state*, i.e., it may or may not represent a quantum state that a system could be in. If a result ket is not a state, then what is it? It may be a linear combination (superposition) of results, computed from operators acting on a superposition of states. It may be represented as a sum (superposition) of basis vectors:

$$\text{Given} \quad |\psi\rangle = \sum_{n=1}^{\infty} c_n |\phi_n\rangle, \quad \text{then} \quad \hat{o}|\psi\rangle = \sum_{n=1}^{\infty} c_n \hat{o}|\phi_n\rangle.$$

In any case, it is a vector in the ket vector-space that contains information of interest.

An operator acts on a ket to produce another ket. Either one or both kets may or may not be "states."

A linear operator acting on a *state* $|\psi\rangle$ produces a superposition of *results* based on the superposition of *states* composing $|\psi\rangle$.

Some references do not properly distinguish between a "ket" and a "state." Some even go so far as to define an "operator" as acting on a "state" to produce another "state." This is wrong.

A linear operator acting on a nonstate ket produces a superposition of new results based on the superposition of prior results composing the given ket.

Recall that in Dirac notation, kets and bras are independent of the representation basis. For Dirac algebra to work, operators must also be independent of representation. Therefore, Dirac operators never have things like $\partial/\partial x$, because that implies a specific representation basis. Instead, Dirac operators are just labels that describe their function, but the actual implementation of an operator in any basis is not specified. That is the beauty of Dirac algebra: much of the tedium of complicated operators is eliminated. The algebra works by universal identities and properties of kets, bras, and operators.

Dirac operators may be written three ways: preceding a ket, between a bra and a ket, and less often, following a bra with nothing to the right:

$\hat{o}|\psi\rangle$ means the operator " $\hat{o}$ " acting on the ket " $|\psi\rangle$."

$\langle\varphi|\hat{o}|\psi\rangle$ means the dot product of $\langle\varphi|$ with the result of " $\hat{o}$" acting on $|\psi\rangle$.

$\langle\psi|\hat{o}$ means the operator acting to the left, $\left(\hat{o}^{\dagger}|\psi\rangle\right)^{\dagger}$, which is a bra.

This last case of a lone operator following a bra is interpreted as the left-action of the operator on the bra. We usually think of operators as acting on the ket to the right, but there is also a way of defining the action of an operator on the bra to the left [16, p. 15m]. Adjoints and left-action are discussed on p. 66.

Here is an example of Dirac algebra using operator algebra and kets and bras. Consider the energy eigenstates (stationary states) of a harmonic oscillator, $|u_0\rangle, |u_1\rangle, |u_2\rangle, ...$, and the lowering operator, $\hat{a}$. What is the result of "$\hat{a}$" on a ket $|u_n\rangle$? In other words, what is $\hat{a}|u_n\rangle$? We can answer that question with Dirac algebra, by starting with an identity for the lowering operator:

$$\text{Given:} \quad \langle u_{n-1}|\hat{a}|u_n\rangle = \sqrt{n}, \quad \forall n, \quad \text{and} \quad \langle u_j|\hat{a}|u_n\rangle = 0, \quad j \neq n-1.$$

This implies that $\hat{a}|u_n\rangle$ is a multiple of $|u_{n-1}\rangle$. Then from the first identity, we must have:

$$\langle u_{n-1}|\hat{a}|u_n\rangle = \sqrt{n}\langle u_{n-1}|u_{n-1}\rangle \quad \text{since} \quad \langle u_{n-1}|u_{n-1}\rangle = 1 \quad \text{(basis functions normalized).}$$

Now "divide" both sides by $\langle u_{n-1}|$:

$$\hat{a}|u_n\rangle = \sqrt{n}|u_{n-1}\rangle.$$

The "divide" is only possible because all other inner products are zero in the "givens." Note that non-Hermitian operators such as this are calculation aids and nothing more.

Composition of Operators: Two operators may be composed, i.e., the first acts on a ket to produce another ket, then the second acts on the result of the first. For example, $\hat{g}\hat{h}|\psi\rangle$ means $\hat{g}$ acts on the result of $\hat{h}$ acting on $|\psi\rangle$. The combination $\hat{g}\hat{h}$ "looks like" multiplying $\hat{g}$ and $\hat{h}$, but it is not. It is the **composition** of $\hat{g}$ on $\hat{h}$. Sometimes, references even call such a composition "$\hat{g}$ times $\hat{h}$," but there may not be any multiplication involved.

For scalar multiplication, such as $ab|\psi\rangle$, where "a" and "b" are complex numbers, then "a" and "b" are, in fact, multiplied. Also, for finite state spaces, where $\hat{g}$ and $\hat{h}$ are matrices, the composition is, indeed, matrix multiplication.

Summary: An operation on a quantum state is not necessarily a quantum state, but the result *is* a function of space. Therefore, it can be represented by a ket (or bra). In other words, the result of an operator on a ket is another ket, e.g., $\hat{o}|\psi\rangle$ is the operator "$\hat{o}$" on the ket $|\psi\rangle$ and is itself a ket. This resulting ket is often not a quantum state (i.e., it does not completely define all the properties of any particle), but it is still a useful function of all space. A linear operator acting on a state $|\psi\rangle$ produces a superposition of results based on the superposition of states which compose $|\psi\rangle$.

In the case of operators for observables, as we have seen, an operator "brings out" some physical property from a quantum state, such as the energy of the particle, or its position, or its momentum. But the computation can only be completed by taking a relevant inner product of a bra with the result of the operator acting on the state.

In some cases, however, an operator converts a state into another state, such as the time evolution operator, a translation operator, or a rotation operator. In other cases, an operator converts a ket into another ket, which we use purely for mathematical convenience. For example, raising and lowering operators for the harmonic oscillator ($\hat{a}$ and $\hat{a}^\dagger$), or for angular momentum ($\hat{J}_+$ and $\hat{J}_-$) are *not* Hermitian. This is a dead giveaway that they are not physically significant: they do not correspond to physical observables (operators for which must be Hermitian). Nonetheless, raising and lowering operators are extremely useful mathematical analytic tools. There are many such relationships between operators and kets that are useful algebraic tools, but which have no physical significance.

2.3.3   Units Revisited

You can almost think of kets and bras as wave-functions (but kets and bras are independent of any representation, such as position or momentum representations). We can, therefore, apply our knowledge of units of wave-function to Dirac notation. We cannot ascribe any units to a ket or a bra by itself, since those units depend on the representation. However, we have seen that the squared-magnitude is dimensionless in any representation, so for a quantum state:

$$\langle \psi | \psi \rangle \quad \text{is dimensionless.}$$

We have also seen that the dimensions of operators are the same in any representation, so operators in Dirac notation have the same units as operators in "function notation:"

$$\hat{x} \qquad \text{has units of m,}$$

$$\langle \psi | \hat{x} | \psi \rangle \qquad \text{has units of m,}$$

$$\langle \psi | \hat{p} | \psi \rangle \qquad \text{has units of kg-m/s, etc.}$$

2.3.4   And Now, the Confusing Part

There is a subtle and confusing ambiguity in common QM notation. The symbol "$\psi(x)$" has two different meanings, in different contexts. First, $\psi(x)$ means a function of space, the whole function. Second, sometimes $\psi(x)$ means the value of ψ at some specific point x, i.e., one specific point of ψ, and *not* the whole function. You can only tell from the context which meaning of the symbol "$\psi(x)$" is intended.

An example of the first meaning is (in our flexible notation):

$$\psi(x) = | \psi \rangle, \quad \text{``}\psi(x)\text{'' means the whole function. A ket is } \textit{always} \text{ a whole function.}$$

$$(2.1)$$

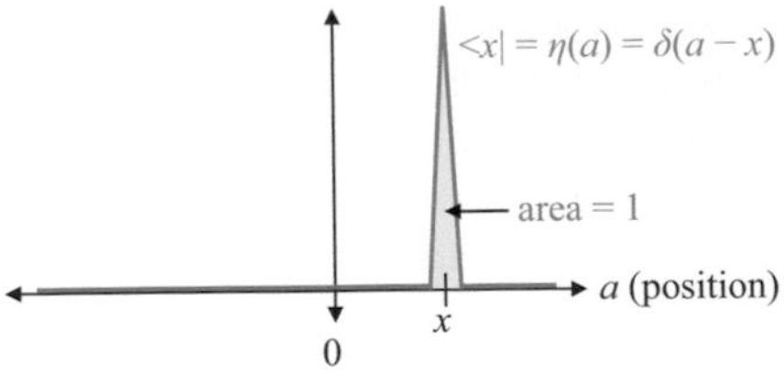

Fig. 2.2 The bra $\langle x|$ for one
particular value of x

Another way to write "the whole function" is just "ψ"; thus, the following are equivalent:

"A particle is in a state, ψ."   (This does not imply any particular basis of ψ.)
"A particle is in a state, $|\psi\rangle$."   (Nor does this.)
"A particle is in a state, $\psi(x)$." (This emphasizes the position representation of ψ.)

An example of the second meaning is

$$\psi(x) = \langle x|\psi\rangle, \quad \text{``}\psi(x)\text{''means the value at some point } x. \qquad (2.2)$$

A dot product is a single complex number, and $\langle x|\psi\rangle$ is the value of ψ at the point x.

"$\langle x|$" is a *variable* bra: for each value of x there is a different bra (a different wave-function). Thus, $\langle x|$ is one of a set of functions of space. It is variable because there are *different spatial functions* for each value of x. "$\langle x|$" is the (conjugate of the) wave-function of a particle completely localized to the position x, specifically, an impulse (Dirac δ-function) at some position "x." Given x, we can write the wave-function for $\langle x|$, call it $\eta(a)$, where we use "a" as the position variable (since "x" is already taken), see Fig. 2.2.

Since $|x\rangle$ is real, the conjugation of the bra has no effect:

$$\langle x| \equiv \eta(a) = \delta(a-x) \qquad \text{for all } a, \ -\infty < a < \infty, \text{ where "}a\text{" is the position variable}$$
(not x).

Then: [16, 1.7.5 p. 52m]:

$$\langle x|\psi\rangle = \int_{-\infty}^{\infty} da\ \delta(a-x)\psi(a) = \psi(x), \qquad \text{the value of } \psi \text{ at position } x.$$

Recall that integrating through an impulse function times some function "picks out" that function at the impulse position; in this case, we pick out ψ at x or $\psi(x)$.

Variable kets and bras are often written with x' as the position, so that x (instead of a) is the usual coordinate axis:

$$|x'\rangle = \delta(x - x') \quad \text{and} \quad \langle x'| = \delta(x - x').$$

In other coordinates, there are other variable bras. In two dimensions, $\langle x, y |$ is a wave-function localized to a point in the plane:

$$\langle x, y | \equiv \eta(a,b) = \delta(a - x) \, \delta(b - y).$$

In polar coordinates, $\langle r, \phi |$ is a wave-function localized to a particular angle. Similarly, $\langle \theta, \phi |$ is a direction in 3D space, though the normalization is tricky in nonrectangular coordinates. In particular, for $\theta > 0$:

$$\langle \theta, \phi | \equiv \eta(c,d) = \frac{1}{\sin \theta} \, \delta(c - \theta) \, \delta(d - \phi) \neq \delta(c - \theta) \, \delta(d - \phi), \quad \theta > 0.$$

(Some references forget the $(1/\sin \theta)$ factor.) See *Funky Mathematical Physics* for more on multidimensional δ-functions in nonrectangular coordinates. (We address more issues of normalization on p. 60.) This concept of multidimensional localized states extends directly to multiparticle states.

2.4 Covering Your Bases: Decomposition

Any wave-function can be written as the (possibly infinite) sum of **basis functions:** defined as a set of functions which can be linearly combined to construct any function in the system of interest (i.e., in the vector-space). This is a common concept in linear system analysis throughout engineering and physics. For example, any (reasonably well-behaved) function can be written as the sum of an infinity of sine waves (a Fourier series or transform). The sine waves are the basis functions, which when weighted with coefficients, sum to compose the original function. In QM, Fourier *transforms* come up for converting between the position-representation and the momentum representation of a wave-function, which is a special case of the more-general concept of changing bases. (Fourier *series* do not come up much in QM.)

2.4.1 Countably Infinite Basis Sets

Let us start with the countably infinite sets of basis functions. **Countably infinite** means we can assign an integer to identify each basis function. There are an infinite number of functions, and an infinite number of integers, so we can count the bases by their corresponding integers. A countably infinite set of functions is also called a **discrete** set. We suppose that a given function can be written as the weighted sum of bases:

$$f(x) = \sum_{n=0}^{\infty} c_n b_n(x) \quad \text{where} \quad f(x), b_n(x), c_n \text{ are complex-valued.}$$

The most common discrete basis functions in QM are energy eigenstates of bound particles. Note that basis functions are most generally complex-valued, though real-valued bases are common. Also, the coefficients c_n are generally complex. A crucial aspect of basis functions is that they are **orthogonal**, i.e., the dot product between any two basis functions is 0:

$$\int_{-\infty}^{\infty} b_i^*(x) b_j(x)\, dx = \langle b_i | b_j \rangle = 0, \quad i \neq j.$$

Basis functions are **normalized** to unit "length" (dot product with itself $=1$):

$$\int_{-\infty}^{\infty} b_i^*(x) b_i(x)\, dx = \langle b_i | b_i \rangle = 1, \qquad \forall\ b_i \text{ in the basis set of functions.}$$

A basis whose functions are both orthogonal and normalized is call **orthonormal**. We can include both properties of orthonormality in one equation:

$$\int_{-\infty}^{\infty} b_i^*(x) b_j(x)\, dx = \langle b_i | b_j \rangle = \delta_{ij} \quad where \quad \delta_{ij} \equiv \text{Kronecker-}\delta = \begin{cases} 1, & \text{if } i = j, \\ 0, & \text{if } i \neq j. \end{cases}$$

Given orthonormality, how can we evaluate the c_n in the previous expansion of $f(x)$? If we take the dot product of $f(x)$ with a basis $b_n(x)$, only one term in the expansion survives:

$$\int_{-\infty}^{\infty} b_n^*(x) f(x)\, dx = \int_{-\infty}^{\infty} b_n^*(x) \left[\sum_{i=0}^{\infty} c_i b_i(x) \right] dx \quad \text{(expanding } f(x) \text{ in bases)}$$

$$= \sum_{i=0}^{\infty} c_i \int_{-\infty}^{\infty} b_n^*(x) b_i(x)\, dx \quad \text{(integral of a sum is the sum of integrals)}$$

$$= c_n \int_{-\infty}^{\infty} b_n(x) b_n(x)\, dx \qquad \text{(because the integral is 0 when } i \neq n)$$

$$= c_n \qquad \text{(because } b_n \text{ is normalized).}$$

Thus, we compute c_n, the coefficient of a basis function b_n in the expansion of $f(x)$, by taking the dot product of b_n with $f(x)$. All the other basis functions in $f(x)$ integrate to zero in the dot product, and only the component of $f(x)$ containing b_n contributes to the dot product. In Dirac notation,

$$c_n = \langle b_n | f \rangle \qquad and \qquad f(x) = |f\rangle = \sum_{n=0}^{\infty} c_n |b_n\rangle.$$

If discrete bases are unfamiliar, another example of a discrete basis set is the Fourier series. (However, Fourier series are not used much in QM. We use them here only because many people are familiar with them.) Fourier series are examples of decomposition into an infinite discrete basis set. In a Fourier series, periodic complex-valued functions are decomposed into a countably infinite number of discrete frequency components, each identified by an integer, n. We say that any (well-behaved) given periodic function can be written as the weighted sum of these bases:

$$\text{Given} \quad f(x) = f(x+L) \qquad \text{where} \quad L \equiv \text{period of } f,$$

$$\text{then} \quad f(x) = \sum_{n=0}^{\infty} c_n \frac{[\cos(nkx) + i\sin(nkx)]}{\sqrt{L}} = \sum_{n=0}^{\infty} c_n \frac{e^{inkx}}{\sqrt{L}}, \quad \text{where} \quad k \equiv \frac{2\pi}{L}.$$

The basis functions are $L^{-1/2}e^{inkx}$, which are orthonormal. To evaluate the c_n, we use dot products with the basis functions:

$$c_n = \int_0^L \frac{1}{\sqrt{L}} \left(e^{inkx}\right)^* f(x)\, dx = \int_0^L \frac{1}{\sqrt{L}} e^{-inkx} f(x)\, dx,$$

which we recognize as the standard definition of the Fourier Series. The limits of integration change to $[0, L]$, because the functions are periodic, and not square-integrable. In this characteristic, they are different from most QM wave-functions, but they illustrate the principle of decomposition into orthonormal basis functions. (Note, though, that in spherical polar coordinates, $\psi(r,\theta,\phi)$ is periodic in ϕ with period 2π.)

The previous integral for c_n (the decomposition integral) is very important. Taking the dot product of a basis function and a given function, $f(x)$, "picks out" the weight of that basis function in the sum of bases that compose $f(x)$. Sometimes this dot product is called correlation: the **correlation** of the basis function b_n with $f(x)$ tells how much of b_n is included in $f(x)$. With complex-valued bases and functions, the correlation is a complex number whose magnitude tells how much of the basis function exists in the given function, and whose angle describes the complex rotation of that basis function component.

An inner product, or correlation, of $\langle b_n | f \rangle$ tells how much of $|b_n\rangle$ is in $|f\rangle$, and with what phase.

2.4.2   Example: Finite Basis Sets

To illustrate decomposition, let us imagine we have a system with only two basis functions: $b_1(x)$ and $b_2(x)$, as shown later. For illustration, these basis functions are real, as is common in QM. You can still decompose an arbitrary complex function into real basis functions, because the coefficients of the basis functions can

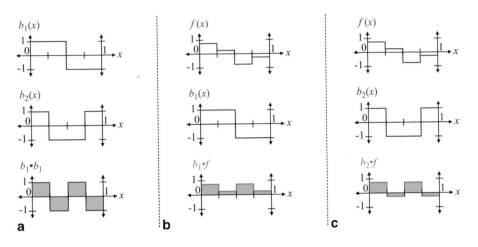

Fig. 2.3 a Basis functions b_1, b_2, and $b_1 \cdot b_2$. **b** f and $f \cdot b_1$. **c** f and $f \cdot b_2$

be complex. However, for simplicity, we decompose a purely real function for this example.

By inspection, we see that the basis functions are normalized, i.e.,

$$\int_{-\infty}^{\infty} |b_1(x)|^2 \, dx = 1 \qquad \text{and} \qquad \int_{-\infty}^{\infty} |b_2(x)|^2 \, dx = 1.$$

And they are orthogonal:

$$\int_{-\infty}^{\infty} b_1^*(x) b_2(x) \, dx = 0 \quad \text{(Fig. 2.3a)}.$$

Now we are given a function composed of some linear combination of b_1 and b_2, i.e.,

$$f(x) = c_1 b_1(x) + c_2 b_2(x) \quad \text{(Fig. 2.3b)}.$$

How can we find the coefficients c_1 and c_2? By decomposition, we find:

$$c_1 = \int_{-\infty}^{\infty} b_1^*(x) f(x) \, dx = \frac{1}{2} \quad \text{(Fig. 2.3b)}.$$

$$c_2 = \int_{-\infty}^{\infty} b_2^*(x) f(x) \, dx = \frac{1}{4} \quad \text{(Fig. 2.3c)}.$$

And, thus:

$$f(x) = \frac{1}{2} b_1(x) + \frac{1}{4} b_2(x).$$

Note that we could choose some different basis functions, say d_1 and d_2, which are normal and orthogonal to each other. We could decompose the same function f into those bases. In QM, we have many choices of bases (basis function sets), and we choose our basis to solve the given problem most easily.

2.4.3 Back to Kets and Bras

It should now be clear why kets and bras are independent of the representation basis. Kets and bras are entire spatial functions, and you can decompose these functions into many different bases. The basis set you choose for decomposition is arbitrary, and does not change the functions themselves. The kets and bras represent the quantum states (entire wave-functions), or the results of operations on wave-functions, which have nothing to do with how you choose to decompose them.

By analogy, the number 11 is a number. Written in base 10, it is "11," representing $10+1$. Written in base 8, it is "13," representing $8+3$. Written in base 4, it is "23," representing $2\cdot4+3$. But it is always the number 11, no matter how we represent it. Its properties do not change: it is always prime; it is always bigger than 10 and smaller than 12, etc.

2.4.4 Continuous (or Uncountable) Basis Sets

We saw previously that some basis functions are countably infinite, and can therefore be numbered, and identified by an integer n. However, other basis functions are uncountably infinite, and cannot be numbered by integers. For example, the position and momentum bases are uncountably infinite. Such basis functions are identified by a real number, rather than an integer, e.g., a real-valued momentum p or a real-valued position x. The momentum basis functions are complex exponentials (sinusoids): Ne^{ikx}, where $k=p/\hbar$. In this case, k can be used as the identifier of the basis function (instead of p), and for each real value of k, there exists a basis function. Between any two values of k, there are an infinite number of other k values and basis functions. Hence, there are an uncountable number of basis functions. Such a set of functions is also called a **continuous** basis. Note that all the basis functions are still orthogonal:

$$\int_{-\infty}^{\infty} \left(e^{ik_1x}\right)^* e^{ik_2x}\, dx = 0, \qquad k_1 \neq k_2.$$

However, we have to refine our definition of "normal." Why? Because the dot product integral for the "length-squared" of each basis function diverges:

$$\int_{-\infty}^{\infty} \left(e^{ikx}\right)^* e^{ikx}\, dx = \int_{-\infty}^{\infty} 1\, dx \to \infty.$$

A similar problem exists for the position basis: the basis functions are Dirac delta functions. Their squared integral diverges:

$$\int_{-\infty}^{\infty} \delta(x)\delta(x)\, dx = \delta(0) \to \infty.$$

We get out of this problem by redefining "normalized." Instead of meaning $\int_{-\infty}^{\infty} |b_k(x)|^2 \, dx = 1$, we say a continuous basis function is normalized if (and only if):

$$\int_{-\infty}^{\infty} b_{k_1}^{*}(x) b_{k_2}(x) \, dx = \delta(k_1 - k_2) \qquad \text{(continuous basis normalization)}.$$

If $k_1 \neq k_2$, the integral is zero, and the basis functions are orthogonal. This normalization scheme is called **delta-function normalization**. Note, though, that the length-squared of a basis function has to equal exactly one $\delta(0)$, and not some multiple of that. It is not good enough that the length-squared go to infinity; it has to go to infinity in the "right" way. When we discuss normalization in more detail later, we will find that:

$$|p\rangle = \frac{1}{\sqrt{2\pi\hbar}} e^{ipx/\hbar}, \quad \text{and} \quad \langle p| = \frac{1}{\sqrt{2\pi\hbar}} e^{-ipx/\hbar} \quad \text{where} \quad \langle p| \equiv \text{variable bra.}$$

Now some Fourier transform mathematics shows that we can write an arbitrary wave-function $\psi(x)$ as a superposition of definite momentum states, using the standard decomposition rule:

$$\psi_p(p) = \langle p | \psi \rangle = \frac{1}{\sqrt{2\pi\hbar}} \int_{-\infty}^{\infty} dx \, e^{-ipx/\hbar} \psi(x).$$

Thus:

The momentum representation of a wave-function is a scaled Fourier transform of the position representation.

2.4.5   Mental Projection: Projection Operators and Decomposition

Recall that $\langle b | \psi \rangle$ is the "amount" of $|b\rangle$ in $|\psi\rangle$, and its phase. In other words, if we were to decompose $|\psi\rangle$ into a superposition of basis kets, the coefficient of $|b\rangle$ in the sum is $\langle b | \psi \rangle$:

$$|\psi\rangle = \langle a | \psi \rangle |a\rangle + \langle b | \psi \rangle |b\rangle + \langle c | \psi \rangle |c\rangle + ..., \quad \text{where} \quad |a\rangle, |b\rangle, |c\rangle ... \text{ are basis kets.}$$

Note that the component of a vector is a vector. Therefore, the actual *component* of the vector $|\psi\rangle$ in the $|b\rangle$ direction is $\langle b | \psi \rangle |b\rangle$ (Fig. 2.4).

We can thus construct a linear operator which takes the component of a given vector along some other vector, say $|b\rangle$:

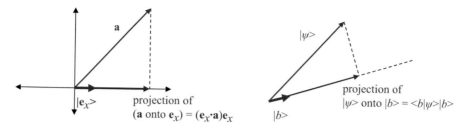

Fig. 2.4 (*Left*) The component (projection) of the vector **a** in the e_x direction is $(e_x \cdot a)e_x$. (*Right*)
The component (projection) of the vector $|\psi\rangle$ in the $|b\rangle$ direction is $\langle b | \psi | b \rangle$

$|b\rangle\langle b|$ is a projection operator such that

$$(|b\rangle\langle b|)|\psi\rangle = |b\rangle\langle b|\psi\rangle = \langle b|\psi\rangle|b\rangle \quad \text{is the projection of } |\psi\rangle \text{ on } |b\rangle.$$

The $|b\rangle\langle b|$ operator looks a little funny, because we usually put the coefficient of $|b\rangle$ in front of $|b\rangle$, but since operators act on kets to the right, we have to put $\langle b|$ on the right of $|b\rangle\langle b|$, so it can match with some target vector $|\psi\rangle$.

Earlier, before we defined Dirac notation, we noted that any operator can be thought of as decomposing the given ket $|\psi\rangle$ into the operator's own eigenfunctions, then multiplying each eigenfunction by its own eigenvalue, and then summing the results. We write this mathematically by projecting $|\psi\rangle$ onto each of the eigenkets, multiplying by the eigenvalues, and summing:

$$\hat{o}|\psi\rangle = \left(\sum_j o_j |o_j\rangle\langle o_j| \right)|\psi\rangle \quad where \quad o_j \equiv \text{eigenvalues}, |o_j\rangle \equiv \text{eigenstates of } \hat{o}.$$

This means any operator (that has eigenvectors and eigenvalues) can be written in terms of its eigenvalues and eigenvectors:

$$\hat{o} = \sum_j o_j |o_j\rangle\langle o_j|.$$

In a continuous basis, for continuous operators (such as position, momentum, energy, and angular momentum), the sum becomes an integral:

$$\hat{o}|\psi\rangle = \left(\int_\infty do\, o|o\rangle\langle o| \right)|\psi\rangle \quad where \quad o \equiv \text{continuous eigenvalues}, |o\rangle \equiv \text{eigenstates of } \hat{o}$$

$$\Rightarrow \qquad \hat{o} = \int_\infty do\, o|o\rangle\langle o|. \qquad \text{E.g.,} \qquad \hat{p} = \int_\infty dp\, p|p\rangle\langle p|.$$

2.4.6   A Funny Operator

Any bra or ket (i.e., wave-function or function of space) can be decomposed into basis functions (or basis kets):

$$|\psi\rangle = \sum_n c_n |b_n\rangle = \sum_n \langle b_n|\psi\rangle |b_n\rangle, \quad where \quad |\psi\rangle \text{ is a ket,}$$

$$c_n = \langle b_n|\psi\rangle, \text{ and } |b_n\rangle \equiv \text{basis kets}$$

We can rewrite the last expression by putting the coefficient of each basis function on the right, and applying parentheses:

$$|\psi\rangle = \sum_n |b_n\rangle\langle b_n|\psi\rangle = \left(\sum_n |b_n\rangle\langle b_n|\right)|\psi\rangle.$$

This is simply decomposing a vector into its components, and then summing all the components, which returns the original vector. Therefore, the parenthesized operator is the identity operator:

$$\sum_n |b_n\rangle\langle b_n| = \mathbf{1}_{op} \quad \text{(the identity operator, aka the completeness operator).}$$

In other words, we can insert a sum over a set of arbitrary basis functions, $\sum_n |b_n\rangle\langle b_n|$, before any ket. That might seem pointless, but it is actually an important tool of Dirac algebra. It works just as well acting to the left on a bra:

$$\langle\psi| = \sum_n \langle\psi|b_n\rangle\langle b_n| = \langle\psi|\left(\sum_n |b_n\rangle\langle b_n|\right).$$

This is also evident because the identity operator is Hermitian. The completeness operator is dimensionless.

This "completeness operator" can also be considered a special case of the prior section on operators "mentally decomposing" a vector into the operator's eigenvectors, multiplying those components by their eigenvalues, and summing the results. Here, the operator is the identity operator; any basis is an eigenbasis of the identity operator; and the eigenvalues are all 1.

This idea also generalizes to continuous bases in a straightforward way: the sum becomes an integral. For example:

$$\int_{-\infty}^{\infty} dx\, |x\rangle\langle x| = \mathbf{1}_{op}.$$

Fig. 2.5 Allowed electron
energies near a proton

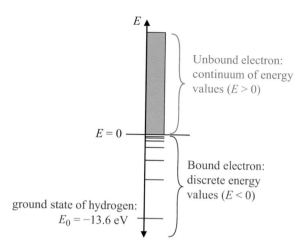

2.4.7 *Countable or Uncountable Does Not Mean Much*

You may be worried that a countably infinite basis set (say, the energy basis) can
compose a function of space, with an uncountably infinite position basis. (If this
does not bother you, skip this section.) But uncountable infinities are "bigger" than
countable ones, so what is the dimensionality of our vector space?

   Let us consider states of a simple harmonic oscillator, in both the energy and
position bases. In the energy basis, the basis functions are the eigenstates of energy
(often labeled $|u_0\rangle, |u_1\rangle, |u_2\rangle, \dots$). Any bound-state wave-function can be com-
posed from the *countably* infinite set of energy eigenstates. In the position representation (basis), the bases (basis functions) are Dirac deltas at all values of x
(position). Any wave-function can be composed from the *uncountably* infinite set of
position impulses. In short, the $|u_n\rangle$ are countably infinite, the $|x\rangle$ is uncountably
infinite, yet both sets are (complete) basis sets for bound states. In fact, the bound
states are a subset of the full vector space spanned by $|x\rangle$, which explains why the
dimension of the function space spanned by $|x\rangle$ is "bigger" than the bound state
space. However, any bound state function can be represented as the sum of energy
bases, or the sum of position bases. Thus, the degree of infinity of the dimension of
a basis is not always significant.

   Finally, consider the energy states of a hydrogen atom: for bound states the electron energies are below the "lip" of the well (usually the lip is set to $E=0$, so this
means for negative electron energies). There are a countably infinite set of such
energy eigenstates ($E_n = Ry/n^2$). But for ionized hydrogen, where the electron is unbound and can be anywhere in space, there are an uncountably infinite set of energy
eigenstates. Thus, the complete basis of energy eigenstates for a (possibly ionized)
hydrogen atom is the union of a countably infinite set and an uncountably infinite
set (Fig. 2.5). C'est la vie.

2.5   Normalize, Schmormalize

We normalize quantum states to make it easier to compute quantities such as probabilities, averages, cross sections, etc. Usually, we normalize our basis functions so that the probability of a given basis in a superposition is given by the squared-magnitude of that basis function's coefficient:

$$\text{Given} \quad |\psi\rangle = c_1|\phi_1\rangle \dots + c_j|\phi_j\rangle + \dots \quad \text{then} \quad \text{Pr(measuring } |\phi_j\rangle) = |c_j|^2 .$$

But this guideline has different implications for different circumstances, and sometimes other criteria are chosen, so there are at least six different normalization methods in common use. We summarize them first, and then describe most of them in more detail. All normalization methods apply to any number of spatial dimensions (not vector-space dimensions); we show some examples as 1D, but they apply just as well to 2D and 3D. Recall that $|\psi|^2 = \psi^*\psi$. Common normalizations are:

1. Bound states usually use wave-function normalization:

$$\langle\psi|\psi\rangle = \int_{-\infty}^{\infty} \psi^*(x)\psi(x)\,dx = 1, \quad \text{or in 3D:} \quad \langle\psi|\psi\rangle = \int_{\infty} \psi^*(\mathbf{r})\psi(\mathbf{r})\,d^3r = 1.$$

2. Unbound states (e.g., free particle states) usually use delta-function normalization:

$$\langle x|x'\rangle = \int_{-\infty}^{\infty} \psi_x^*(a)\psi_{x'}(a)\,da = \delta(x-x'), \text{ or } \langle\mathbf{r}|\mathbf{r}'\rangle = \int_{\infty} \psi_\mathbf{r}^*(\mathbf{a})\psi_{\mathbf{r}'}(\mathbf{a})\,d^3a = \delta^3(\mathbf{r}-\mathbf{r}').$$

3. Box normalization:

$$\langle\psi|\psi\rangle = \int_{Length} \psi^*(x)\psi(x)\,dx = 1, \quad \text{or} \quad \int_{Volume} \psi^*(\mathbf{r})\psi(\mathbf{r})\,d^3r = 1 .$$

4. Radial wave-functions, and the like: $|R(r)|^2$ is the radial part of the 3D probability density function $\text{pdf}_{volume}(\mathbf{r})$; it is *not* the 1D $\text{pdf}_{radius}(r)$. The $\text{pdf}_{radius}(r) = r^2|R(r)|^2$. Therefore [8, 4.31 p. 138], $\int_0^\infty r^2 R^*(r)R(r)\,dr = 1$.

5. For perturbation theory, with orthonormal basis state $\langle n|n\rangle = 1$, we normalize the perturbed state $|n'\rangle$ such that $\langle n|n'\rangle = 1$, which means $|n'\rangle = |n\rangle + \sum_{j \neq n} c_j|j\rangle$ [1, 11–17 p. 227b].

6. In quantum field theory, energy normalization is often used, because it is Lorentz covariant. For example: [from Pes & Sch, 2.36 p. 23m], $\int_{Volume} \bar{\psi}\psi\,d^3r = 2E$.

We now discuss some of these further.

2.5.1   Wave-Function Normalization

Wave-function normalization might be the most "obvious" normalization method: simply make the probability density, $\text{pdf}(\mathbf{r}) = |\psi(\mathbf{r})|^2$. Bound states use this, because the particle is localized to a region of space, and ψ is square-integrable. The probability of finding the particle *somewhere* in space is 1, so we must have:

$$1 = \int_{-\infty}^{\infty} dx \; \mathrm{pdf}(x) = \int_{-\infty}^{\infty} dx \left| \psi(x) \right|^2, \qquad \text{or in 3D:} \quad \int_{\infty} d^3 r \left| \psi(\mathbf{r}) \right|^2 = 1.$$

In higher dimensions, when the wave-function is separable, there is a PDF for each wave-function coordinate. For example, if $\psi(x,y,z) = X(x)Y(y)Z(z)$, then each of X, Y, and Z are separately normalized:

$$\int_{-\infty}^{\infty} \left| X(x) \right|^2 dx = 1, \qquad \int_{-\infty}^{\infty} \left| Y(y) \right|^2 dy = 1, \qquad \int_{-\infty}^{\infty} \left| Z(z) \right|^2 dz = 1.$$

Physically, this means the probability of measuring the particle's x-coordinate *somewhere* between $-\infty$ and ∞ is 1, therefore $\mathrm{pdf}_x(x)$ must be normalized. Similarly for y and z, which imply proper normalization for the complete wave-function,

$$\int_{\infty} d^3 r \left| \psi(\mathbf{r}) \right|^2 = \iiint_{\infty} \left| X(x)Y(y)Z(z) \right|^2 dx \, dy \, dz$$

$$= \int_{-\infty}^{\infty} \left| X(x) \right|^2 dx \int_{-\infty}^{\infty} \left| Y(y) \right|^2 dy \int_{-\infty}^{\infty} \left| Z(z) \right|^2 dz = 1.$$

See the radial function section later for spherical coordinates.

2.5.2   Delta-Function Normalization

Delta-function normalization is commonly used for idealized states which are completely localized, (in either position or momentum), and for scattering states. For example, the state $|x\rangle$ is an idealized state where the particle location is exactly x. Though this is physically impossible, it *can* serve as a mathematical basis for expressing ket vectors. Similarly, $|p\rangle$ is an idealized state (also physically impossible) where the particle momentum is exactly p. Such states cannot be normalized with wave-function normalization, and are therefore sometimes called "unnormalizable."

The key difference between the so-called unnormalizable localized states, such as $|x\rangle$, and a normalized discrete state $|a\rangle$ is that, when used in an inner product, the former produces a probability *density*, whereas the latter produces an absolute *probability*. For example:

$$\left| \langle x | \psi \rangle \right|^2 = \left| \psi(x) \right|^2 \qquad \text{is a probability } density, \text{ but}$$

$$\left| \langle a | \psi \rangle \right|^2 \qquad \text{is a } probability.$$

In other words, the ket $|x\rangle$ is the limiting case of a discrete, normalizable state where the width of the wave-function (Δx) is made arbitrarily small, but furthermore, the result is divided by $\sqrt{\Delta x}$, so that the squared magnitude of an inner product produces a probability per-unit-distance, i.e., a probability density:

$$\text{Define }\ R(x) \text{ such that }\ \begin{cases} R(x)=0, & x<0, x>\Delta x \\ R(x)=1/\sqrt{\Delta x}, & 0\le x\le \Delta x \end{cases}$$

$$\Rightarrow \qquad \langle R|R\rangle = \int_{-\infty}^{\infty} |R(x)|^2\, dx = 1, \qquad \text{for all } \Delta x.$$

$$\text{Then }\quad |x'\rangle \equiv \lim_{\Delta x\to 0} \frac{R(x-x')}{\sqrt{\Delta x}} = \lim_{\Delta x\to 0} \begin{cases} \dfrac{1}{\Delta x} & 0\le (x-x')\le \Delta x \\ 0 & \text{elsewhere} \end{cases} = \delta(x-x').$$

The last line is a fundamental definition of the delta function. This limit process is rigorous, and there is no mathematical "fudging." Confusion arises primarily because this limit process is not often (ever?) explained.

In contrast to localized states, idealized unbound states have "large" probabilities to be outside *any* arbitrarily large region of space. For example, momentum eigenstates have all the wave-function moving in one single direction, **p** (or in 1D, p), and hence cannot be localized. Idealizing from experiment, we know they can be written as:

$$|p\rangle = \psi_p(x) = Ne^{ipx/\hbar}.$$

Such a wave-function has uniform probability over all space, which means that any finite amplitude N cannot be wave-function normalized. However, we can consider a momentum eigenstate as a wave-function comprising a narrow range of momenta, and take the limit as that "width" goes to zero. This leads to Dirac δ-functions. Also, two different momenta should be orthogonal. Thus, we must find N such that:

$$\langle p'|p\rangle \equiv N^2 \int_{-\infty}^{\infty} dx \left(e^{ip'x/\hbar}\right)^* e^{ipx/\hbar} = N^2 \int_{-\infty}^{\infty} dx\, e^{i(p-p')x/\hbar}$$

$$= \delta(p-p') \qquad (\text{a unit magnitude } \delta)$$

$$\Rightarrow \qquad \langle p|p\rangle = \delta(0) = \infty.$$

Recall that a delta function can be multiplied by a weight; it is still infinite (at 0), but a different "size" infinite. We must find N discussed previously to make the weight of the eigenstate δ-function be 1. From Fourier analysis, we know:

$$\int_{-\infty}^{\infty} e^{ikx}\, dx = 2\pi \delta(k) \qquad \Rightarrow$$

$$\langle p' | p \rangle = N^2 \int_{-\infty}^{\infty} dx\, e^{i(p-p')x/\hbar}. \qquad \text{Let} \qquad y = \frac{x}{\hbar}, \quad dx = \hbar\, dy :$$

$$\langle p' | p \rangle = \hbar N^2 \int_{-\infty}^{\infty} dy\, e^{i(p-p')y} = 2\pi\hbar N^2 \delta(p-p') \qquad \Rightarrow$$

$$N = \frac{1}{\sqrt{2\pi\hbar}}, \qquad | p \rangle = \frac{1}{\sqrt{2\pi\hbar}} e^{ipx/\hbar}, \quad \text{and} \quad \langle p | = \frac{1}{\sqrt{2\pi\hbar}} e^{-ipx/\hbar}.$$

In general, we can compare the "sizes" of δ-functions by their weights. For example, suppose we have a superposition of two momenta: maybe $1/3$ of p_1 and $\sqrt{8}/3$ of p_2:

$$\psi(x) = \frac{1}{3} N e^{ip_1 x/\hbar} + \frac{\sqrt{8}}{3} N e^{ip_2 x/\hbar}. \qquad \text{Then}$$

$$\Pr(\text{momentum is } p_1) = \left|\frac{1}{3}\right|^2 = \frac{1}{9}, \quad \text{and} \quad \Pr(\text{momentum is } p_2) = \left|\frac{\sqrt{8}}{3}\right|^2 = \frac{8}{9}.$$

We can evaluate this formally as a space integral:

$$\Pr(\text{momentum is } p_1) = \frac{|\langle p_1 | \psi \rangle|^2}{\langle p_1 | p_1 \rangle \langle \psi | \psi \rangle} = \frac{\left|\int_{-\infty}^{\infty} dx\, \frac{1}{\sqrt{2\pi\hbar}} e^{-ip_1 x/\hbar} \psi(x)\right|^2}{\delta(0)\delta(0)}$$

$$= \frac{|(1/3)\delta(0)|^2}{\delta^2(0)} = \frac{1}{9}.$$

The inner product $\langle p_1 | \psi \rangle$ is infinite, $c\delta(0)$, but we can still use the coefficient c of such a δ-function to compute probability, just as we would use the coefficient of any other basis function. We "cancel" the infinity of $\delta(0)$ from both the numerator and denominator. Note that our earlier definition of the δ-function as a limit makes such a cancellation perfectly justified.

Position eigenstates also use δ-function normalization, and can be tricky. Recall that a position eigenstate $|x'\rangle$ is zero everywhere, and infinite at the position x' (see Sect. 2.3.4 earlier). So its wave-function is a δ-function, *and* it is δ-function normalized. This is consistent, because:

$$|x'\rangle = \psi_{x'}(x) = \delta(x-x').$$

$$\langle x' | x' \rangle = \int_{-\infty}^{\infty} dx\, \delta(x-x')\delta(x-x'). \quad \text{But} \quad \int_{-\infty}^{\infty} dx\, \delta(x-x')f(x) = f(x'), \ \ \forall\, f(x),$$

$$\text{So we let } f(x) = \delta(x-x') \qquad \Rightarrow \qquad \langle x' | x' \rangle = \delta(x'-x') = \delta(0) = \infty.$$

The inner product is a unit δ-function, as required by δ-function normalization.

Note also that this is consistent with our decomposition rules: because $|x'\rangle = \psi_{x'}(x) = \delta(x - x')$ we can write an arbitrary wave-function $\psi(x)$ as an inner product:

$$\langle x'|\psi\rangle = \int_{-\infty}^{\infty} dx\, \delta(x - x')\,\psi(x) = \psi(x'), \quad \text{or more commonly} \quad \langle x|\psi\rangle = \psi(x).$$

A variant of delta function normalization is often used in scattering with momentum eigenstates. Scattering analysis usually drops the prefactor, so:

$$|p\rangle \equiv \psi_p(x) = e^{ipx/\hbar}.$$

In this book, we call this "unit amplitude delta-function normalization." This is also said to be normalizing to one particle per unit volume, because evaluating the normalization integral over a finite volume yields:

$$\int_V \left|e^{ipx/\hbar}\right|^2 d^3r = \int_V d^3r = V.$$

However, it is probably simpler to say that we are simply choosing our normalization such that the (implicit) prefactor of "1" before $e^{ipx/\hbar}$ represents one particle. Scattering is generally analyzed one particle at a time. Claiming "one particle per unit volume" might suggest a multiparticle system, which would be a very different quantum mechanical system.

2.5.3   Box Normalization

Box normalization is good for things like scattering: it makes it easy to compute incident flux density (particles/sec/m^2). Box normalization declares that the particle must be inside some finite region of space, usually rectangular. Therefore, the probability of finding the particle *somewhere* in the box is 1. Generally, we take the size of the box to be known, and very large, so its boundaries are far from the region of interest, and do not affect the result. The exact volume of the box is never important, because it always cancels out of any observable. This is equivalent to taking the limit as the box size goes to infinity. So:

Box normalization is just wave-function normalization within a large box.

As an example, assume a particle moves in a momentum eigenstate in the z direction:

$$|p\rangle = \psi(x, y, z) = Ce^{ipz/\hbar}, \quad \text{and} \quad \text{Pr(particle in } dx\, dy\, dz) = |\psi(x, y, z)|^2\, dx\, dy\, dz$$

$$\iiint_{Vol} |\psi|^2\, dx\, dy\, dz = 1 \quad \Rightarrow \quad C = \frac{1}{\sqrt{Vol}}.$$

Note that there is no longer a need for the $1/\sqrt{2\pi\hbar}$ normalization used for a free particle. The final result is general for any state:

$$\text{pdf}(x,y,z) = |\psi(x,y,z)|^2.$$

Note that in 1D or 2D, "volume" is replaced by "length" or "area." The result is similar:

$$\text{pdf}(x) = |\psi(x)|^2 \qquad \text{and} \qquad \text{pdf}(x,y) = |\psi(x,y)|^2.$$

The 1D case is the famous "particle-in-a-box."

2.5.4   Funny Normalization of Radial Wave-Functions (and the Like)

For 3D bound-state or box-normalized wave-functions, $\psi(r)$ (in any coordinate system) gives the 3D probability density (per unit volume), such that:

$$\text{Pr}(\text{particle in volume } d^3r) = |\psi(\mathbf{r})|^2\, d^3r \quad \Rightarrow \quad \text{pdf}(\mathbf{r}) = |\psi(\mathbf{r})|^2.$$

Recall that the 3D wave-functions of spherically symmetric potentials, e.g., atomic electron states, separate into three parts [18, 3.6.5 p. 96]:

$$\psi_{nlm}(r,\theta,\phi) = R_{nl}(r)Y_{lm}(\theta,\phi)$$

$$\text{where} \quad Y_{lm}(\theta,\phi) \equiv \begin{cases} (-1)^m \sqrt{\dfrac{(2l+1)}{2}\dfrac{(l-m)!}{(l+m)!}}P_{lm}(\cos\theta)\dfrac{e^{im\phi}}{\sqrt{2\pi}}, & m \geq 0, \\[4ex] \sqrt{\dfrac{(2l+1)}{2}\dfrac{(l-|m|)!}{(l+|m|)!}}P_{l|m|}(\cos\theta)\dfrac{e^{im\phi}}{\sqrt{2\pi}}, & m < 0, \end{cases} \quad (2.3)$$

$P_{lm}(x)$ is the associated Legendre function,

$$l = 0,1,2..., \qquad m = -l, -l+1, ... l-1, l.$$

The previous sign convention (also called "phase convention," because a negative sign is the same as a complex phase of π) is called the "Condon–Shortley phase." With it [24, 3.6.7 p. 97],

$$Y_{lm}^{*}(\theta,\phi) = (-1)^m\, Y_{l,-m}(\theta,\phi).$$

This is somewhat odd, since the coefficients of $m \geq 0$ alternate in sign with m, but those for $m < 0$ are all positive.

To find out how the individual components R_{nl} and Y_{lm} normalize, we write the probability equation as components in $R_{nl}(r)$ and $Y_{lm}(\theta, \phi)$:

$$\text{pdf}_{volume}(r, \theta, \phi) = \left| R_{nl}(r) \right|^2 \left| Y_{lm}(\theta, \phi) \right|^2$$

$$\Rightarrow \quad \iiint_{\infty} r^2 dr \, \sin\theta \, d\theta \, d\phi \, \left| R_{nl}(r) \right|^2 \left| Y_{lm}(\theta, \phi) \right|^2 = 1.$$

Separating: $\left(\int_0^\infty dr \, r^2 \left| R_{nl}(r) \right|^2 \right) \left(\int_0^\pi \sin\theta \, d\theta \int_0^{2\pi} d\phi \left| Y_{lm}(\theta, \phi) \right|^2 \right) = 1.$

Since we may want to know, say, the 1D radial distribution, independent of angle, we would like the radial integral and the combined (θ, ϕ) integral each to separately equal 1. That is, the probability of finding the particle at *some* radius between 0 and ∞ is 1, and separately, the probability of finding the particle at *some* angular position (θ, ϕ) is also 1. Hence:

$$\int_0^\infty dr \, r^2 \left| R_{nl}(r) \right|^2 = 1 \quad \Rightarrow \quad \text{pdf}_{radius}(r) = r^2 \left| R_{nl}(r) \right|^2.$$

Note that $\text{pdf}_{radius}(r)$ is *not* $|R(r)|^2$. $R(r)$ is the function which we multiply the angular function by to get the 3D *volume* probability density, which is different than the number we square to get the 1D $\text{pdf}_{radius}(r)$. The r^2 appears in $\text{pdf}_{radius}(r)$ because when we look only at the radius, we must take into account that:

(volume covered by some interval dr) $\alpha \ r^2 \, dr$.

There is no 4π in the radial normalization integral.

For a state with $l=0$, the *state* is spherically symmetric (not just the potential). All angles are equally likely, so for our wave-function we have:

$$\psi(\mathbf{r}) = \frac{1}{\sqrt{4\pi}} R(r) \quad \text{so that} \quad \iiint_{\infty} r^2 dr \, \sin\theta \, d\theta \, d\phi \left| \psi(\mathbf{r}) \right|^2$$

$$= 4\pi \int_0^\infty r^2 dr \left| R(r) \right|^2 = 1,$$

where *both* $\psi(\mathbf{r})$ and $R(r)$ are normalized as previously.

For the angles θ and ϕ, the $Y_{lm}(\theta, \phi)$ are normalized such that

$$\int_0^\pi d\theta \int_0^{2\pi} d\phi \, \sin\theta \left| Y_{lm}(\theta, \phi) \right|^2 = 1.$$

For the spherically symmetric state with $l=0$, all solid angles are equally likely, i.e., the solid angular probability density is a constant. This implies that $Y_{00}(\theta, \phi) =$ constant, and to be normalized:

Fig. 2.6 The solid angle
(area on unit sphere) per
unit θ is larger for θ near the
equator than near the poles

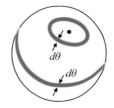

$$\int_0^\pi d\theta \int_0^{2\pi} d\phi \sin\theta \left| Y_{00}(\theta,\phi) \right|^2 = 4\pi \left| Y_{00}(\theta,\phi) \right|^2 = 1 \quad \Rightarrow \quad Y_{00}(\theta,\phi) = \frac{1}{\sqrt{4\pi}}, \quad \text{and}$$

$$\psi_{n00}(\mathbf{r}) = R(r)Y_{00}(\theta,\phi) = \frac{1}{\sqrt{4\pi}}R(r).$$

By convention, we choose Y_{00} to be real, for simplicity.

Often, we leave the θ and ϕ components bunched as $Y_{lm}(\theta,\phi)$, but it is worth noting their separate properties. We separate the θ and ϕ normalizations with the same method we used to separate $R(r)$: First, we consider the $e^{im\phi}$ factor in the $Y_{lm}(\theta,\phi)$ previously. The probability of finding the particle at *some* azimuth ϕ between 0 and 2π is 1:

$$\int_0^{2\pi} d\phi\, C^2 \left| e^{im\phi} \right|^2 = 1 \quad \Rightarrow \quad C = \frac{1}{\sqrt{2\pi}}, \quad \text{and} \quad \text{pdf}_\phi(\phi) = \frac{1}{2\pi} \text{ per radian.}$$

This is why the spherical harmonics were written in Eq. (2.3) with the $1/\sqrt{2\pi}$ factor near the $e^{im\phi}$ factor.

For any spherically symmetric potential, the ϕ probability density is uniform: the particle is equally likely to be found at any azimuth.

Finally, the θ density follows the (appropriately normalized) associated Legendre function, $P_{lm}(\cos\theta)$:

$$\int_0^\pi \sin\theta\, d\theta\, N^2 \left| P_{lm}(\cos\theta) \right|^2 = 1 \quad \Rightarrow$$

$$\text{pdf}_\theta(\theta) = \sin\theta\, N^2 \left| P_{lm}(\cos\theta) \right|^2, \quad |N| \equiv \sqrt{\frac{(2l+1)}{2}\frac{(l-|m|)!}{(l+|m|)!}},$$

where the Condon–Shortly phase convention determines the sign of N. Thus, the PDFs for ϕ and θ are each separately normalized. Note that the PDF for θ depends on both the quantum numbers l and m. Even when $l=0$, which has a uniform *solid* angle density, θ has a nonuniform density: θ is more likely to be near the equator than the poles, because there is more area covered per unit θ near the equator (Fig. 2.6).

[Some references refer to the P_{lm} as "associated Legendre polynomials," even though they are not, in general, polynomials. The correct term is "associated Legendre functions." The *ordinary* Legendre polynomials $P_l = P_{l0}$ are, indeed, polynomials.]

2.6   Adjoints

The "adjoint" of an operator is another operator. Adjoints are an essential part of Dirac algebra, being used in proofs, and allowing the crucial concept of a **self-adjoint operator**: an operator which is its own adjoint (aka "Hermitian" operator). All quantum operators corresponding to observable properties of systems are self-adjoint. Self-adjoint operators have real eigenvalues, which are required since observable properties are real. If you are comfortable with matrices, you may want to preview the section on finite-dimensional adjoints in the Matrix Mechanics chapter, since adjoints are easier to understand in finite dimensions.

We usually think of operators as acting to the right on a ket. However, operators can also be thought of as acting to the left on a bra. (In matrix mechanics, these correspond to (respectively) premultiplying a vector (ket) by a matrix, and postmultiplying a row vector (bra) by a matrix.) Recall that, when acting to the right, an operator produces a new ket from a given ket. Since we can alternatively think of operators as acting to the left, we can also say that an operator produces a new bra from a given bra. However, in general, the effect of an operator acting to the left on a bra is different than the effect of the operator acting to the right on a ket. We now show that we can define the left action of an operator to make inner products associative.

This section assumes you understand the dual nature of kets and bras: the "dual" of a vector (ket) $|\psi\rangle$ is a bra, written $\langle\psi|$. The bra is also sometimes called the "adjoint" of the ket. (This definition applies to both wave mechanics and matrix mechanics. When written in Dirac notation, all the formulas apply equally well to wave and matrix mechanics, which illustrates again the utility of Dirac notation.) For this section, we define a notation to help distinguish this dual-conjugate (relating kets to or from bras) from operator adjoints (following [16]). In many references, the † symbol is used for both. For clarity, we define:

$$|\psi\rangle^{\dagger} \equiv |\psi\rangle^{DC} = \langle\psi|, \qquad \text{and} \qquad \langle\psi|^{\dagger} \equiv \langle\psi|^{DC} = |\psi\rangle.$$

For example,

$$|\psi\rangle = \psi(x) \qquad \Rightarrow \qquad |\psi\rangle^{DC} \equiv \langle\psi| = \psi^{*}(x).$$

In a 2D vector space:

$$|\psi\rangle = \begin{pmatrix} a \\ b \end{pmatrix} \qquad \Rightarrow \qquad |\psi\rangle^{DC} \equiv \langle\psi| = \begin{pmatrix} a^{*} & b^{*} \end{pmatrix}.$$

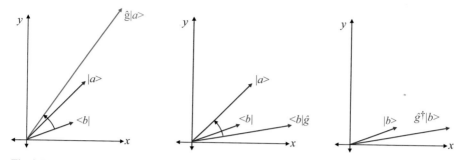

Fig. 2.7 Example of inner product of vectors with an operator. (*Left*) The operator acts to the right. (*Middle*) The operator acts to the left. (*Right*) The adjoint acting to the right

Note that for discrete vectors and matrices, † means "complex-conjugate-transpose," whereas * means just "complex-conjugate."

2.6.1   Continuous Adjoint Operators

We now take an example, simpler than QM, to illustrate the concept of an adjoint operator. Consider a set of 2D vectors in the x-y plane (arrows in space). We define the inner product as the usual dot-product (product of the parallel components of the two vectors). We define an operator $\hat{g}$ that rotates a vector $10°$ counter-clockwise and doubles its magnitude (Fig. 2.7).

In this simplified example, there is no distinction between a bra and a ket; they are both vectors in a plane, so the vector $\langle a|$ is the same as $|a\rangle$. We can take the inner product of two arbitrary vectors and $\hat{g}$:

$$\langle b|\hat{g}|a\rangle \qquad \text{(a number)}.$$

We already defined $\hat{g}$'s action to the right (Fig. 2.7, *left*). We *define* its action to the left as that required to keep the inner product invariant, whichever way we choose to think of $\hat{g}$'s action (to the right or left). This means that $\hat{g}$ acting to the left rotates the vector clockwise (CW) (opposite to how it acts to the right), but still doubles the size of the vector (Fig. 2.7, *middle*). In this way, the inner product remains unchanged. Essentially, we have made the previous inner product associative:

$$\langle b|(\hat{g}|a\rangle) = ((\langle b|\hat{g})|a\rangle) \qquad \text{(associative)}.$$

Note that the action of $\hat{g}$ to the right completely defines how it must act to the left, to keep every inner product invariant.

We define the "adjoint" of $\hat{g}$, written $\hat{g}^\dagger$, as the operator which acts to the right the way $\hat{g}$ acts to the left. In this example, $\hat{g}^\dagger$ rotates the vector $10°$ CW, and doubles its size (Fig. 2.7, *right*). Note that the adjoint is *not* the inverse of the original operator, but its definition depends on the left-action of the given operator, which is defined by our requirement that the inner product be invariant whichever side the operator acts on. (There is a special class of operator in which the adjoint *is* the inverse operator. Such operators are called **unitary**. We discuss these later.)

Similarly in QM, we *define* the action of an operator to the left to be the action that leaves every inner product unchanged whether we think of the operator acting to the right or left:

$$\langle a|\hat{O}|b\rangle = \underbrace{\langle a|\left(\hat{O}|b\rangle\right)}_{\text{acting right}} \equiv \underbrace{\left(\langle a|\hat{O}\right)|b\rangle}_{\text{acting left}}.$$

(In matrix mechanics, this equality follows necessarily from the associativity of matrix multiplication.) Thus, the action of an operator to the right fully defines its action to the left. In general, the left-action of an operator may be different than the right-action. In QM, we have the additional complexity that a bra is different than a ket, so that given an operator $\hat{O}$, we write:

$$\underbrace{\langle a|\hat{O}}_{\text{acting left}} \neq \underbrace{\left(\hat{O}|a\rangle\right)^{DC}}_{\substack{\text{acting} \\ \text{right}}} \qquad\qquad \text{(in general).}$$

Given an operator $\hat{O}$, there is a related operator called the adjoint of $\hat{O}$, which can be defined as:

> The **adjoint** of an operator acts to the *right* in the same way the original operator acts to the *left*.

In other words, the adjoint of an operator is that operator which, when acting to the right, produces a ket from a ket, in the same way that the original operator acting to the left produces a bra from a bra. The adjoint of an operator, $\hat{O}$, is written $\hat{O}^\dagger$, and pronounced "o dagger." Mathematically:

$$\hat{O}^\dagger|\psi\rangle \equiv \left(\langle\psi|\hat{O}\right)^{DC}. \qquad\qquad (2.4)$$

Recalling that $\langle a|b\rangle = \langle b|a\rangle^*$, this definition implies the adjoint satisfies, for all kets $|\phi\rangle$ and $|\psi\rangle$:

$$\langle\psi|\hat{O}|\phi\rangle = \left(\langle\psi|\hat{O}\right)|\phi\rangle = \langle\phi|\hat{O}^\dagger|\psi\rangle^*.$$

(Some references use this equation as the definition of operator adjoint.) By letting $\hat{O} \to \hat{O}^\dagger$, swapping $|\psi\rangle$ and $|\phi\rangle$, and complex conjugating, we learn about the adjoint of the adjoint:

$$\langle\psi|\hat{O}|\phi\rangle = \langle\phi|\hat{O}^\dagger|\psi\rangle^* = \langle\psi|\hat{O}^{\dagger\dagger}|\phi\rangle \qquad \Rightarrow \qquad \hat{O}^{\dagger\dagger} = \hat{O}.$$

We thus see that the adjoint operator also satisfies the reverse of its definition: the adjoint operates to the *left* the way the original operator acts to the *right*.

In wave-mechanics notation, the adjoint satisfies, for all functions $\psi(x)$:

$$\hat{O}^\dagger \psi(x) = \left(\psi(x)\hat{O}\right)^*.$$

The identity $\langle\psi|\hat{O}|\phi\rangle = \langle\phi|\hat{O}^\dagger|\psi\rangle^*$ becomes:

$$\int_{-\infty}^{\infty} \psi^*(x)\hat{O}\,\phi(x)\,dx = \int_{-\infty}^{\infty}\left[\hat{O}^\dagger\psi(x)\right]^*\phi(x)\,dx = \left(\int_{-\infty}^{\infty}\phi^*(x)\left[\hat{O}^\dagger\psi(x)\right]dx\right)^*.$$

for all vectors in the vector-space (e.g., **square-integrable** complex functions $\phi(x)$ and $\psi(x)$). Recall that square integrable functions are those whose squared-magnitude integrates over all space to a finite value. For example, in one dimension:

$$\langle\psi|\psi\rangle \equiv \int_{-\infty}^{\infty}|\psi|^2 dx = \int_{-\infty}^{\infty}\psi^*(x)\psi(x)\,dx < \infty \qquad \text{(is finite)}.$$

Square-integrability is important to the definitions and uses of QM adjoints. For example, the momentum operator is only Hermitian in the space of square-integrable functions. (In some disciplines, square-integrable functions are called "finite energy." We avoid that term here, because square-integrability has nothing to do with particle energy.)

We will see that in some cases, the adjoint of an operator is itself! Such an operator is called **self-adjoint**, or **Hermitian**, and satisfies:

$$\hat{h}|\psi\rangle = \left(\langle\psi|\hat{h}\right)^{DC} \qquad \text{(self-adjoint operator } \hat{h}\text{)}.$$

It can be shown that all operators for observable properties of a quantum system must be self-adjoint.

As examples of adjoints in action, we use the familiar harmonic oscillator lowering operator $\hat{a}$, and the spin lowering operator $\hat{s}$. We have:

$$\hat{a}|2\rangle = \sqrt{2}|1\rangle \quad \Rightarrow \quad \langle 2|\hat{a}^\dagger = \sqrt{2}\langle 1|, \quad \text{and}$$

$$\hat{s}_-|2\;2\rangle = 2|2\;1\rangle \quad \Rightarrow \quad \langle 2\;2|\hat{s}_-^\dagger = \langle 2\;2|\hat{s}_+ = 2\langle 2\;1|.$$

2.6.2 Evaluating Adjoints

How do we find the adjoint of an operator? To evaluate an adjoint, we usually first express the operator in some basis, typically the position basis (x-representation). All of the following rules for evaluating adjoints follow directly from the definition of adjoints given previously. These few rules will get you through almost all adjoints:

1. Multiplication by any complex-valued function of x:

$$z(x)^\dagger = z^*(x).$$

Note that multiplication by a constant is a special case of a (constant valued) function of x, so

$$(c)^\dagger = (c^*), \qquad \text{and in particular, multiplication by } i, (i)^\dagger = -i.$$

Note also that multiplication by a real-valued function of x is an important case because the conjugate of a real number is itself. Hence, multiplication by a real-valued function of x is self-adjoint, aka Hermitian. For example,

$$x^{n\dagger} = x^n \qquad \text{(where } n \text{ is any constant)},$$
$$\cos(x)^\dagger = \cos(x), \text{ etc.}$$

2. Addition of operators: $\qquad \left(\hat{A} + \hat{B}\right)^\dagger = \hat{A}^\dagger + \hat{B}^\dagger$.

3. Composition of two operators: $\left(\hat{A}\hat{B}\right)^\dagger = \hat{B}^\dagger \hat{A}^\dagger$.

4. Differentiation: $\qquad\qquad \left(\dfrac{\partial}{\partial x}\right)^\dagger = -\dfrac{\partial}{\partial x} \qquad$ (for square-integrable functions).

This last adjoint is calculated from integration by parts: $\int U\, dV = UV - \int V\, dU$. Given a dot product:

$$\int_{-\infty}^{\infty} \phi^*(x) \frac{\partial}{\partial x} \psi(x)\, dx,$$

integration by parts lets us exchange the derivative of ψ in favor of the derivative of ϕ:

Let $\qquad U = \phi^*(x), \qquad dU = \dfrac{\partial}{\partial x}\phi^*(x), \qquad\qquad dV = \dfrac{\partial}{\partial x}\psi(x)\, dx.$

Then $\qquad V(x) = \int dV = \int \dfrac{\partial}{\partial x}\psi(x)\, dx = \psi(x).$

Therefore,

$$\int_{-\infty}^{\infty} \phi^*(x) \frac{\partial}{\partial x} \psi(x)\, dx = UV - \int V\, dU = \left[\cancel{\phi^*(x)\psi(x)}\right]_{-\infty}^{\infty} - \int_{-\infty}^{\infty} \psi(x) \frac{\partial}{\partial x}\phi^*(x)\, dx$$

$$= 0 - \int_{-\infty}^{\infty} \psi(x) \frac{\partial}{\partial x}\phi^*(x)\, dx$$

$$= \int_{-\infty}^{\infty} \left[-\frac{\partial}{\partial x}\phi(x)\right]^* \psi(x)\, dx \qquad \Rightarrow \qquad \left(\frac{\partial}{\partial x}\right)^\dagger = -\frac{\partial}{\partial x}.$$

Note that square integrability requires that $\phi(x)$ and $\psi(x)$ both go to zero at both $\pm$ infinity, so the UV term is zero. UV vanishes only for well-behaved functions that go to zero at the infinities, which is required for the adjoint of $\partial/\partial x$ to be well-defined.

Thus, we see that $\partial/\partial x$ is *not* Hermitian, but $\partial^2/\partial x^2$ *is* Hermitian, by rules 4 and 3. This is critical, since Hamiltonians usually contain $\partial^2/\partial x^2$ terms, and the Hamiltonian is a Hermitian operator for the observable energy.

Momentum Example: The momentum operator is Hermitian. In the x-representation, $\hat{p} = -i\hbar\dfrac{\partial}{\partial x}$. This is the composition of two operators: $(-i\hbar)$ and $\partial/\partial x$. So the adjoint is:

$$\hat{p}^\dagger = \left(-i\hbar\frac{\partial}{\partial x}\right)^\dagger = \left(\frac{\partial}{\partial x}\right)^\dagger (-i\hbar)^\dagger = \left(-\frac{\partial}{\partial x}\right)(i\hbar) = -i\hbar\frac{\partial}{\partial x} = \hat{p}.$$

Hence, the momentum operator is Hermitian, as it must be to correspond to an observable property.

2.6.3 Adjoint Summary

We have shown that we can define a left-action of an operator, in terms of its right-action, as that required to preserve the value of all inner products regardless of which way we think of the operator acting. In other words, we make inner products $\langle a|\hat{o}|b\rangle$ associative. The left action then defines the "adjoint" operator, which acts to the right the way the original operator acts to the left. With respect to adjoints, there are two special classes of operators: "Hermitian" operators are self-adjoint: they are their own adjoint operator. "Unitary" operators have an adjoint which is also the operator inverse. We have much to say about these two special classes of operators throughout the rest of this book.

2.7 The WKB Approximation

Since few QM bound states can be computed exactly, approximation methods are very important for computing practical quantities. In particular, knowing the quantized energy levels for the bound states of a given potential is quite important, because it gives us the radiation spectrum. For bound states with moderate-to-high energies, the WKB approximation provides a good estimate of quantized energy levels [18, p. 277]. (WKB can also be used to estimate tunneling probabilities but we do not address that here [18, p. 278].) WKB is named for Gregor Wentzel, Hans Kramers, and Léon Brillouin, three of many people who described the method early on. WKB goes by other acronyms, as well.

The WKB approximation can be used for any 1D problem, including a separated variable from a 2D or 3D problem, such as the radial equation from a central potential. The energy spectrum is our concern here; we do not focus on approximating the wave-function itself. WKB also introduces the semiclassical approximation for momentum, which is used in many situations.

For WKB to apply, the potential energy function which binds the particle must be either smooth near the classical turning points or essentially infinite at a hard edge (more later). We take the following approach:

1. Review of qualitative consideration of high energy wave-functions.
2. Understand the wave-function cycles in the bowl of a bound state.
3. Understand the wave-function near the edges (turning points).
4. Match the bowl solution to the edge solutions, to arrive at the quantization condition.
5. From the quantization condition, we determine the quantized energies for integer n, $E(n)$.
6. Examine the validity conditions: when is WKB a valid approximation?

High Energy Wave-Functions: We consider again Fig. 1.12, a particle-in-a-box, with a potential step at the bottom. The first time we considered it, we noted that $\psi(x)$ is larger amplitude, and longer wavelength (smaller k) on the left, where the particle's local speed is "slow" compared to the right.

For energy levels and spectra from WKB, we are mostly concerned with the wavelength of ψ. Therefore, we compute a local value of k based on the local momentum:

$$\frac{p(x)^2}{2m} = \frac{(\hbar k(x))^2}{2m} = E - V(x) \qquad \Rightarrow \qquad k(x) = \frac{\pm\sqrt{2m(E-V(x))}}{\hbar} \text{ (rad/m)}. \tag{2.5}$$

The stationary state $\psi(x)$ is real, and is everywhere a superposition of positive and negative momenta. The real-valued sinusoids oscillate according to $|k(x)|$. However, we will consider only the complex-valued right-moving component of $\psi(x)$, so we can take $k(x) > 0$ from now on. Because $\psi(x)$ oscillates, we can write it as a product of a real-valued amplitude and a real-valued phase:

$$\psi(x) = A(x)e^{i\phi(x)} \qquad A(x), \phi(x) \text{ real}.$$

Thus, we can say $\psi(x)$ has a phase, $\phi(x)$, at each point. Note that as usual, the energies are quantized by the boundary conditions on ψ, in this case, that $\psi(x) = 0$ at the edges. For ψ to be zero at both edges ($x=0$ and $x=c$), and if $A(x)$ is smooth, the phase must satisfy:

$$\phi(c) - \phi(0) = n\pi, \qquad where \quad n \text{ is an integer quantum number}.$$

Thus, the allowed energies, i.e., the spectrum, of the system are those energies that make n an integer.

For WKB, we mostly consider the phase of ψ, but to later determine the validity of approximation, we briefly reiterate its amplitude characteristics. On the left of the step-bottom box in Fig. 1.12, the particle moves "slowly" and on the right it moves "quickly." The particle "spends more time" on the left, so the amplitude of ψ is larger (more likely to find the particle there). Conversely, the particle "spends less

Fig. 2.8 A bound particle in a potential with energy suitable for the WKB approximation

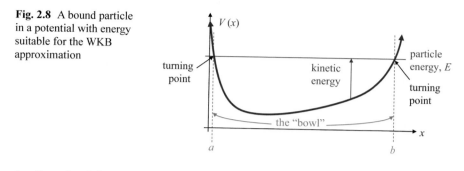

time" on the right, so the amplitude of ψ is smaller (less likely to find the particle there).

With those considerations in mind, we proceed to WKB: given a 1D potential, $V(x)$, such as Fig. 2.8, find the energy levels of high-energy bound states.

For the approximation to hold, $V(x)$ must vary "slowly" compared to the wavelength $\lambda = 2\pi/k$. We will examine this requirement more closely at the end.

Wave-Function Cycles in a Bowl: Recall the harmonic equation:

$$\psi''(x) + k^2 \psi(x) = 0 \qquad where \quad k \equiv \text{spatial-frequency} = \text{constant}.$$

The solutions are sinusoids of spatial frequency k, rad/m. If we now let k be a function of x that varies slowly, we expect the solutions to be approximately sinusoids with a position-dependent frequency $k(x)$. Therefore, in the middle region of the potential (the "deep bowl"), we approximate the spatial frequency (wave-number) from Eq. (2.5) as a function of x:

$$k(x) = \frac{\sqrt{2m(E - V(x))}}{\hbar}, \qquad where \quad E \text{ is the total particle energy.}$$

$\psi(x)$ oscillates in the bowl, with approximate spatial frequency $k(x)$. $k(x)$ is the rate of change with distance of the phase ϕ of $\psi(x)$. We focus on this phase because it is the phase change of ψ that will determine the quantization condition. By this approximation, the total phase change of ψ in the bowl would be:

$$k(x) \equiv \frac{d\phi(x)}{dx} \quad \Rightarrow \quad \Delta\phi = \int_a^b dx \, k(x) = \int_a^b dx \, \frac{\sqrt{2m(E - V(x))}}{\hbar}. \tag{2.6}$$

The wave-function near the edges: Now consider the wave-function near a turning point (the right one, $x = b$). The bowl approximation is good so long as $V(x)$ varies little in one wavelength of ψ. For short wavelengths, $V(x)$ more easily meets this condition, i.e., so long as the kinetic energy of ψ is "large." Near the turning point, $k(x)$ goes to zero, and the wavelength gets arbitrarily long. Hence, the "slow" variation of $V(x)$ cannot be satisfied near the turning point, and we must investigate this region in more detail.

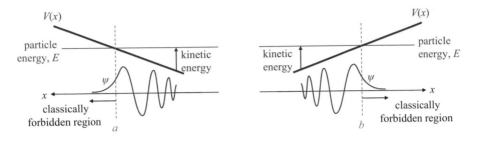

Fig. 2.9 The wave-function near left and right turning points

Qualitatively, we expect the wave-function near an edge to look something like Fig. 2.9.

The frequency increases and amplitude decreases inside the bowl (increasing kinetic energy). Outside the bowl, in the classically forbidden region, ψ decreases rapidly. We approximate the potential as a straight line near the turning point, since smooth functions are well approximated by their first derivative in a small region. This allows us to approximate the Schrödinger equation near (say) the right turning point:

$$V(x) \approx E + V'(b)(x-b) \quad where \quad V'(b) \equiv \left.\frac{\partial V}{\partial x}\right|_{x=b} \quad \text{is the slope of the potential at } b.$$

Then: $E\psi(x) = -\frac{\hbar^2}{2m}\psi''(x) + V(x)\psi(x) \approx -\frac{\hbar^2}{2m}\psi''(x) + \left[E + V'(b)(x-b)\right]\psi(x).$

$E\psi(x)$ cancels from both sides. Putting the rest in standard differential equation form, we have:

$$\psi''(x) - \frac{2mV'(b)(x-b)}{\hbar^2}\psi(x) = 0. \tag{2.7}$$

This is of the form of a standard equation, whose (bounded) solution is a special function called the Airy function, Ai(x) (which can be written as Bessel functions). (The "i" here is part of the two-character abbreviation for Airy; it is not $\sqrt{-1}$). In general, the special function Ai(y) is defined as the bounded solution to the linear differential equation:

$$\psi_c''(y) - y\psi_c(y) = 0 \quad\quad \text{(standard equation satisfied by Ai(y)).}$$

Comparing this to the harmonic equation, we expect that if $y < 0, \psi_c(y)$ will oscillate, and if $y > 0, \psi_c(y)$ will decay rapidly. Indeed, Ai(y) looks like Fig. 2.10 (*blue*). It oscillates when $y < 0$, and decays rapidly for $y > 0$.

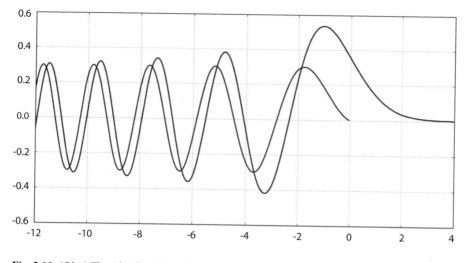

Fig. 2.10 (*Blue*) The Airy function, Ai(y). (*Red*) (0.3)sin[2/3(−x)³ᐟ²]

A "scale and offset" change of variables from x to y gets us from the Schrödinger equation to this standard equation. To find such a change of variables, we write the Schrödinger Eq. (2.7) in simpler form:

$$\psi''(x)-(c)(x-b)\psi(x)=0 \qquad where \quad c \equiv \frac{2mV'(b)}{\hbar^2}.$$

We apply a scale change of the form $x = fy + b$, where f and b (the turning point) are constants:

$$x(y) = fy + b, \qquad\qquad x'(y) = f,$$
$$\psi_c(y) \equiv \psi(x(y)), \qquad\qquad \psi_c'(y) = \psi'(x(y))x'(y) = f\psi'(x(y)),$$
$$\psi_c''(y) = f^2\psi''(x(y)).$$

$$\psi''(x)-(c)(x-b)\psi(x)=0 \quad \rightarrow \quad \frac{\psi_c''(y)}{f^2}-cfy\psi_c(y)=0,$$

$$or \quad \psi_c''(y)-cf^3 y\psi_c(y)=0.$$

To achieve standard form, we must choose f such that $cf^3=1$ or $f=c^{-1/3}$. Then:

$$y = \frac{x-b}{f} = \left(\frac{2mV'(b)}{\hbar^2}\right)^{1/3}(x-b) \qquad\qquad \Rightarrow$$

$$\psi(x) = \psi_c(y) \approx N\,\mathrm{Ai}\left[\left(\frac{2mV'(b)}{\hbar^2}\right)^{1/3}(x-b)\right] \qquad where \quad N \text{ is for normalization.}$$

This result is approximate because we approximated $V(x)$ as a straight line through the turning point.

As noted earlier, we focus on the phase of ψ, and not on its amplitude. The Airy function has an asymptotic form far to the left of zero [8, 8.20 p. 173]:

$$\psi_c(y) \sim \text{Ai}(y) \underset{\substack{y \text{ big} \\ \text{negative}}}{\rightarrow} B(y) \sin\left(\frac{2}{3}(-y)^{3/2} + \frac{\pi}{4}\right)$$

where $B(y)$ is some amplitude function   [8, 8.20 p. 173].

Considering only the phase (the argument of the sine function) far to the left, we have:

$$\phi(y) \rightarrow \frac{2}{3}(-y)^{3/2} + \frac{\pi}{4}, \qquad\qquad y \text{ far to the left of zero.} \qquad (2.8)$$

Match the Bowl Solution to the Edge Solutions: In the simple particle-in-a-box (with infinite potential walls), the quantization condition is that there must be an integral number of half-wave-lengths between the walls of the box. This is required for the wave-function to be continuous, and to be exactly zero outside the box. This quantization condition determines the allowed energies of a particle in the box. We will use a similar requirement to find the quantization condition for a particle in a "bowl."

We have seen that the deep-bowl approximation fails near the turning points, and we cannot use $k(x) \approx \sqrt{2m(E - V(x))}/\hbar$. However, let us ignore this failure for a moment: we consider the potential to be a straight line, but naively use the bowl approximation all the way to the turning point. In that case, at the right turning point, we define $\phi(b)=0$. The phase accumulated from $x<b$ to b is:

$$\Delta\phi(x) = \int_x^b dx'\, k(x') = \int_x^b dx' \, \frac{\left[2mV'(b)(b-x')\right]^{1/2}}{\hbar}$$

$$= \frac{\left[2mV'(b)\right]^{1/2}}{\hbar} \frac{2}{3}(b-x)^{3/2} = \frac{2}{3}(-y)^{3/2}.$$

Compare this to the asymptotic phase of the exact Airy solution, previous Eq. (2.8), and the red curve in Fig. 2.10: far to the left of zero, the difference between the phase from the Airy solution and our naive integral is just $\pi/4$ radians. In the graph, moving left from the turning point ($y=0$), the Airy function reaches a given phase (a given value of $\Delta\phi(x)$) "sooner" than the sine function, because it has a "head start" of $\pi/4$ radians.

Now note that the left turning point is just a mirror image of the right, with the same $\pi/4$ phase difference from the naive integral. Therefore, the total phase shift between the turning points is:

$$\phi(b) - \phi(a) = \frac{\pi}{4} + \frac{\pi}{4} + \int_a^b dx \, \frac{\sqrt{2m(E-V(x))}}{\hbar} = n\pi. \quad \text{(quantization condition).} \quad (2.9)$$

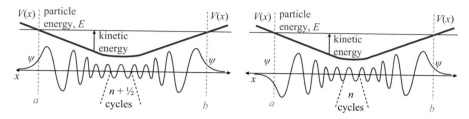

Fig. 2.11 The stationary state wave-function must fit an exact number of half-cycles between the sinusoidal limits of the Airy functions at each edge

Similar to the particle-in-a-box, for the wave-function to fit between the turning points, with the phase needed to match the Airy functions at each side, the total phase shift between the turning points must be a multiple of half cycles, i.e., a multiple of π. This is the quantization condition, which determines the allowed energies. "n" is the quantum number. Note that the phase shift need not be a multiple $(2n+1)\pi$ because the Airy function on either edge can be negated and still be a solution (Fig. 2.11, *right*).

From the Quantization Condition, Find the Allowed Energies: To find the quantized energy levels, we first note that the turning points a and b are themselves functions of the particle energy, E:

$$V\big(a(E)\big) = E, \qquad V\big(b(E)\big) = E \qquad \Rightarrow \qquad a(E) = V^{-1}(E), \quad b(E) = V^{-1}(E).$$

For a potential with bound states, $V^{-1}(E)$ is necessarily a multivalued "function," providing the two values $a(E)$ and $b(E)$. We can now define the naive phase integral, Eq. (2.6), as a function of E:

$$\Phi(E) \equiv \int_{a(E)}^{b(E)} dx\, \frac{\sqrt{2m\big(E - V(x)\big)}}{\hbar}.$$

For a given integer quantum number $n \geq 1$, its energy E_n must satisfy the quantization condition Eq. (2.9):

$$\Phi(E_n) = \left(n - \frac{1}{2}\right)\pi.$$

In principle, we can invert the function $\Phi(E)$ to find the allowed energies for (integer) quantum numbers, n:

$$E_n = \Phi^{-1}\left[\left(n - \frac{1}{2}\right)\pi\right].$$

In practice, $\Phi(E)$ may be hard or impossible to invert in closed form. Indeed, $\Phi(E)$ itself may be hard or impossible to write in closed form. (Some references use $(n+\frac{1}{2})$ instead of $(n-\frac{1}{2})$ in the previous equation, which means that n starts at 0

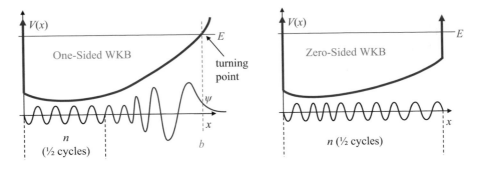

Fig. 2.12 (*Left*) A 1-side WKB approximation. (*Right*) A zero-sided WKB approximation

instead of 1; however, we use the "minus" form for consistency with the one-sided case later.)

The energy spacing (= radiation spectrum) is then well approximated by the energy difference between two adjacent quantum numbers n:

$$\Delta E = e_{n+1} - E_n \approx \frac{dE(n)}{dn},$$

where $E(n)$ is the continuous function for E_n. The derivative approximation gets better for larger n.

Note that there is also a one-sided WKB approximation, if only one edge has a smooth turning point (Fig. 2.12, *left*) [18, p. 277]. $\psi(a)$ must then go to zero (as with a particle-in-a-box), and there is no $\pi/4$ phase shift on the left.

For the one-sided case, we get only one phase shift of $\pi/4$, so the quantization condition is [16, 2.4.50 p. 108]:

$$\Phi(E_n) = \left(n - \frac{1}{4}\right)\pi \qquad \Rightarrow \qquad E_n = \Phi^{-1}\left[\left(n - \frac{1}{4}\right)\pi\right]. \qquad (2.10)$$

There is even a zero-sided WKB approximation, if both sides are "hard" (Fig. 2.12, *right*). For the zero-sided case, we have simply:

$$\Phi(E_n) = n\pi \qquad \Rightarrow \qquad E_n = \Phi^{-1}(n\pi).$$

(Aside: Interestingly, the WKB approximation leads directly to the "old" (i.e., wrong) Wilson–Sommerfeld quantization rule [8, p. 177, 14, p. 107]. The Schrödinger equation did not exist yet, and physicists believed that classical mechanical action was quantized to integer multiples of h. When n is large, the phase shift of $\pi/2$ or $\pi/4$ has negligible effect on the energy *spacing*, and the energy spacing satisfies the Wilson–Sommerfeld quantization rule.)

How fast can $V(x)$ vary? We now consider under what conditions the WKB approximation is valid. This puts to use several QM and mathematical concepts, illustrating a practical application for understanding the meaning of the wave-function. Finally, we interpret the approximation conditions physically, and describe why they fail for a particle-in-a-box.

We have noted that $V(x)$ must vary "slowly," which we must quantify. To highlight the oscillatory nature of $\psi(x)$ in the bowl, we rewrite the time-independent Schrödinger equation to look more like the harmonic equation ($\psi(x)$ oscillates in space, but since it is a stationary state, we are not concerned with its time dependence):

$$\psi''(x) + \frac{2m}{\hbar^2}(E - V(x))\psi(x) = 0 \quad \text{or} \quad \psi''(x) + \frac{p_{cl}^{\,2}(x)}{\hbar^2}\psi(x) = 0 \quad \text{(exact)}$$

$$\text{where} \quad p_{cl}(x) \equiv \sqrt{2m(E - V(x))}.$$

$p_{cl}(x)$ is the momentum of a classical particle of energy E at position x. This equation is exact. Experience has shown that we can express our approximation conditions conveniently in terms of p_{cl}. Note that for a "particle-in-a-box," $V(x)$ is constant, the equation is exactly harmonic, and the solutions are exact sinusoids. For both WKB and particle-in-a-box, the boundary conditions on the sinusoidal phase determine the quantization condition.

To determine when the WKB approximation is valid, we use the self-consistency method: we assume the approximation is valid and derive what conditions must be true for that assumption to hold. In particular, as we did earlier, we write $\psi(x)$ in the bowl as:

$$\psi(x) = A(x)e^{i\phi(x)}, \qquad \text{and} \qquad \phi'(x) \equiv k(x) \approx p_{cl}(x)/\hbar.$$

To substitute for $\psi(x)$ into the previous form of the Schrödinger equation, we must evaluate $\psi''(x)$. We first find $\psi'(x)$, and abbreviate $p \equiv p_{cl}(x)$:

$$\psi'(x) = A'(x)e^{i\phi(x)} + A(x)i\phi'(x)e^{i\phi(x)} = A'(x)e^{i\phi(x)} + iA(x)\frac{p}{\hbar}e^{i\phi(x)}.$$

You might think that, if $A'(x) << A(x)p/\hbar$, we can drop the first term. However, this is not generally true, because when we take the second derivative, we will introduce another factor of $\phi'(x) \sim p$, which increases the "power" of p in that term. In other words:

Just because a term is small, does not insure that its derivative is small.

Therefore, we keep all terms in the second derivative:

$$\psi''(x) = A''(x)e^{i\phi(x)} + iA'(x)\frac{p}{\hbar}e^{i\phi(x)}$$

$$+ iA'(x)\frac{p}{\hbar}e^{i\phi(x)} + iA(x)\frac{p'}{\hbar}e^{i\phi(x)} + iA(x)\frac{p}{\hbar}i\phi'(x)e^{i\phi(x)}$$

$$= A''(x)e^{i\phi(x)} + 2iA'(x)\frac{p}{\hbar}e^{i\phi(x)} + iA(x)\frac{p'}{\hbar}e^{i\phi(x)} - \frac{p^2}{\hbar^2}A(x)e^{i\phi(x)}.$$

The last term is what will satisfy the Schrödinger equation, $\psi''(x) + \frac{p^2}{\hbar^2}\psi(x) = 0$, and therefore, all other terms must be small compared with it for the approximation to be valid. The $e^{i\phi(x)}$ factor cancels from all terms, leaving:

$$A''(x) + 2iA'(x)\frac{p}{\hbar} + iA(x)\frac{p'}{\hbar} \ll \frac{p^2}{\hbar^2}A(x). \tag{2.11}$$

Now note that we can scale $|A(x)|$ in terms of p. As noted in Eq. (8.5), Chap. 8, the probability of finding a particle in an interval dx is inversely proportional to its speed, and in QM, is also proportional to $|A(x)|^2$. Therefore,

$$|A(x)|^2 \sim \frac{1}{v} \sim \frac{1}{p} \quad \Rightarrow \quad A(x) \sim p^{-1/2}, \quad A'(x) \sim -\frac{1}{2}p^{-3/2}p',$$

$$\text{and} \quad A''(x) \sim \left(\frac{3}{4}p^{-5/2}(p')^2 - \frac{1}{2}p^{-3/2}p''\right).$$

We have omitted the normalization for $A(x)$, but our condition for the approximation to be valid, Eq. (2.11), is linear in $A(x)$, and therefore the normalization factor cancels from all terms. Substituting these results for $A(x)$, $A'(x)$, and $A''(x)$ into our approximation condition Eq. (2.11), we get:

$$\left(\frac{3}{4}p^{-5/2}(p')^2 - \frac{1}{2}p^{-3/2}p''\right) - ip^{-3/2}p'\frac{p}{\hbar} + i\left(p^{-1/2}\right)\frac{p'}{\hbar} \ll \frac{p^2}{\hbar^2}p^{-1/2}.$$

Multiplying through by $\hbar^2 p^{-3/2}$, and rearranging slightly:

$$\frac{3}{4}\hbar^2\frac{(p')^2}{p^4} - \frac{1}{2}\hbar^2\frac{p''}{p^3} \ll 1.$$

The magnitudes of each term on the left must be $\ll 1$. The numeric coefficients are of order 1, so we can drop them. Furthermore, the first term is a perfect square, so we write as such. Our final approximation conditions are (restoring the full notation for $p_{cl}(x)$) [cf. 8, 8.15 p. 172]:

$$\left(\hbar\frac{p_{cl}'(x)}{p_{cl}^2(x)}\right)^2 \ll 1, \quad \text{and} \quad \hbar^2\frac{p_{cl}''(x)}{p_{cl}^3(x)} \ll 1 \quad \text{(approximation conditions)}.$$

These are the conditions for the "semiclassical" approximation to apply, i.e., $k(x) \approx p_{cl}(x)/\hbar$. Sometimes the first condition is written without the square, since if $X \ll 1$, the $X^2 \ll 1$. We retain the square so the two conditions are directly comparable. They must be satisfied everywhere in the bowl, except near the turning points. Near the turning points, the potential must be approximately straight until $p_{cl}(x)$ satisfies the previous conditions. Then the Airy function solution handles the region near the turning point where the previous conditions are not met. Note that we have just introduced a new approximation condition that $V(x)$ must be "straight enough" near the turning points, but we have not quantified that condition. We do not address that further here.

What do the approximation conditions mean physically? The condition that $\hbar p'/p^2$ must be small is saying something about the rate-of-change of momentum compared to inverse momentum or wavelength. Therefore, it is reasonable to express this condition entirely in terms of distance and wavelength. We recognize that [2, 36.19 p. 112]:

$$1 \gg \hbar \frac{p'}{p^2} = \hbar \left| \frac{d}{dx} p^{-1} \right| = \cancel{\hbar} \frac{d}{dx} \frac{\lambda}{2\pi \cancel{\hbar}} \qquad \text{(using } p = \frac{2\pi \hbar}{\lambda}\text{)}$$

$$\Rightarrow \qquad \frac{d\lambda}{dx} \ll 2\pi.$$

That is, the rate of change of wavelength (in meters per meter) must be $\ll 2\pi$. (We retain the 2π, since it is of order 10, rather than 1.) Multiplying both sides of the inequality by λ, we can answer "how much can the wavelength change from one cycle to the next?"

$$\frac{d\lambda}{dx}\lambda \approx (\Delta\lambda \text{ in one cycle}) \ll 2\pi\lambda.$$

In other words, the change in wavelength, $\Delta\lambda$, during one cycle must be small compared to the wavelength of that cycle.

Comparison to Particle-in-a-Box: It is instructive to consider what happens to $\psi(x)$ when we smoothly deform a bowl potential into a hard-walled particle-in-a-box potential. Consider a one-sided WKB bowl, with a turning point on the right (Fig. 2.13, *left*). Let the bottom of the bowl be flat [$V(x)$=constant], until near the right edge, where $V(x)$ rises to become a straight line.

Our quantization condition for a one-sided WKB approximation is Eq. (2.10), i.e., E must satisfy:

$$\Phi(E) = \left(n - \frac{1}{4}\right)\pi \qquad \text{(one-sided WKB approximation)}.$$

Now we deform $V(x)$ so that the right edge gets steeper and steeper (above right). The potential approaches a particle-in-a-box, for which the quantization condition is:

$$\Phi(E) = n\pi \qquad \text{(particle-in-a-box)}.$$

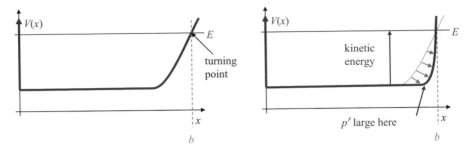

Fig. 2.13 (*Left*) A one-side WKB approximation. (*Right*) Deforming the potential to a particle-in-a-box

Somehow, in the transition from one-sided WKB to particle-in-a-box, we lost the $\pi/4$ phase. We must have violated one or both of the WKB approximation conditions. In fact, when $V(x)$ approaches particle-in-a-box, p changes rapidly near the right knee of $V(x)$, quickly going from finite to zero. When p changes rapidly, it means p' is large, which violates the condition $\hbar p'/p^2 <<1$. Hence, we cannot deform the WKB approximation into a hard-walled potential. Fortunately, we know that the solution at a hard wall requires simply that $\psi(x)$ at the wall be zero.

2.8 Long-Distance Operators

This section attempts to clarify some subtleties of spatial operators.

The position and momentum bases (basis functions) are parameterized by continuous observables: position and momentum. Between any two eigenvalues (position or momentum) there are an infinite number of basis functions and eigenvalues, i.e., there is a continuum of basis functions. (In fact, there are an uncountably infinite set of basis functions.) *Observable* operators in continuous bases are simple multiplier or differential operators, e.g., in the position basis:

$$\hat{p} = \frac{\hbar}{i}\frac{\partial}{\partial x}, \qquad \hat{p}^2 \equiv \hat{p}\hat{p} = -\hbar^2\frac{\partial^2}{\partial x^2}, \qquad \hat{x} = x \qquad \text{(position basis)}.$$

For these simple operators, at any given point, the only contribution to the value of the operation on a spatial function comes from that point, or at most, a differentially small neighborhood of that point. Distant points do not contribute, and therefore distant places in the spatial function do not interact.

Mathematical operators (as opposed to observable operators) do not necessarily have such simple differential forms. We now consider some important operators: space translation, rotation, time evolution, and parity. Later, we will use the concepts of rotations, angular momentum, and generators, developed here, to deduce the quantization rules for angular momentum, without the need to solve the Schrödinger equation.

2.8.1 Introduction to Generators

The 1D **space translation operator** produces a function which is the given function shifted in space by a constant distance a:

$$\phi(x) \equiv \hat{T}(a)\psi(x) = \psi(x-a)$$

$$\text{where} \quad \hat{T}(a) \equiv \text{translation by distance } a \text{ (not kinetic energy)}.$$

$\hat{T}(a)$ is really a family of operators, parameterized by the translation distance, a, where a is a constant. $\hat{T}(a)$ is easily defined implicitly, as previously, but there is no *simple* mathematical operation that can shift a function in space by a finite amount. However, we can write $\hat{T}(a)$ as the composition of an infinite number of *infinitesimal* translations. For finite a:

$$\hat{T}(a) = \lim_{N\to\infty} \left[\hat{T}\left(\frac{a}{N}\right) \right]^N \quad \text{where} \quad a \equiv \text{finite, constant displacement distance.}$$

$$(2.12)$$

Conveniently, there *is* a simple operator for an *infinitesimal* translation, by a distance dx:

$$\psi(x) \to \psi(x-dx) = \psi - \frac{\partial \psi}{\partial x}dx = \left(1 - dx\frac{\partial}{\partial x}\right)\psi = \left(1 - \frac{i}{\hbar}\hat{p}\,dx\right)\psi$$

$$\Rightarrow \qquad \hat{T}(dx) \equiv 1 - \frac{i}{\hbar}\hat{p}\,dx.$$

Substituting this into Eq. (2.12), the formula for finite translation distance a, we have:

$$\hat{T}(a) = \lim_{N\to\infty} \left[1 - \frac{i}{\hbar}\left(\frac{a}{N}\right)\hat{p} \right]^N = \exp(-ia\hat{p}/\hbar).$$

Notice that the previous limit yields the exponential of an operator. This is in analogy with numbers, where the same limit yields the exponential of a number:

$$\lim_{N\to\infty}\left(1 + \frac{x}{N}\right)^N = e^x.$$

As we saw earlier with the time evolution operator:

The exponential of an operator is also an operator.

So with exponentials, we *can* write an operator equation for $\hat{T}(a)$ (as we could for, say, $\hat{p}$). When an operator (or transformation), such as $\hat{T}$, is "generated" by the exponential of a number times another operator $(\hat{P})$, we say that $\hat{p}$ is the **generator** of $\hat{T}$. In words, "momentum is the generator of space translation." Note that the translation distance, a, is a free parameter used to generate a particular translation in the family of translation operators.

In this case, the value of $\hat{T}(a)\psi(x)$ is *not* determined solely by a differential neighborhood of x; it is determined by values of $\psi(\cdot)$ that are "far" (i.e., a finite distance) from x. However, $\hat{T}(a)\psi(x)$ *is* determined by a unique differential neighborhood, just not the local neighborhood. Therefore in this case, there is still no interaction between separated neighborhoods; each neighborhood is still independent of all others. With other, more abstract operators, there may be interactions between separated neighborhoods.

Note that $\hat{T}(a)$ preserves the magnitude (or "norm") of its operand, say $\psi(x)$, i.e.,

$$\text{Let}\quad \phi(x) \equiv \hat{T}(a)\psi(x). \quad\text{Then}\quad M^2 \equiv \int_{-\infty}^{\infty} \phi^*(x)\phi(x)\,dx = \int_{-\infty}^{\infty} \psi^*(x)\psi(x)\,dx.$$

An operator which preserves the magnitude of its operand is called **unitary**. It is easy to show that the exponential of any Hermitian operator is unitary.

The previous discussion generalizes straightforwardly to three dimensions, for a translation vector **a**, giving:

$$\hat{T}(\mathbf{a}) = e^{i\mathbf{a}\cdot\hat{\mathbf{p}}/\hbar} \quad\text{where}\quad \mathbf{a}\cdot\hat{\mathbf{p}} \equiv a_x\hat{p}_x + a_y\hat{p}_y + a_z\hat{p}_z.$$

Rotation Operators: A similar situation exists for $\hat{R}_z(\alpha)$, the operator for rotation about the z-axis. Starting with a rotation by an infinitesimal angle, $d\phi$:

$$\psi(r,\theta,\phi) \to \psi(r,\theta,\phi - d\phi) = \psi - \frac{\partial\psi}{\partial\phi}d\phi = \left(1 - d\phi\frac{\partial}{\partial\phi}\right)\psi.$$

$$L_z = \frac{\hbar}{i}\frac{\partial}{\partial\phi} \quad\Rightarrow\quad \hat{R}_z(d\phi) = 1 - \frac{i}{\hbar}\hat{L}_z\,d\phi.$$

Thus, $\hat{L}_z$ is the generator of rotations about the z-axis. For finite angle, α:

$$\hat{R}_z(\alpha) = \lim_{N\to\infty}\left[\hat{R}_z\left(\frac{\alpha}{N}\right)\right]^N = \lim_{N\to\infty}\left[1 - \frac{i}{\hbar}\left(\frac{\alpha}{N}\right)\hat{L}_z\right]^N = \exp\left(-i\hat{L}_z\alpha/\hbar\right).$$

For rotation about an arbitrary unit-vector $\hat{\mathbf{n}} \equiv \dfrac{\alpha}{|\alpha|}$, by $|\alpha|$ radians:

$$\hat{R}(\boldsymbol{\alpha}) = \exp\left(-i\boldsymbol{\alpha}\cdot\hat{\mathbf{L}}/\hbar\right) \quad\text{where}\quad \boldsymbol{\alpha} \equiv \alpha_x\hat{L}_x + \alpha_y\hat{L}_y + \alpha_z\hat{L}_z$$

$$\equiv \text{rotation parameter (\textit{not} a vector).}$$

We see that the angular momentum vector operator is the generator of rotations about any axis. Much more on this later.

As before, $\hat{R}(\boldsymbol{\alpha})$ is unitary, since rotating a function does not change its "magnitude". Be careful: $\boldsymbol{\alpha}$ is *not* a vector; it is the product of an angle times a unit vector (rotation axis):

$$\boldsymbol{\alpha} \equiv |\boldsymbol{\alpha}|\,\hat{\mathbf{n}} \qquad where \qquad |\boldsymbol{\alpha}| \equiv \text{the angle to rotate by.}$$

$\boldsymbol{\alpha}$ is just three parameters describing the rotation. Since finite rotations do not commute, we *cannot* decompose $\boldsymbol{\alpha}$ into x, y, and z "components," and the composition of two rotations, $\boldsymbol{\alpha}_1$ and $\boldsymbol{\alpha}_2$, is *not* $\boldsymbol{\alpha}_1 + \boldsymbol{\alpha}_2$. This latter equation would have to be satisfied for $\boldsymbol{\alpha}_1$ and $\boldsymbol{\alpha}_2$ to be vectors. For more information, see *Quirky Mathematical Physics Concepts*.

As with space translation, each point of the rotated spatial function is determined by only a single point of the original function, and there is no interaction between separated points.

Time Evolution: In contrast, as we have seen, operators such as time evolution, $\hat{U}(t)$, introduce true *interactions* between multiple (in fact, all) distant points of the wave-function:

$$\psi(t,x) \equiv \hat{U}(t)\,\psi(0,x) = \left[\psi(0,x) \text{ evolved by a time } t\right].$$

The value of $\hat{U}(t)\,\psi(x)$ at a point x depends not only on the value of $\psi(\cdot)$ at x, but also on the values of $\psi(\cdot)$ at all other points. Hence, $\hat{U}(t)$ is not a "local operator."

Earlier, we derived the time evolution operator (for time-independent Hamiltonians) from the simple form of time evolution for energy eigenstates. Here, we derive it directly from the Schrödinger equation, and so show that the Hamiltonian is the generator of time translation, i.e., time evolution. Using the same approach as for space-translation and rotation, and assuming a time-independent Hamiltonian (we drop the argument "x" from ψ for notational simplicity, i.e., $\psi(t) \equiv \psi(t,x)$):

$$\psi(t+dt) = \psi(t) + \frac{\partial}{\partial t}\psi(t)\,dt, \quad \text{but from the Schrodinger equation:} \quad \frac{\partial}{\partial t}\psi(t) = \frac{\hat{H}}{i\hbar}\psi(t).$$

$$\psi(t+dt) = \psi(t) + \frac{\hat{H}}{i\hbar}dt\,\psi = \left(1 - \frac{i}{\hbar}\hat{H}dt\right)\psi \qquad \Rightarrow \qquad \hat{U}(dt) = 1 - \frac{i}{\hbar}\hat{H}dt.$$

Then for finite Δt:

$$\hat{U}(\Delta t) = \lim_{N \to \infty}\left[\hat{U}\left(\frac{\Delta t}{N}\right)\right]^N = \lim_{N \to \infty}\left[1 - \frac{i}{\hbar}\hat{H}\frac{\Delta t}{N}\right]^N \equiv \exp\left(-i\hat{H}\Delta t/\hbar\right).$$

The space translation and rotation operators are unique to continuous bases, such as position or momentum. In contrast, we will see that the time evolution operator, and this derivation of it, work for both continuous and discrete bases, because the Schrödinger equation has the same form for continuous and discrete bases. The quantum state previously, $\psi(t)$, could be either a wave-function, or a discrete vector.

It is interesting to note that in more advanced QM, in particular the Dirac equation describing relativistic fermions, the Hamiltonian is *defined* as the generator of time evolution. It is *not* the total energy. This is perhaps similar to some cases in classical Hamiltonian mechanics, where the Hamiltonian is not total energy, but is still the classical generator of time evolution.

2.8.2   Parity

Another example of a nonlocal operator, the **parity operator** $\hat{P}$, is easily defined implicitly:

$$\phi(x) \equiv \hat{P}\psi(x) = \psi(-x).$$

Here again, $\hat{P}$ is not "local;" i.e., a distant point affects the value at a given point, x, but as with space translation and rotation, only a single distant neighborhood (in this case, a single distant *point*) affects the result at x. ($\hat{P}$ can be defined explicitly as a special case of the dilation operator, but this is outside our scope.)

2.9   Vector Spaces and Hilbert Spaces

Linear vector spaces are used throughout science and engineering. QM organizes cleanly into vector spaces, so it will make your life much easier if you understand them. Vector spaces also form a foundation for the future study of group theory. However, for spatial wave-functions alone, you probably do not need all of this section, because you can work most of the mathematics without it. However, it is critically important to understand the concept of a "zero vector" (defined later), and its distinction from the number 0. For example, the zero vector is an essential part of how the Dirac algebra derives energy quantization in the harmonic oscillator.

For angular momentum states, and other finite-dimensional quantum states, you really *need* to understand vector spaces, including the zero vector. Similarly to the harmonic oscillator, the zero vector is essential to deriving the requirement of half-integer multiples of $\hbar$ for all angular momenta.

Vector spaces derive from the common notion of vectors in elementary physics: such a vector is two or three real values, representing a magnitude and direction. The vectors are drawn in either a 2D space (the space of a plane) or a 3D space (the space of volumes). The number of real values in such a vector is the **dimension** of the vector: 2D vectors have two real numbers (x, y). 3D vectors have three real numbers (x, y, z).

Such physically based vector spaces have some important properties, which we detail later. But it turns out that many types of mathematical elements share important properties with physical vector spaces. Thus, it becomes very useful to define *general* vector spaces, called simply **vector spaces**, which are both more general and more abstract than spatial vectors in physical space. We describe vector spaces as follows:

- Overview: a vector space comprises fields, groups, and more.
- Fields.
- Groups.
- Vector spaces, plus inner products, and operators.

2.9.1   Vector Space Overview

Briefly, a **vector space** comprises a "field" of scalars and a "group" of vectors. (We define "field" and "group" shortly.)

> A vector space has the minimum properties required to allow solving simultaneous linear equations both for unknown scalars and unknown vectors.

For example, a system of equations in unknown scalars,

$$\left.\begin{array}{l} \gamma|x\rangle + \delta|y\rangle = a|z\rangle \\ \gamma|u\rangle + \delta|v\rangle = b|w\rangle \end{array}\right\} \quad where \quad \left\{\begin{array}{l} a,b, \equiv \text{given constants, } |x\rangle,|y\rangle,|z\rangle,|u\rangle,|v\rangle,|w\rangle \\ \qquad\quad \equiv \text{given vectors,} \\ \gamma,\delta \equiv \text{unknown scalars,} \end{array}\right.$$

can be solved for the unknown scalars α and δ. Similarly, a system of equations in unknown vectors,

$$\left.\begin{array}{l} a|\alpha\rangle + b|\beta\rangle = c|x\rangle \\ d|\alpha\rangle + e|\beta\rangle = f|y\rangle \end{array}\right\} \quad where \quad \left\{\begin{array}{l} a,b,c,d,e,f \equiv \text{given constants, } |x\rangle,|y\rangle \\ \qquad\qquad\quad \equiv \text{given vectors,} \\ |\alpha\rangle,|\beta\rangle \equiv \text{unknown vectors,} \end{array}\right.$$

can be solved for the unknown vectors $|\alpha\rangle$ and $|\beta\rangle$.

A vector space also defines scalar multiplication of a vector, with a distributive property of scalar multiplication over vector addition.

QM vector spaces have two additional characteristics: they define a dot product between two vectors, and they define linear operators which act on vectors to produce other vectors. This allows us to solve more complex equations, "operational equations," such as:

$$\hat{E}|\psi\rangle = \hat{H}|\psi\rangle \quad \text{where} \quad \hat{E}, \hat{H} \text{ are given operators, and } |\psi\rangle \equiv \text{unknown vector.}$$

For example,

$$\hat{E} \equiv i\hbar\frac{\partial}{\partial t}, \qquad \hat{H} \equiv -\hbar^2\frac{\partial^2}{\partial x^2} + V(x) \qquad \rightarrow \qquad i\hbar\frac{\partial}{\partial t}\psi(t,x)$$

$$= \left[-\hbar^2\frac{\partial^2}{\partial x^2} + V(x)\right]\psi(t,x).$$

Operational equations are more complicated than simple linear combination equations, and lead to families of solutions, rather than a unique solution. The Schrödinger equation is an operational equation. In wave-mechanics, the Schrödinger equation is a partial differential equation in time and space, but in matrix mechanics, the Schrödinger equation is a discrete vector equation that is differential in time only.

In summary, QM vector spaces have the following properties:

Vector space	
Field of scalars	Group of vectors
Scalars form a commutative group under addition (+), (closure, associativity, identity, inverses)	Vectors form a commutative group under addition (+), (closure, associativity, identity, inverses)
Scalars, excluding 0, form a commutative group under multiplication (·)	
Distributive property of scalar (·) over +	
Scalar multiplication of a vector produces another vector	
Distributive property of scalar multiplication over vector +	
Additional QM features	
Conjugate bilinear dot product produces a scalar from two vectors	
Linear operators act on vectors to produce other vectors	

Note that scalar addition is a *different* operation than vector addition.

2.9.2 Field of Scalars

Before we get to (general) vector spaces, we need to understand mathematical "fields." A mathematical field is a set of elements (scalars) and two operators which follow the rules of linear algebra, i.e., you can solve simultaneous linear equations with them. For example, the set of real numbers, and the operators "+" (addition) and "·" (multiplication) constitute a field (called the "field of real numbers"). Any set of N (linearly independent) simultaneous equations and N unknowns can be solved in the field of real numbers. In general, a field can have either a finite or infinite number of elements. QM is only concerned with the infinite field of complex numbers. Therefore, in QM, the operators "+" and "·" are the familiar arithmetic operations of complex addition and multiplication. (For a general field, they can be anything that meets the following criteria.)

A **field** has the following defining properties (sometimes called "axioms" of a field):

1. A set of elements $(a, b, c, \ldots)$, either finite, countably infinite, or uncountably infinite.
2. Two operators, which we call "+" and "·," satisfying:
   closure: "$a+b$" and "$a \cdot b$" are also elements of the field;
   commutativity: "$a+b$"="$b+a$" and "$a \cdot b$"="$b \cdot a$," and
   associativity: $(a+b)+c=a+(b+c)$ and $(ab)c=a(bc)$,
   When writing formulas, the "·" may be omitted: $ab \equiv a \cdot b$.
3. An additive identity, called "0": $a+0=a$.
4. Additive inverses: $a+(-a)=0$.
5. A multiplicative identity called "1": $a \cdot 1=a$.
6. Multiplicative inverses, except for 0: $a \cdot (a^{-1})=1$.
7. Distributivity of "·" over "+": $a \cdot (b+c)=a \cdot b+a \cdot c$.

In QM, the field of interest is the field of complex numbers.

[It may be interesting to note that the elements form a mathematical "group" under the operator "+,"and the elements excluding 0 form a group under the operator "·." Groups are described later.]

[Note: we often speak of physical "fields" that are totally different from this kind of mathematical field: a physical "field" is a function of space, typically a scalar or vector function of space. For example, the electric potential field is a scalar function of space, $V(\mathbf{r})$. The magnetic induction field is a vector function of space, $\mathbf{B}(\mathbf{r})$. This meaning of "field" is unrelated to a mathematical "field."]

2.9.3   Group of Vectors

In contrast to the field of scalars, which in QM is always the field of complex numbers, QM "vectors" come in different types. Some are finite-dimensional vectors (say, angular momentum), some are discrete (countably) infinite vectors (e.g., the basis vectors of bound states), and some are uncountably infinite (e.g., unbound states of particles in space). In all cases, vectors can be added, and constitute a mathematical "group."

A mathematical **group** is a set of elements with the following defining properties (sometimes called "axioms" of a group):

1. A set of elements $(\mathbf{v}, \mathbf{w}, \mathbf{y}, \mathbf{z}, \ldots)$, either finite, countably infinite, or uncountably infinite.

2. An operator, here called "+," satisfying:
   closure: "$\mathbf{v}+\mathbf{w}$" is also an element in the group, and
   associativity: $(\mathbf{v}+\mathbf{w})+\mathbf{y}=\mathbf{v}+(\mathbf{w}+\mathbf{y})$.
3. There is a group additive identity, called the **zero element** (in QM, the **zero vector**, written here as $\mathbf{0}_v$), such that: $\mathbf{0}_v+\mathbf{w}=\mathbf{w}$.
4. All elements have an **inverse** such that: $\mathbf{v}+\mathbf{v}^{-1}=\mathbf{0}_v$

Note that vector operator "+" is a *different* operation than the scalar "+." To distinguish "+" as scalar addition or vector addition, simply look at the operands.

If a group has the additional property of commutativity of addition, then it is called a **commutative group** (aka an **abelian** group). The vectors of a vector space form a commutative group. Mathematically, vectors are abstract entities that can be almost anything. In QM, they can be wave-functions, spinors, or other discrete vectors.

> In QM, the kets form a commutative group, and are the vectors of the vector space.

Bras form another group, and are the vectors of another vector space, the "dual" vector space, i.e., the bra-space is dual to the ket-space. More on this shortly.

2.9.4   Scalar Field + Vector Group + Scalar Multiplication = Vector Space

A vector space combines the field of scalars and the group of vectors in a particular way. Specifically, a **vector space** has the following properties (sometimes called "axioms" of a vector space):

1. A *field* of elements (a, b, c,...). The field operators "+" and "·" apply to the field elements. In a vector space, the field elements are called **scalars**. In QM, the scalars are the field of complex numbers.
2. A *commutative group* of vectors ($\mathbf{v}$, $\mathbf{w}$, $\mathbf{y}$, $\mathbf{z}$,...). The vectors are not necessarily related to the field elements (scalars) at all; they can be almost anything. In particular, a vector is *not* necessarily a list of scalars. The vectors form a commutative (aka abelian) group under addition, as defined previously.
3. A scalar multiplication operator "·" such that

   "$a \cdot \mathbf{v}$" is also a vector in the space,   ("a" is an element of the field).

   Scalar multiplication may omit the "·" symbol: $a\mathbf{v}\equiv a \cdot \mathbf{v}$. Scalar multiplication by 0 yields $\mathbf{0}_v$:

   $$0 \cdot \mathbf{w}=\mathbf{0}_v.$$

4. Associativity of scalar multiplication with field multiplication such that:

   $$a \cdot (b \cdot \mathbf{v})=(a \cdot b) \cdot \mathbf{v}.$$

5. Scalar multiplication distributes over vector addition:

$$a \cdot (\mathbf{v} + \mathbf{w}) = a \cdot \mathbf{v} + a \cdot \mathbf{w}.$$

6. The field multiplicative identity "1" is also the scalar multiplicative identity:

$$1 \cdot \mathbf{v} = \mathbf{v}.$$

A vector space allows you to solve simultaneous linear equations with scalars as the coefficients of vectors, i.e., you can use linear algebra on vector spaces. From the definition of "field," with just the scalars, you can solve ordinary linear equations such as:

$$\left. \begin{aligned} a_{11}x_1 + a_{12}x_2 + \ldots a_{1n}x_n &= c_1 \\ a_{21}x_1 + a_{22}x_2 + \ldots a_{2n}x_n &= c_2 \\ \vdots \qquad \vdots \qquad \vdots \\ a_{n1}x_1 + a_{n2}x_2 + \ldots a_{nn}x_n &= c_n \end{aligned} \right\} \text{equivalent to } \mathbf{ax} = \mathbf{c}.$$

All the usual methods of linear algebra work to solve the previous equations: Cramer's rule, Gaussian elimination, etc.

With the whole vector space, you can solve simultaneous linear *vector* equations for unknown scalars or unknown vectors, such as

$$\left. \begin{aligned} a_{11}\mathbf{v}_1 + a_{12}\mathbf{v}_2 + \ldots a_{1n}\mathbf{v}_n &= \mathbf{w}_1 \\ a_{21}\mathbf{v}_1 + a_{22}\mathbf{v}_2 + \ldots a_{2n}\mathbf{v}_n &= \mathbf{w}_2 \\ \vdots \qquad \vdots \qquad \vdots \\ a_{n1}\mathbf{v}_1 + a_{n2}\mathbf{v}_2 + \ldots a_{nn}\mathbf{v}_n &= \mathbf{w}_n \end{aligned} \right\} \text{equivalent to } \mathbf{av} = \mathbf{w}.$$

> The same methods of linear algebra work just as well to solve vector equations as scalar equations.

A vector space has a **dimension**, which is the number of basis vectors needed to represent any vector in the space. In wave mechanics, the wave-function vector space has square-integrable functions as its vectors. It is infinite dimensional: we need an infinite number of basis vectors (in this case, basis functions) to construct an arbitrary function in the space; i.e., no finite number of basis functions can compose every possible wave-function.

In contrast, angular momentum vector spaces are finite dimensional. The spin-1/2 vector-space is 2D. In general, the space of angular momentum quantum number j is $(2j+1)$ dimensional. For example, finite-dimensional vectors may be written as:

$$2\text{D}: \ |\alpha\rangle = \begin{bmatrix} a \\ b \end{bmatrix} \qquad\qquad 3\text{D}: \ |\beta\rangle = \begin{bmatrix} a \\ b \\ c \end{bmatrix}.$$

(Kets are written as column vectors, while bras are written as row vectors.)

Hilbert Spaces: Note that a vector space is not required to have a dot product defined. Without such, there is no way to measure the magnitude of a vector, or the angle between two vectors. Such a space is a **nonmetric** vector space. However, all QM vector spaces have dot products.

There is no general agreement among mathematicians on the precise definition of a Hilbert Space, much less agreement between mathematicians and physicists. However, in QM, a **Hilbert space** is a vector space, with the following additional properties:

1. There is a dot product operator "·" between any two vectors, yielding a scalar (an element of the field), i.e., a complex number. It need not be commutative, and in QM it is *not* (as described earlier and later):

$$\mathbf{v} \cdot \mathbf{w} = a, \qquad \text{or in Dirac notation,} \qquad \langle v | w \rangle = a .$$

The dot product of a vector with itself is its squared magnitude, aka its squared **length**:

$$\mathbf{v} \cdot \mathbf{v} = \mathbf{v}^2 = |\mathbf{v}|^2 = \langle v | v \rangle = [\text{length}(\mathbf{v})]^2.$$

2. The dot product is **conjugate bilinear** over scalar multiplication and vector addition:

$$\mathbf{v} \cdot \mathbf{w} = (\mathbf{w} \cdot \mathbf{v})^* \qquad \text{or in Dirac notation:} \qquad \langle v | w \rangle = \langle w | v \rangle^*.$$
$$\mathbf{v} \cdot (a\mathbf{w}) = a(\mathbf{v} \cdot \mathbf{w}),$$

and therefore $\quad (b\mathbf{v}) \cdot \mathbf{w} = b^*(\mathbf{v} \cdot \mathbf{w}).$

Also, the dot-product distributes over vector addition:

$$\mathbf{v} \cdot (\mathbf{w} + \mathbf{y}) = \mathbf{v} \cdot \mathbf{w} + \mathbf{v} \cdot \mathbf{y}.$$

3. The dot product of the zero vector is always zero: $\mathbf{0}_v \cdot \mathbf{w} = \mathbf{w} \cdot \mathbf{0}_v = 0$ (the scalar 0).

In wave mechanics, the zero vector wave-function is a function that is everywhere zero: $\mathbf{0}_v \equiv \psi_0(\mathbf{r}) = 0$. In discrete spaces, $\mathbf{0}_v$ has all zero components, e.g., in two or three dimensions:

$$2\text{D}: \ \mathbf{0}_v = \begin{bmatrix} 0 \\ 0 \end{bmatrix} \qquad\qquad 3\text{D}: \ \mathbf{0}_v = \begin{bmatrix} 0 \\ 0 \\ 0 \end{bmatrix}.$$

In QM, the zero vector is sometimes called the "null ket." Note that the ket written "$|0\rangle$" is *not* the zero vector, it is the (normalized) ground state of a particle or system. The zero vector cannot be normalized and cannot be a quantum state.

A Hilbert space is a **metric** vector space: the metric is the dot product. In QM, the scalars are the field of complex numbers. The vectors are kets, e.g., in wave mechanics, complex-valued functions of space. The QM dot product for 1D wave-functions was defined earlier as $\mathbf{v} \cdot \mathbf{w} \equiv \int_{-\infty}^{\infty} v^*(x)w(x)\, dx$. In three dimensions the dot product is $\mathbf{v} \cdot \mathbf{w} \equiv \iiint_{\infty} v^*(\mathbf{r})w(\mathbf{r})\, d^3\mathbf{r}$, where $\mathbf{r}$ covers all physical 3D space. The dot product is sometimes called an **inner product**.

In summary, in physics, an infinite-dimensional vector space with an inner product defined is usually called a Hilbert space.

Be careful to distinguish the dimension of real space in which a QM wave-function exists, from the dimension of the Hilbert space of wave-functions. QM wave-functions are often simplified into one dimension: motion confined to a line, or two dimensions, and motion confined to a plane. Most generally, wave-functions cover all three dimensions of real space. In all of those cases, though, the wave-function is a vector in an infinite-dimensional mathematical Hilbert space.

Looking ahead to matrix mechanics, the spin state of a single spin-1/2 particle exists in a two-dimensional mathematical vector space over the field of complex numbers. Kets are pairs of complex numbers, which are coefficients in some basis, e.g.,

$$|\chi\rangle = \begin{bmatrix} a \\ b \end{bmatrix} = a|z+\rangle + b|z-\rangle \quad \text{where} \quad a, b \equiv \text{complex components,}$$

$$|z+\rangle, |z-\rangle \equiv \text{basis vectors.}$$

Other finite-dimensional vector spaces are also used in QM, and will be discussed later.

2.9.5   Observations on Vector Spaces

Most mathematicians demand that a Hilbert space be infinite dimensional; physicists are not always so demanding. However, to be clear, we use the term "vector space" for finite-dimensional spaces, and reserve "Hilbert space" for infinite-dimensional spaces. Also, most mathematicians demand that a Hilbert space be "complete," which means the limit of any convergent infinite sum of vectors is also a vector in the space. The QM definition of Hilbert space meets this requirement, but it is not important in this work. Do not worry about it.

Finally, note that the set of *normalized* quantum states is *not* a vector space, because arbitrary linear combinations of states are *not* normalized. This is not a problem in QM, because all equations for quantum states are formed in such a way that the normalization works out properly in the end. In some cases, we deliberately use different normalizations for convenience, but the final results can always be normalized, and the vector mathematics is rigorous.

Chapter 3
Introduction to Scattering

Scattering is important for several reasons: lots of physics was discovered by scattering one thing off another. Rutherford and his students discovered atomic nuclei by scattering alpha particles off gold foil; Davisson and Germer discovered the wave nature of electrons by scattering them off a nickel block, thus ushering in the era of quantum mechanics. Today, the most advanced quantum theories are tested by scattering (aka "colliding") particles off each other, and measuring the angles and energies of particles in the shower of debris that results. (Scattering theory is also called "collision theory.") Furthermore, 1D scattering essentially includes the topic of tunneling, which is of great practical importance. Tunneling has many experimental applications, such as the scanning tunneling microscope (STM), and everyday practical uses, such as high-frequency tunnel diodes.

The mathematics for scattering is covered in most books, but the big picture preceding the mathematics is often skipped. We start here with the big picture, before delving into a few selected computations. Our goal is to convey the concepts, so that the presentations in standard texts are more accessible. ([8, Chap. 23] provides an unusually clear introduction to scattering.)

Recall that a "stationary" state has properties that do not change with time; however, it is not "static." For example, a particle in a pure momentum eigenstate is "stationary," but it is also moving (it has momentum). A particle in an orbital angular momentum eigenstate is also stationary, but "moving": it is revolving around the center.

In the following analysis, we use a variant of delta-function normalization that is often used in scattering with momentum eigenstates. We drop the prefactor, so

$$|p\rangle \equiv \psi_p(x) = e^{ipx/\hbar} \quad \text{(unit magnitude delta-function normalization)}.$$

In other words, we are simply choosing our normalization such that the (implicit) prefactor of "1" before $e^{ipx/\hbar}$ represents one particle. This can also be considered normalizing to one particle per unit volume, but scattering is generally analyzed one incident particle at a time.

We present a five-step program for scattering, in increasing order of complexity:

1. 1D scattering: Solving Schrödinger's equation.
2. Three-dimensional (3D) classical scattering: defining cross sections.

E. L. Michelsen, *Quirky Quantum Concepts*, Undergraduate Lecture Notes in Physics, DOI 10.1007/978-1-4614-9305-1_3, © Springer Science+Business Media New York 2014

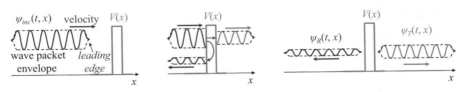

Fig. 3.1 Not to scale. (*Left*) The particle wave-packet approaches from the left. (*Middle*) The particle interacts with the barrier potential, and is partially transmitted and partially reflected. (*Right*) The wave-function has split into spatially separate transmitted and reflected parts

3. 3D quantum scattering: Solving Schrödinger's equation and defining quantum cross sections.
4. Born approximation: good for mild scattering (particle energy >> scattering potential).
5. Partial wave expansion, including the low-*l* approximation (good for particle wavelength >> target size).

Alternatively, one can view scattering as a time-dependent perturbation problem, and use Fermi's Golden Rule [1, p. 252b], but we do not discuss that development here.

This whole chapter assumes you are familiar with solving the time-independent Schrödinger equation, wave-packets. The 3D scattering sections assume understanding angular momentum, spherical Bessel functions, and solutions to the SE (Schrödinger equation) in spherical coordinates.

3.1 1D Quantum Scattering

3.1.1 1D Scattering: Finite Width Barrier

Suppose we have a scattering potential, $V(x)$, which is localized (Fig. 3.1, *left*). This potential is the "target." The force between the target and incident particle is short range, as indicated by the potential being zero outside a small region. Therefore, the particle is a free particle outside the range of the target. This assumption is also important in 3D scattering.

A *single* particle is incident from the left, with a very well-defined momentum, so it is described by a wave-packet that is very wide (narrow $\Delta p \Rightarrow$ wide Δx). The particle is "free" until it hits the target. A realistic wave-packet is much wider than the target width, so the diagram is not to scale. The wave-packet comprises waves of a narrow range of momenta (a superposition of momenta), and can be normalized. However, since the spread in momentum is very small, we approximate it as a momentum eigenstate, and use unit-amplitude δ-function normalization (the wave function is nearly a δ-function in the momentum basis).

> The particle energy may be above or below the target height. Either way, some of the wave-function is reflected, and some is transmitted through the target.

The particle interacts with the target ("collides," Fig. 3.1, *middle*). This interaction splits the wave-packet into two parts: a transmitted part that continues past the target and a reflected part that moves back to the left.

Classically, we are used to thinking of scattering as a *dynamic* process: a particle enters, collides, and leaves. In QM, however, the process can be computed as *stationary* (steady-state). Because the wave-packet is wide, when its center reaches the target, the packet envelope is nearly constant in space and time, but the wave-function phase changes with time. In other words, we have (essentially) a stationary quantum state, with the incident part of the wave-function moving to the right. The leading and trailing edges of the wave-packet are far from the barrier. Therefore, to a good approximation, Schrödinger's time-independent equation applies. Note that "stationary" does not mean "static." The particle is always moving, but during the collision, far from the wave-packet edges, the wave-function amplitudes do not significantly change with time. Therefore, the incident particle wave-function is closely approximated by that of a definite momentum:

$$\psi_{inc}(t,x) = e^{ipx/\hbar}e^{-iEt/\hbar} = e^{ikx}e^{-i\omega t} \qquad where \qquad p = \hbar k, \text{ and } E = \hbar\omega.$$

This is where all the mathematics comes in: we solve Schrödinger's time-independent equation and find the amplitudes for the transmitted piece and the reflected piece. We have to match the wave-function value and slope at both interfaces (front and back of barrier). Most standard references do this, so we do not discuss it here (see for example, [8, Chap. 4]).

After the collision is done, the particle has a probability to be found on the left, and moving left, and some probability to be found on the right, and moving right (Fig. 3.1, *right*). That is, it has a probability to be reflected and a probability to be transmitted.

Things to note: scattering amplitudes, and therefore the transmission and reflection probabilities, depend on p, the incident particle momentum, or equivalently on k, the incident particle spatial frequency (wave number). Furthermore:

In 1D barrier scattering, the energy (and therefore momentum, spatial frequency, and speed) is the same on both sides of the barrier. Therefore, the widths of the incident, transmitted, and reflected wave-packets are all the same.

If we take the amplitude of the incident wave-packet as unity (unit amplitude δ-function normalization), then the probabilities of transmission and reflection are simply:

$$|\psi_{inc}|^2 \equiv 1 \quad \Rightarrow \quad \Pr(\text{transmission}) = |\psi_T|^2, \text{ and } \Pr(\text{reflection}) = |\psi_R|^2.$$

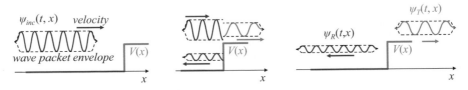

Fig. 3.2 Not to scale. (*Left*) The particle approaches from the left. (*Middle*) The particle interacts with the potential. (*Right*) The wave-function has split into separate transmitted and reflected parts, whose wave-packets are different lengths

3.1.2   1D Scattering: Step Potential

A step potential is similar to a finite-width barrier, but has three important differences. For definiteness, we assume a step-increase in potential, but similar arguments apply to a step-decrease. We consider only the case where the incident particle energy is greater than the potential step, so a fraction of the particle is transmitted forever beyond the step. (If the particle had less energy than the step, it would bounce off, with an exponentially decaying probability of being past the step.)

Differences from finite width barrier scattering are: (1) there is only one interface, so we only have to match the wave-function once (instead of twice); (2) the energy, momentum, spatial frequency (k), and speed of the transmitted wave-packet is lower; and therefore (3) the transmitted wave-packet is narrower. Here is why:

We can view the collision as the incident packet interacting with the target, and during the time of collision, generating a transmitted wave-packet and a reflected wave-packet. Each of the transmitted and reflected packets grows with time as the incident packet feeds into the step. Both the transmitted and reflected packets are generated for the same amount of time: the duration of the collision. But the transmitted packet moves more slowly (less momentum), so it grows more slowly, and reaches a smaller width than the reflected packet. In fact, the width of the *transmitted* packet is smaller by the same fraction that its speed is smaller than the incident packet:

$$w_T = \frac{speed_T}{speed_{inc}} w_{inc} = \frac{p_T}{p_{inc}} w_{inc} = \frac{k_T}{k_{inc}} w_{inc}$$

where w_{inc}, $w_T \equiv$ incident and transmitted packet-widths.

p_{inc}, $p_T \equiv$ incident and transmitted momenta

k_{inc}, $k_T \equiv$ incident and transmitted.

The *reflected* packet has the same width, energy, momentum, spatial-frequency, and speed as the incident packet, just as with a finite barrier. This affects our transmission and reflection probabilities:

$$\text{Pr(transmission)} = \int_{w_T} dx \, |\psi_T|^2 = w_T |\psi_T|^2, \quad \text{and similarly,}$$

$$\text{Pr(reflection)} = \int_{w_R} dx \, |\psi_R|^2 = w_R |\psi_R|^2 \qquad \text{(for a wide, flat packet).}$$

We (again) choose our normalization so that the incident amplitude is unity. Also, without loss of generality, we choose the incident width to be unity. Our probabilities of transmission and reflection are:

$$|\psi_{\text{inc}}|^2 \equiv 1 \quad \Rightarrow \quad \text{Pr(transmission)} = \frac{w_T}{w_R} |\psi_T|^2 = \frac{k_T}{k_R} |\psi_T|^2,$$

$$\text{and} \qquad \text{Pr(reflection)} = |\psi_R|^2.$$

The final result is independent of the incident packet width. This differing transmit/ receive packet-width phenomenon is specific to inelastic scattering, where the "outgoing" particle has different energy than the incident particle; it does not apply to 1D barrier scattering or 3D elastic scattering, because the scattered wave is the same "width" as the incident wave.

3.2 Introduction to 3D Scattering

The basic idea of 3D classical and quantum scattering is that we send a uniform spray of particles at a target, and they bounce off in all kinds of directions. However, we consider each incident particle individually, assuming no interaction between incident particles.

By "uniform spray" we mean a fixed number of particles per second per unit area: a uniform flux density. A **flux** is particles per second (N/s); **flux density** is particles per second per area (N/s/area). Flux density is also called "luminosity," especially in experimental physics. (Many references use the term "flux" to mean "flux density.") The incident beam is wide, much wider than the target, so its exact width is not a factor in the scattering. In realistic quantum scattering, the incident particles are far enough apart that they do not interact.

Before considering 3D *quantum* scattering, we start with an overview of *classical* scattering, which is the foundation on which we build.

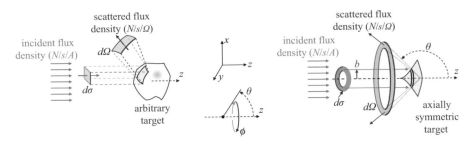

Fig. 3.3 (*Left*) Classical scattering off an arbitrary target. (*Right*) Classical scattering off an axially symmetric target. "*b*" is the classical impact parameter

3.2.1   3D Hard-Target Classical Scattering

"Hard target" scattering has rigid, impenetrable particles and a rigid, impenetrable target. (This situation does not exist at the quantum level.) Therefore, an incident particle either hits or misses the target. We consider the following topics in classical hard-target scattering:

- Incident particles.
- Scattered particles.
- Differential cross section.
- Total cross section.
- Measuring differential cross section.
- Axially symmetric targets.
- Reduction to one body.

Incident particles   Because the incident flux density is constant over the *x-y* plane, the flux through any area normal to the flux is just area times flux density (Fig. 3.3):

$$\text{flux} \equiv \text{particles/s} = \left(\text{flux\_density}\right)\left(\text{area}\right) \qquad \text{or}$$

$$F = JA \qquad\qquad \textit{where} \quad J \equiv \text{flux density} = \text{particles/s/area}.$$

A **hard-target** is a lump of hard stuff which is impenetrable, but has no effect outside the target boundary. (We discuss "soft-targets" later.) The scattering is usually elastic: the target is fixed (heavy) and acquires negligible energy; therefore the scattered particles have the same energy as the incident particles. In classical mechanics, the cross section of a hard-target is literally the cross-sectional area of the target perpendicular to the flux, measured in, say, cm^2. By analogy, in quantum scattering, the *effective* area capturing flux is called a **cross section**.

Scattered particles   The scattered particles are measured differently. Far from the target, it appears to be a point source of scattered particles. Since particles are conserved, the total outward flux through a large sphere centered on the (effectively point-source) target, including particles which "miss" the target and continue past it, equals the incident flux. For a differential solid angle $d\Omega$ measured far from the target, the flux per steradian is constant along the radius away from the target. Therefore, for *scattered* particles, the flux density is measured as particles per second per solid angle (particles/s/sr), rather than particles/s/area.

Differential cross section Consider Fig. 3.3, *left*. Each infinitesimal area, $d\sigma$, scatters into an infinitesimal solid angle, $d\Omega$, in the direction (θ, ϕ).

$$\text{particles/s into } d\sigma = \text{particles/s out of } d\Omega.$$

For infinitesimal $d\sigma$, if we vary its size, the $d\Omega$ into which it scatters will vary proportionately: if we double $d\sigma$, we will double $d\Omega$. In other words, in an infinitesimal region around any direction (θ, ϕ), the ratio $d\sigma/d\Omega$ is constant. We here call it $s(\theta, \phi)$:

$$s(\theta,\phi) \equiv \frac{d\sigma}{d\Omega} \quad \text{(units of area/sr).} \tag{3.1}$$

Generally, though, we think in the reverse direction: we can control the size $(d\Omega)$ and position (θ, ϕ) of particle detectors counting scattered particles. Then we can ask (and measure): how big a $d\sigma$ contributes particles to the given $d\Omega$. Since the incident flux density j is uniform, we do not care *where* $d\sigma$ is, we only care how big it is. In general, the size (and position) of the incident cross section, $d\sigma$, that fills a given small solid angle, $d\Omega$, varies in different scattered directions, i.e., it is a function of (θ, ϕ).

$$J_{inc} \, d\sigma = \text{flux out of } d\Omega$$

where $J_{inc} \equiv$ incident flux density (particles/s/area).

Then: $d\sigma = \dfrac{\text{flux out of } d\Omega}{J_{inc}} \equiv s(\theta,\phi) \, d\Omega.$

The function $s(\theta, \phi)$ is (somewhat inappropriately) called the **differential cross section**. Of course, the "differential cross section" is literally just $d\sigma$, and is proportional to $d\Omega$, but this misnomer is universally used. Griffiths [9, p. 395b] notes that $s(\theta, \phi)$ is a derivative, not a differential, which we see from Eq. (3.1). The units of differential-cross section are cm^2/steradian, or just cm^2, because the steradian (like the radian) is equivalent to dimensionless.

Note that more than one location for $d\sigma$ could scatter into the same solid angle, $d\Omega$; this is only possible because far from the target, it appears to be a point source.

> The concept of multiple incident regions scattering into the same solid angle is critical for quantum scattering.

In this case, the differential cross section is simply the sum of all small incident areas that scatter into the given $d\Omega$.

> The differential cross section does not care which region of the incident beam the particles come from; it tells only how much total incident area scatters into a solid angle in a given direction.

Total cross section The **total cross section** is the cross-sectional area of the incident flux that is scattered (in any direction). Therefore, we can find the total cross section by integrating $d\sigma$:

$$\sigma = \int_{\text{sphere}} \frac{d\sigma}{d\Omega}\, d\Omega = \int_0^\pi d\theta \int_0^{2\pi} d\phi\, s(\theta,\phi)\,\sin\theta \qquad where \quad d\Omega = d\theta\sin\theta\, d\phi.$$

Measuring differential cross section We measure differential cross sections exactly the same way for all scattering (hard classical, soft classical, and quantum). Send in a known number of particles/cm^2 toward a target. Put small detectors around the target in all directions. Measure the detection rate in each direction, knowing the solid angle $d\Omega$ covered by each detector. Then simply divide:

$$d\sigma = \frac{\text{particles/s}}{j_{inc}} \left(\text{in cm}^2\right), \qquad \frac{d\sigma}{d\Omega} \equiv s(\theta,\phi) = \frac{d\sigma}{\text{solid\_angle\_of\_detector}}.$$

Note that the detection rate (particles/s) of any detector is proportional to the incident flux density; therefore, the flux density cancels, and the cross sections $d\sigma$ are independent of flux density; $d\sigma/d\Omega$ is a function solely of a single incident particle (including its energy) and the target.

Axially symmetric targets Many targets are axially symmetric (Fig. 3.3, *right*). In particular, spherically symmetric targets are also axially symmetric. Because of this symmetry, there is no ϕ dependence. Then the differential cross section is a function of θ only:

$$\frac{d\sigma}{d\Omega} = s(\theta) \quad \text{(axially symmetric target)}.$$

We must still include ϕ when integrating to find total cross section, but it integrates trivially to 2π:

$$\sigma = \int_0^\pi d\theta\, s(\theta)\sin\theta \int_0^{2\pi} d\phi = 2\pi\int_0^\pi d\theta\, s(\theta)\sin\theta \qquad \text{(axially symmetric target)}.$$

Reduction to one-body The targets we have described so far are fixed (typically macroscopic) targets, such as the nickel block Davisson and Germer scattered electrons off, or the gold foil Rutherford and his students scattered alpha particles off. However, it is quite common to study collisions of two particles which are both moving, and both moved by the collision. There is then no "fixed" target. Conveniently, if the interaction of two particles is described by a potential function only of their separation, they can be reduced to an equivalent one-body dynamic system, with a potential fixed in the new "space." Then our fixed-target scattering model can still be used. This is exactly the same "reduction to one-body" that is used in classical mechanics to study orbits in central potentials, such as gravity or electric forces. Reduction to one-body works just as well in QM as in classical mechanics [1, pp. 169–172, 8, pp. 193–195], because the Schrödinger equation applies to any set of generalized coordinates. You can see that this is true for reduction to one-body by direct substitution of the transformed coordinates into the Schrödinger equation for the original coordinates. We use this same reduction to one-body when analyzing the hydrogen atom: the proton and electron are reduced to a single (hypothetical) particle in a fixed potential.

We reduce to one-body by first converting the two-body scattering system into the center-of-mass coordinates (aka center-of-momentum, or COM system), using the usual formulas for reduced mass and coordinates:

$$m_1, \mathbf{r}_1, m_2, \mathbf{r}_2, V(\mathbf{r}_1 - \mathbf{r}_2) \qquad \rightarrow \qquad m \equiv \left(\frac{1}{m_1} + \frac{1}{m_2} \right)^{-1},$$

$$\mathbf{r} \equiv \mathbf{r}_1 - \mathbf{r}_2, \qquad V(\mathbf{r}) \equiv V(\mathbf{r}_1 - \mathbf{r}_2).$$

With scattering, though, there is one more complication: we compute angles in the COM frame, but experimentally measure them in our laboratory frame. So we must be able to convert angles between the two frames. This conversion is straightforward (though slightly tedious), so we defer to references such as [8, Chap. 23, 18, pp. 111–114].

3.2.2 Soft-Target Classical Scattering

Soft-targets are targets with finite potentials, outside the hard boundaries of the target, that interact with the incident beam. For example, a hard charged sphere has an infinite potential inside the sphere, but a soft electrostatic potential, outside the sphere, that interacts with a charged incident beam. Similarly, a planet has a hard surface and a soft gravitational potential that scatters incident space dust. Soft-targets need not have any well-defined boundaries ("hard" boundaries).

> In soft-target scattering, the scattering cross section may be larger than the hard part of the target cross section, because the target's potential reaches outside the hard part. A soft-target is completely defined by its potential field.

For example, a charged solid blob scatters incident charges far beyond the blob's boundary; a planet scatters incident mass far beyond its physical surface (cross section).

Hard-target scattering may be considered the limiting case where a soft-target potential is infinite inside the target boundary, and zero outside.

3.2.3 Wave Scattering: Electromagnetic and Quantum Waves

Electromagnetic (EM) waves have no charge or mass, so they are not scattered by static fields. They are scattered by dielectric targets (electrically polarizable), magnetically susceptible targets, or conducting targets. EM scattering introduces two new twists over particle scattering:

- Energy takes the place of particles: the flux is power or energy/time; incident flux-density is power/area, aka **intensity**, such as Joule/s/m^2 = W/m^2. The scattered flux density is therefore power/solid-angle, such as W/steradian.
- Interference: waves scattered into a given solid angle from more than one small incident area interfere: the total wave amplitude is the coherent sum of all the contributing wave amplitudes. The power is proportional to the square of the total wave amplitude.

For EM wave scattering, one considers an incident wave of uniform intensity (power density). The differential cross section $d\sigma$ is defined as the incident area needed to provide the power scattered into a small solid angle $d\Omega$.

The scattering cross section can be larger or smaller than the physical cross section, e.g., stronger dielectrics have larger cross sections than weak ones.

Quantum scattering is wave-function scattering, a form of wave scattering, so interference is important. In wave scattering in general, and in QM in particular, *every* $d\Omega$ in every direction gets contributions from *many* areas of the incident beam, because of diffraction. This is why we do not care *where* $d\sigma$ is: in wave scattering, $d\sigma$ is spread out over a wide area, includes interference effects, and cannot be localized.

Spherical targets and spherical waves If the scattering target is spherically symmetric, the scattered wave is a spherical wave, but usually not spherically symmetric. A spherical wave is one which can be written in the form:

$$\psi(r, \theta, \phi) = \frac{e^{ikr}}{r} f(\theta, \phi) \qquad where \quad f(\theta, \phi) \text{ is generally complex (spherical wave).}$$

$f(\theta, \phi)$ imparts direction-dependent phase and amplitude shifts to the wave. In scattering, we are concerned with outbound (expanding) spherical waves, and not inbound waves. Following any ray away from the origin, the e^{ikr}/r factor imposes regular phase shifts at spatial frequency k, and an amplitude drop off which (when squared) conserves outbound particle flux. In the case where $f(\theta, \phi)$ is approximately real (or has any constant complex phase), the zero crossings of both the real and imaginary parts of ψ would be spheres, but ψ's amplitude may vary with angle (θ, ϕ). [We prove the optical theorem later, which mandates that $f(\theta, \phi)$ be complex.]

3.3   3D Quantum Scattering

3.3.1   3D Quantum Scattering Overview

Quantum scattering combines wave-scattering concepts with quantum probabilities. Quantum scattering is wave-function scattering, so interference is important. In QM, each small incident area has a probability of scattering into most any solid angle. Conversely, a given solid angle usually has contributions from a large (or infinite) incident area, i.e., a given $d\Omega$ has probabilities of coming from a large set of $d\sigma$'s. The overall QM $d\sigma$ for a given $d\Omega$ is a weighted sum of every possible infinitesimal incident area da, each weighted by its amplitude for scattering into $d\Omega$, and including the interference of all such contributors. Recall that:

> We define (and measure) cross sections by counting particles.

In particle-target quantum scattering, we usually consider the target fixed, and it absorbs negligible energy from the incident beam. (As noted earlier, two-body collisions can be converted to this form.) We assume the interaction force is short-range (defined soon). Therefore, outside the range of the scattering potential, the outbound waves have (essentially) the same kinetic energy as the inbound waves, and the scattering is elastic. This means the particle is a free particle outside the range of the interaction force. The outbound wave is proportional to the incident wavefunction, which is a plane wave, written as:

$$\psi_{inc}(z) = Be^{ikz} \quad where \quad B \equiv \text{normalization factor.}$$

From far away, the target appears to be a point source of scattered waves, so the scattered wave fronts must be spherical, i.e., proportional to $\exp(ikr)$. We can deduce the mathematical form by considering conservation of particle flux (sometimes called "conservation of probability"). In steady state, the particle flux *into* any thick (or thin) spherical shell must equal the flow *out*. In general, through any area in space, flux = flux–density times area, and flux-density equals volume-density times velocity:

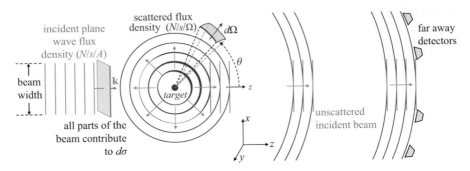

Fig. 3.4 Quantum scattering: each small incident area scatters in all directions; the contribution to a given solid angle is the incident area times the amplitude for scattering into that solid angle

$$F = Ja = \rho v a = |\psi_{out}|^2 \frac{\hbar k}{m} a$$

where $m \equiv$ mass of particle, $k \equiv$ spatial frequency,

$$a \equiv \text{area},$$

$$v = \frac{p}{m} = \frac{\hbar k}{m} \equiv \text{quantum velocity of particle.}$$

Therefore, the scattered wave-function $\psi_{out}(\mathbf{r})$ must decrease as $1/r$, so that $|\psi_{out}|^2$ decreases as $1/r^2$, and the total flux through spheres of any radius is constant:

$$\text{flux} = v \int_{\text{sphere}} |\psi_{sc}(r,\theta,\phi)|^2 r^2 \, d\Omega = v \int_{\text{sphere}} \left| \frac{e^{ikr}}{r} \right|^2 r^2 \, d\Omega = const.$$

The outbound wave-function ψ_{sc} is thus a spherical wave (but not spherically symmetric). ψ_{sc} is also a solution to Schrödinger's equation with $V = 0$, since V is zero outside the range of the scattering potential. Therefore, placing the origin of our coordinates at the target, we define a scattering amplitude $f(\theta, \phi)$:

$$\psi_{sc}(r,\theta,\phi) = f(\theta,\phi) B \frac{e^{ikr}}{r}, \qquad \text{and}$$

$$\psi(x,y,z) = \psi_{inc}(z) + \psi_{sc}(r,\theta,\phi) = B\left[e^{ikz} + f(\theta,\phi)\frac{e^{ikr}}{r} \right]. \qquad (3.2)$$

$f(\theta, \phi)$ is implicitly a function of the incident momentum (or k), as well, but we omit k for brevity, as is conventional. $f(\theta, \phi)$ has units of $m^{-1/2}$. B is arbitrary (it will cancel from all calculations), so we take $B = 1$ for simplicity (unit amplitude delta-function normalization). We have written the stationary-state wave-function $\psi(x, y, z)$ in the conventional manner on the RHS, which mixes both rectangular coordinates (z), and spherical (r, θ, ϕ) coordinates.

We show later that $d\sigma/d\Omega = |f(\theta, \phi)|^2$. Then the total cross section is found by integrating over all angles, just as for classical scattering:

$$\sigma = \int_{\text{sphere}} \frac{d\sigma}{d\Omega} \, d\Omega = \int_0^\pi d\theta \int_0^{2\pi} |f(\theta,\phi)|^2 \, \sin\theta \, d\phi \quad \text{where} \quad d\Omega = d\theta \sin\theta \, d\phi.$$

We can define "hard" quantum scattering as the limit of a potential that is infinite inside its boundary, and zero outside. But,

> Unlike classical scattering, the hard-target quantum scattering cross section is *not* equal to the physical target cross-sectional area.

Scattering calculations may include many conditions on the incident and outbound particles, such as their spin, polarization, etc. The differential cross section $d\sigma/d\Omega$ for a given set of conditions is a statistical description of quantum scattering, and defines everything there is to know about such scattering.

Inelastic scattering You sometimes hear about "inelastic" quantum scattering: this refers to scattering where the outbound particles are different than the inbound particles, and may have different total mass. For example, photons may be radiated, and carry away energy and momentum. Similarly, in condensed matter, scattering may radiate phonons, which carry away energy and momentum. Inelastic scattering includes the case where particles are "absorbed" into the target, such as photon absorption or nuclear capture. It also includes cases where the target changes state, and may *give* energy to the incident particle, *increasing* its outbound energy. Whenever the outbound kinetic energy may be different than the inbound kinetic energy, the scattering is **inelastic**.

Inelastic scattering can be modeled by a potential $V(\mathbf{r})$ which is complex [18, Sect. 20, p. 129], much like EM wave propagate in an absorbing medium with a complex wave-vector $\mathbf{k}$, but we do not address inelastic scattering further.

3.3.2 3D Scattering Off a Target: Schrödinger's Equation

We usually think of scattering as a dynamic, time-dependent process: the incident particle flies in from the left, crashes into the target, and bounces off (scatters) in some direction. But in QM, the incident wave-packet ("beam") is very broad in x and y, compared to the target, and very long in z (Fig. 3.5). However, the beam is narrow compared to the positions of the detectors. As in 1D barrier scattering, the scattering is dominated by the middle of the wave-packet. Hence, the incident and scattered waves are essentially stationary in time (but not static), and we solve the

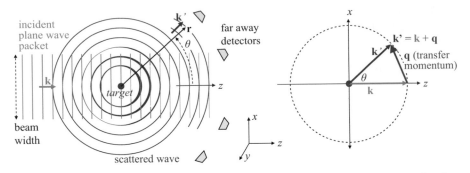

Fig. 3.5 Quantum scattering: a spherically symmetric target scatters into spherical waves, but they are not spherically symmetric. In practice, detectors are much farther away than shown

time-*in*dependent Schrödinger equation for energy eigenstates that include an incident wave, the scattering potential, and a scattered wave.

Finding an exact energy eigenstate is impossible for most scattering potentials. However, there are two cases of interest where approximations work well: when the scattering potential is small compared to particle energy (Born approximation), and when the particle wavelength is large compared to the target size (low *l* approximation). We will examine these special cases after the general scattering description.

We use a common notation: **k** is the incident wave vector; **k'** is a representative scattered wave vector, in the direction (θ, ϕ). (Notation varies in the literature.)

In classical scattering, with localized point particles, the impact parameter is a key input to the scattering effect (Fig. 3.3, *right*). But in QM, a single incident particle is spread over broad, plane waves, which sample all impact parameters from zero to essentially ∞. Therefore, in QM scattering, there is no single impact parameter. However, though the beam is *wide* compared to the target, the beam is *narrow* compared to the detector distance, and some fraction of the incident beam passes through undeflected [18, Fig. 21, p. 115]. This part interferes with the scattered spherical waves, but at large distances, this interference occurs only very near to $\theta = 0$.

In this section, we discuss elastic scattering of spinless, structureless, particles off heavy (immovable), short-range targets. We allow for asymmetric targets, or axially symmetric targets, until the method of partial waves, which requires spherically symmetric targets. We stress physical reasoning, and identify the symmetries throughout the discussion.

As shown earlier, this approach can also be used for particle–particle scattering, using reduction to one-body. After reduction to one-body, the target is fixed in space, i.e., infinitely heavy. For heavy (immovable) targets, conservation of momentum tells you that the target acquires negligible energy, so the scattered particle leaves with the same kinetic energy it started with. This means:

$$|\mathbf{k}| = |\mathbf{k}'| = k, \quad \mathbf{k} = k\hat{\mathbf{z}}, \quad \mathbf{k}' = k\hat{\mathbf{n}}(\theta, \phi), \quad \mathbf{q} \equiv \mathbf{k}' - \mathbf{k} \quad \text{(the momentum transfer)}.$$

(Some references define $\mathbf{q}$ with the opposite sign.) From the law of cosines, we find the magnitude of $\mathbf{q}$:

$$|\mathbf{q}|^2 \equiv q^2 = k^2 + k^2 - 2k^2 \cos\theta = 2k^2(1 - \cos\theta).$$

This is sometimes written $q = 2k\sin(\theta/2)$, found from the right triangle made by bisecting the angle θ.

We consider a hypothetical set of detectors, spread over all θ and ϕ, which measure the momentum and direction of the scattered particle. We define $\hbar\mathbf{q}$ as the momentum acquired by the particle (from the target) during scattering; the target must then acquire momentum of $-\hbar\mathbf{q}$ during scattering. Note that $\mathbf{q}$ changes only the direction of momentum, since the magnitude is fixed by conservation of kinetic energy (elastic collision). Also, the maximum magnitude of $\mathbf{q} = \mathbf{k}' - \mathbf{k}$ is $2k$ (when the particle bounces straight back along the $-z$-axis), so $\mathbf{q}$ is of order $\mathbf{k}$.

A target is modeled as a potential energy, $V(\mathbf{r})$, which may be either attractive (negative) or repulsive (positive) at each point. We assume here that the target has finite reach, and thus the words "far" from the target, or "large r," both mean radii beyond which the potential is (essentially) zero.

Note that the restriction to short-range implies this analysis is not suitable for Coulomb scattering. The Coulomb potential drops as $1/r$, and is therefore too long range for this approach. The quantum analysis for Coulomb scattering requires particular care, though it reduces to the classical result for nonidentical particles [18, pp. 138–141]. We do not consider Coulomb scattering any further.

3.3.3 When is a Potential "Short Range?"

For our 3D scattering derivations, we required that the potential be "short range," so that the asymptotic wave-function be an outgoing spherical wave: $\psi_{sc}(r \to \infty) \sim e^{ikr}/r$. This is an asymptotic solution to the Schrödinger equation for a free particle ($V(r) = 0$). But what does it mean for a potential to be "short range?" More importantly, what kind of potentials have outgoing spherical waves as their (large r) asymptotic solution? The answer is simple (though not necessarily obvious): the *scattered* particle must have finite energy at large distances from the target.

As before, we choose our zero of potential such that $V(r \to \infty) = 0$, or more formally:

$$\lim_{r \to \infty} V(r) = 0.$$

But this alone is insufficient to insure a short range potential. A common counterexample is the Coulomb potential: $V(r) \sim 1/r$. This satisfies the previous limit, but we show later that it is still a "long range" potential.

A potential is **short range** if the potential energy due to a spherical wave function is finite. This is equivalent to saying the potential energy can be made negligible beyond some finite radius.

To explore the implications of this, recall that the potential energy of a particle is given by:

$$PE = \int_\infty \psi^*(\mathbf{r})\, V(r)\psi(\mathbf{r})\, d^3r.$$

For a scattered wave, $\psi^*\psi \sim 1/r^2$. For a long incident wave-packet, $\psi_{sc}(r)$ extends all the way from the target ($r=0$) to essentially infinite r. We substitute $1/r^2$ for $\psi^*\psi$, and rewrite the integral in terms of the coordinate r, so $d^3r \rightarrow 4\pi r^2 dr$:

$$\int_0^\infty V(r)\frac{1}{r^2} 4\pi r^2\, dr < \infty \qquad \Rightarrow \qquad \int_0^\infty V(r)\, dr < \infty \text{ (is finite).}$$

In other words, $V(r)$ must be integrable (in the 1D sense). To connect this with the idea of "short range," note that for $V(r)$ to be integrable, it must be true that the particle's potential energy outside some r_0 can be made arbitrarily small by choosing a sufficiently large r_0:

$$\lim_{r_0 \rightarrow \infty} \left(\int_{r_0}^\infty V(r)\, dr \right) = 0.$$

Therefore, we can estimate the "size" of a given target, r_0, by choosing the amount of the particle's potential energy we are willing to neglect, ΔV, and estimating r_0 such that:

$$\int_{r_0}^\infty |\psi_{sc}(r)|^2 V(r)\, 4\pi r^2\, dr = 4\pi \int_{r_0}^\infty V(r)\, dr \; <\sim \Delta V.$$

Then kr_0 is the number of radians of wave-function spanned by the target. (If $kr_0 \ll 2\pi$, then the target spans only a small fraction of a wavelength.)

If $V(r)$ is integrable and monotonically decreasing outside some r_0 (as realistic potentials are), it must satisfy [cf. [15], 8.62, p. 248]

$$V(r \rightarrow \infty) <\sim \frac{1}{r^{1+\varepsilon}} \qquad where \quad \varepsilon > 0 \qquad \text{(short range potential).}$$

Note that we can apply this definition even to nonspherically symmetric potentials, so long as it is satisfied for all angles (θ, ϕ).

Beware that it is common to say that to be short range, $V(r)$ must drop "faster" than $1/r$. This can be misleading, because $1/(r+1) < 1/r$, but is still not integrable.

[One way to write such a relation is: $1/(r+1) \sim 1/r$ for large r; in other words, they are of the same "scale."]

Finally, note that:

For some approximations to hold, $V(r)$ must drop "faster" than $1/r^2$, which is more stringent than the general requirement for a "short-range" potential.

We discuss some such situations later.

3.3.4 Differential Cross Section from Scattering Amplitude

We now show the remarkably simple connection between the differential cross section and scattering amplitude: $d\sigma/d\Omega = |f(\theta, \phi)|^2$, as in Eq. (3.2). We show this by comparing the particle flux intercepted by an effective area $d\sigma$ to the scattered flux per steradian in a solid angle $d\Omega$ around a direction (θ, ϕ).

Classically, the scattered particles follow a radius away from the target. Quantum mechanically, the scattered wave function $\psi_{sc}(r, \theta, \phi)$ propagates outward along all radii. Figure 3.4 shows that the scattered wave's flux *per steradian* $J_{sc}(\theta, \phi)$, for large r, is a function of the spherical angles (θ, ϕ), but is independent of r:

$$\psi_{sc}(r, \theta, \phi) = f(\theta, \phi)\frac{e^{ikr}}{r} \quad \Rightarrow \quad \left|\psi_{sc}(r, \theta, \phi)\right|^2 = \frac{|f(\theta, \phi)|^2}{r^2}.$$

$$J_{sc}(r, \theta, \phi) = vr^2 \left|\psi_{sc}^2\right| = v\left|f(\theta, \phi)\right|^2 \quad \text{(particle/s/sr)}.$$

Now consider the outward flux in some $d\Omega$. There is an effective area $d\sigma$ which captures the same amount flux, and satisfies:

$$J_{inc}\, d\sigma = J_{sc}(\theta, \phi)\, d\Omega \quad \Rightarrow \quad \left|\psi_{inc}(z)\right|^2 v\, d\sigma = v\left|f(\theta, \phi)\right|^2 d\Omega, \quad \text{or} \quad \frac{d\sigma}{d\Omega} = \left|f(\theta, \phi)\right|^2,$$

since $|\psi_{inc}|^2 = 1$. The inbound and outbound velocities are the same, because the particle has the same kinetic energy (elastic scattering). Note that if we chose ψ to have a different amplitude, B, then B would also appear in $f(\theta, \phi)$, and so cancel from both sides. This confirms that our choice of normalization ($B=1$) is legitimate.

Note that if the target is axially symmetric (also included in spherical symmetry), then there will be no ϕ dependence in ψ_{sc}, so $f(\theta, \phi)=f(\theta)$. However, even for a spherically symmetric target, the scattered wave is *not* generally spherically symmetric: it has a θ dependence.

3.3.5   The Born Approximation

Finding an exact energy eigenstate is impossible for most scattering potentials. However, if the potential is small compared to the energy of the incident particle, then we expect the scattering amplitude ψ_{sc} to be small, and we can treat it as a perturbation. We then follow a common route: expand the perturbation in orders of the small magnitude of $V(\mathbf{r})$, and approximate a solution with a mathematical device. This device is called the Born approximation, and is most commonly used to first order (sometimes called the First Born approximation). However, you can apply it iteratively to get more accuracy.

> The Born approximation uses a Green's function, but not in a way that leads to an exact solution.

It is based on the standard Green's function method, though, so a basic understanding Green's functions is essential to understanding the Born approximation [see *Quirky Mathematical Physics Concepts*]. The Born approximation does not require any symmetry in $V(\mathbf{r})$.

Recall that a Green's function allows us to solve for an unknown scalar function $\psi(\mathbf{r})$ in an equation of this form:

$$L\{\psi(\mathbf{r})\} = s(\mathbf{r}) \quad where \quad L \equiv \text{linear operator}, \ s(\mathbf{r}) \text{ is called the "source"}. \quad (3.3)$$

To use Green's functions, we consider an infinitesimal volume element at a point $\mathbf{r}_s$ as a δ-function of magnitude $s(\mathbf{r}_s)d^3r$. We find a solution (with boundary conditions) for a single, unit-magnitude δ-function at $\mathbf{r}_s$:

$$L\{G(\mathbf{r};\mathbf{r}_s)\} = \delta^3(\mathbf{r} - \mathbf{r}_s).$$

By linearity, the solution for the δ-function of magnitude $s(\mathbf{r}_s)d^3r$ is then $s(\mathbf{r}_s)d^3r$ $G(\mathbf{r}, \mathbf{r}_s)$. But our source comprises many such δ-functions. Since L is linear, we add each of their contributions (i.e., we integrate). Thus, given L, and boundary conditions on $\psi(\mathbf{r})$, we can use a Green's function to find the solution:

$$\psi(\mathbf{r}) = \int_{source} G(\mathbf{r};\mathbf{r}_s)\, s(\mathbf{r}_s)\, d^3r_s.$$

Rewriting Schrödinger's equation similar to the Green's function form, Eq. (3.3), allows us to use a Green's function method to approximate the solution. The incident particle momentum $\hbar k$ is a given (thus also implying the energy E_k), so the time-independent Schrödinger equation is:

$$E_k \psi(\mathbf{r}) = \left(-\frac{\hbar^2}{2m}\nabla^2 + V(\mathbf{r})\right)\psi(\mathbf{r}) \quad \Rightarrow \quad \left(\frac{\hbar^2}{2m}\nabla^2 + E_k\right)\psi(\mathbf{r}) = V(\mathbf{r})\psi(\mathbf{r})$$

$$where \quad E_k \equiv \text{the particle energy} = p^2/2m = (\hbar k)^2/2m$$

$$V(\mathbf{r}) \equiv \text{scattering potential (target)}.$$

This is almost of the form we can solve with a Green's function, where $V(\mathbf{r})\psi(\mathbf{r})$ acts like the source function $s(\mathbf{r})$. The key difference from the standard Green's function equation is that our unknown function, $\psi(\mathbf{r})$, appears on *both* sides of the equation. Nonetheless, we use a Green's function satisfying the usual equation:

$$\left(\frac{\hbar^2}{2m}\nabla^2 + E_k\right)G_k(\mathbf{r}-\mathbf{r}_s) = \delta^3(\mathbf{r}-\mathbf{r}_s).$$

The subscript k on G_k reminds us that the Green's function depends on the particle momentum (or equivalently, on its energy), i.e., each incident momentum k has its own Green's function.

Since the operator and boundary conditions are translation invariant (the BCs for a wave-function are that it goes to zero at infinity), the Green's function is also translation invariant ("portable"), i.e., it is a function only of $(\mathbf{r}-\mathbf{r}_s)$. Finally, the general Green's function is not unique (add any solution to the homogeneous solution to generate another Green's function), so we must apply boundary conditions to fully specify it. In scattering, an incident wave hits the target, and some of it scatters outward from the target. Therefore, we choose the unique Green's function comprising only outgoing waves. Outgoing waves from a point source (the δ-function) are spherically symmetric, so our Green's function is a function of only the magnitude $|\mathbf{r}-\mathbf{r}_s|$. Using Fourier transforms, we find [from [16], 19.4.17, p. 538]:

$$G_k(\mathbf{r}-\mathbf{r}_s) = -\frac{m}{2\pi\hbar^2}\frac{e^{ik|\mathbf{r}-\mathbf{r}_s|}}{|\mathbf{r}-\mathbf{r}_s|}. \tag{3.4}$$

Call the incident plane wave $\psi_{inc}(\mathbf{r})=\exp(ikz)$. Our stationary state solution can then be expanded in orders of the small magnitude of $V(\mathbf{r})$:

$$\psi(\mathbf{r}) = \psi_{inc}(\mathbf{r}) + \psi_{sc}(\mathbf{r}) = \psi_{inc}(\mathbf{r}) + \underbrace{\psi^{(1)}(\mathbf{r}) + \psi^{(2)}(\mathbf{r}) + \ldots}_{\psi_{sc}(\mathbf{r})}$$

$$where \quad \psi^{(1)} \text{ is } 1^{st} \text{ order in } V, \quad \psi^{(2)} \text{ is } 2^{nd} \text{ order in } V, \ldots.$$

Plugging into our Green's function equation for $\psi(\mathbf{r})$, Eq. (3.4), we get (exactly):

$$\psi(\mathbf{r}) = \int_{target} d^3r_s \, G(\mathbf{r}-\mathbf{r}_s)\,V(\mathbf{r}_s)\psi(\mathbf{r}_s)$$

$$= \int_{target} d^3r_s \, G(\mathbf{r}-\mathbf{r}_s)\,V(\mathbf{r}_s)\left(\psi_{inc}(\mathbf{r}_s) + \psi^{(1)}(\mathbf{r}_s) + \psi^{(2)}(\mathbf{r}_s) + \ldots\right) \quad \text{(exact)}.$$

Fig. 3.6 Large r limit for
distant observation points

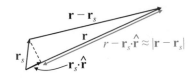

Then the only first-order perturbation is the first term of the integrand:

$$\psi^{(1)}(\mathbf{r}) = \int_{\text{target}} d^3r_s \, G_k(\mathbf{r} - \mathbf{r}_s) V(\mathbf{r}_s) \psi_{inc}(\mathbf{r}_s) = -\frac{m}{2\pi\hbar^2} \int_{\text{target}} d^3r_s \, \frac{e^{ik|\mathbf{r}-\mathbf{r}_s|}}{|\mathbf{r}-\mathbf{r}_s|} V(\mathbf{r}_s) \, e^{i\mathbf{k}\cdot\mathbf{r}_s}.$$
(3.5)

We now simplify this further. As always in scattering, our detectors observe the wave far from the target, where $\mathbf{r}_s$ is small. For such a large $\mathbf{r}$, we take the limit (Fig. 3.6):

$$|\mathbf{r} - \mathbf{r}_s| \underset{\text{large } r}{\longrightarrow} r - \mathbf{r}_s \cdot \hat{\mathbf{r}}.$$

In the denominator, this goes to r, but we must keep both terms in the exponent in the numerator (to get a first order result). Also, because the scattered wave is spherical, the momentum $\mathbf{k}'$ of the scattered particle at $\mathbf{r}$ is in the direction of $\mathbf{r}$ (Fig. 3.5). So:

$$\mathbf{k}' = k\hat{\mathbf{r}} \qquad \text{and} \qquad k\mathbf{r}_s \cdot \hat{\mathbf{r}} = \mathbf{k}' \cdot \mathbf{r}_s.$$

Using both of these in our equation for $\psi^{(1)}$, Eq. (3.5) (note that $\mathbf{r}$ is fixed in these integrals):

$$\psi^{(1)}(\mathbf{r}) \underset{\text{large } r}{\longrightarrow} -\frac{m}{2\pi\hbar^2} \int_{\text{target}} d^3r_s \, \frac{e^{ikr} e^{-ik\mathbf{r}_s \cdot \hat{\mathbf{r}}}}{r} V(\mathbf{r}_s) \, e^{i\mathbf{k}\cdot\mathbf{r}_s}$$

$$= -\frac{m}{2\pi\hbar^2} \frac{e^{ikr}}{r} \int_{\text{target}} d^3r_s \, e^{-i\mathbf{k}'\cdot\mathbf{r}_s} V(\mathbf{r}_s) \, e^{i\mathbf{k}\cdot\mathbf{r}_s} \qquad (3.6)$$

$$= -\frac{m}{2\pi\hbar^2} \frac{e^{ikr}}{r} \int_{\text{target}} d^3r_s \, e^{-i(\mathbf{k}'-\mathbf{k})\cdot\mathbf{r}_s} V(\mathbf{r}_s) \qquad \text{(large } r\text{)}.$$

We recognize this as essentially the 3D Fourier transform in the wave-vector $(\mathbf{k}' - \mathbf{k})$, which we define as $\mathbf{q}$. In other words, to find the scattering amplitude in the direction of $\mathbf{k}'$, we evaluate the Fourier transform of $V(\mathbf{r})$ for the magnitude and (different) direction $\mathbf{q}$. By inspection of the previous equation, we find $f(\theta, \phi)$:

$$\psi^{(1)}(r, \theta, \phi) = f(\theta, \phi)\frac{e^{ikr}}{r} \quad \Rightarrow \quad f(\theta, \phi) = \frac{-m}{2\pi\hbar^2} \int_{\text{target}} d^3r_s \, e^{-i\mathbf{q}\cdot\mathbf{r}_s} V(\mathbf{r}_s)$$
(3.7)

$$\mathbf{q} \equiv \mathbf{k}' - \mathbf{k}, \quad \text{and} \quad \mathbf{k}' = k\hat{\mathbf{r}} = k\hat{\mathbf{n}}(\theta, \phi).$$

Recognizing the Fourier transform is helpful primarily because we can draw on a vast body of knowledge of the properties of Fourier transforms, and apply it to quantum scattering. For example, the maximum magnitude frequency component of the transform accessible to a particle of momentum $\mathbf{k}$ is $q = -2\mathbf{k}$ (for $\theta = \pi$). This means that small scale structure of the target $< \sim 1/2k = \lambda/4\pi$ in size does *not* contribute to the scattering amplitude. This is consistent with the general result that we cannot resolve a feature of a size much smaller than the wavelength of the wave used to "see" the feature.

For spherically symmetric targets, no matter what the direction of $\mathbf{q}$, we can integrate out the angles of the first Born approximation. This provides a quantitative condition for validity of the approximation, and also illustrates how an infinite potential at the origin is not always fatal. We choose angular coordinates ϕ_q and θ_q, aligned along the $\mathbf{q}$ axis (instead of the usual z axis). Then integrate over ϕ_q and θ_q [16, 7.2.4, p. 386]:

$$\mathbf{q} \cdot \mathbf{r}_s = qr \cos \theta_q, \qquad where \quad q \equiv |\mathbf{q}| = 2k \sin(\theta/2), \qquad and$$

$$f(\theta, \phi) = \frac{-m}{2\pi\hbar^2} \int_0^{2\pi} d\phi_q \int_0^\infty dr_s \, r_s^2 \, V(r_s) \int_0^\pi d\theta_q \sin\theta_q \, e^{-iqr_s \cos\theta_q}$$

$$= \frac{-m}{2\pi\hbar^2} (2\pi) \int_0^\infty dr_s \, r_s^2 \, V(r_s) \left[\frac{1}{iqr_s} e^{-iqr_s \cos\theta_q} \right]_{\theta_q=0}^{\pi}$$

$$= \frac{-m}{\hbar^2} \int_0^\infty dr_s \, r_s \, V(r_s) \left[\frac{e^{iqr_s} - e^{-iqr_s}}{iq} \right]$$

$$= \frac{-2m}{\hbar^2 q} \int_0^\infty dr_s \, r_s \sin(qr_s) \, V(r_s) \qquad\qquad \text{(spherically symmetric target).}$$

Note that near zero, the integrand $\sim r_s^2 \, V(r_s)$, which tames any blow up of $V(r \to 0)$ up to $1/r^2$. Therefore, even most potentials that blow up at 0 can be suitable for the Born approximation (if they meet the later conditions).

When is the Born approximation valid? This is a tricky question that we cannot completely answer here. We now show that it can be valid at both low and high energies. Certainly, if $|\psi^{(1)}/\psi_{inc}| \ll 1$ everywhere, then the approximation is valid. With our normalization, $|\psi_{inc}| = 1$ everywhere, and $|\psi^{(1)}(\mathbf{r})|$ is near maximum at $r = 0$, so the approximation holds if [cf. [16], 7.2.12, p. 388]:

$$\psi^{(1)}(r = 0) \ll 1 \qquad\qquad \text{(satisfies Born approximation).}$$

We can quantify this from the Born approximation near the origin, Eq. (3.5) [not the form for large r, Eq. (3.6)], since we seek $\psi^{(1)}(r \to 0)$:

$$\psi^{(1)}(\mathbf{r} = \mathbf{0_v}) = -\frac{m}{2\pi\hbar^2} \int_{\text{target}} d^3 r_s \frac{e^{ik|\mathbf{0_v} - \mathbf{r_s}|}}{|\mathbf{0_v} - \mathbf{r_s}|} V(\mathbf{r_s}) e^{i\mathbf{k} \cdot \mathbf{r_s}}$$

$$= -\frac{m}{2\pi\hbar^2} \int_{\text{target}} d^3 r_s \frac{e^{ikr_s}}{r_s} V(\mathbf{r_s}) e^{ikr_s \cos\theta} \tag{3.8}$$

Here again, for spherically symmetric targets, we can integrate out the angles ϕ and θ [16, 19.4.44]:

$$\psi^{(1)}(r = 0) = -\frac{m}{2\pi\hbar^2} \int_0^{2\pi} d\phi \int_0^\infty dr_s \, r_s^2 \frac{e^{ikr_s}}{r_s} V(r_s) \int_0^\pi \sin\theta \, d\theta \, e^{ikr_s \cos\theta}$$

$$= \frac{m}{2\pi\hbar^2}(2\pi) \int_0^\infty dr_s \, r_s \, e^{ikr_s} V(r_s) \left[\frac{e^{ikr_s \cos\theta}}{ikr_s} \right]_{\theta=0}^\pi \tag{3.9}$$

$$= -\frac{2m}{\hbar^2 k} \int_0^\infty dr_s \, e^{ikr_s} \sin kr_s \, V(r_s) \qquad \text{(spherically symmetric)}.$$

Note that even if $V(r_s \to 0)$ blows up as $1/r_s$, the $\sin kr_s \sim kr_s$, and keeps the integral finite, so the Born approximation can still be valid if $|\psi^{(1)}(r=0)| \ll 1$.

At low particle energy, $kr_0 \ll 1$, $\sin kr_s \to kr_s$, $\exp(ikr_s) \to 1$, and $\psi^{(1)}(r=0)$ further simplifies [16, 19.4.45]:

$$\psi^{(1)}(r = 0) \approx -\frac{2m}{\hbar^2} \int_0^\infty dr_s \, r_s \, V(r_s) \qquad \text{(low energy)} \tag{3.10}$$

This is independent of incident energy (or k). Thus for low energy, the condition for validity of the Born approximation becomes, Eq. (3.10) $\ll 1$ (independent of energy), where the constant depends only on the particle mass and the scattering potential, $V(r)$.

At high energy (large k), $\psi^{(1)}(r=0)$ becomes small for two reasons: first, k appears in the denominator of $\psi^{(1)}(r=0)$, Eq. (3.9). Second, if we rewrite the integrand slightly,

$$\psi^{(1)}(r = 0) = -\frac{2m}{\hbar^2 k} \int_0^\infty dr_s \, e^{ikr_s} \frac{e^{ikr_s} - e^{ikr_s}}{2i} V(r_s)$$

$$= -\frac{m}{i\hbar^2 k} \int_0^\infty dr_s \left(e^{2ikr_s} - 1 \right) V(r_s) \tag{3.11}$$

we see that the first term in the integrand oscillates at frequency k. This makes it cancel more completely as k gets larger, reducing the magnitude of $\psi^{(1)}(r=0)$. Thus, the Born approximation gets progressively better at higher energies [16, p. 389t].

Alternatively, with the usual choice of our zero of energy such that $V(\mathbf{r} \to \infty)=0$, we can directly compare $V(\mathbf{r})$ to $E_k=(\hbar k)^2/2m$. We expect the Born approximation to be valid when $V(\mathbf{r}) << E_k$ everywhere. However, in practice $V(\mathbf{r})$ often has small regions where it is large (or even modeled as infinite), yet the resulting $\psi^{(1)}$ is still appropriately small (as shown previously), so this test is not always useful. Other considerations may improve the region of validity; see [18, p. 325b] and [20, pp. 544–545] for more information.

Observations on the Born approximation   Note the following features of the Born approximation:

- The particle mass multiplies every condition for the validity of the Born approximation. Therefore:

> $\psi^{(1)}$ is proportional to the particle mass, so heavier particles are less often amenable to the Born approximation than lighter ones.

- For a given $V(\mathbf{r})$, the scattering amplitude $f(\theta, \phi)$ [Eq. (3.7)] is a function of $\mathbf{q} = \mathbf{k}' - \mathbf{k}$ only, not of $\mathbf{k}$, θ, or ϕ separately [16, p. 388].
- For low energy (small k and q):

$$f(\theta, \phi) = \frac{-m}{2\pi\hbar^2} \int_{\text{target}} d^3 r_s\, e^{-i\mathbf{q}\cdot\mathbf{r}_s}\, V(\mathbf{r}_s) = \frac{-m}{2\pi\hbar^2} \int_{\text{target}} d^3 r_s\, V(\mathbf{r}_s),$$

which is independent of θ and ϕ [cf. 16, p. 388]. In other words, the scattering is spherically symmetric. This is the limiting case where all frequency components of the Fourier Transform of $V(\mathbf{r})$ greater than 0 are negligible. The whole target, being much smaller than the wavelength of the particle, cannot be resolved at all, and appears to be just a point. We will see later, in the partial wave expansion, that this is equivalent to s-wave scattering ($l=0$), which is spherically symmetric.

- For spherically symmetric potentials, the scattering amplitude is approximately real [16, p. 388]. This is at odds with the optical theorem (discussed in detail later), which states that the scattering cross section is proportional to the imaginary part of $f(\theta=0)$. The resolution of this conflict is that the cross section is second order in the size of the scattering potential, and is too small to be computed with the Born approximation [6, p. 400b].

Higher order Born approximations   From our derivation of $\psi^{(1)}$, we see that it is first order in $V(\mathbf{r})$. The expansion of $\psi(\mathbf{r})$ into orders of $V(\mathbf{r})$ shows that, having found $\psi^{(1)}$, we can compute $\psi^{(2)}$, which is second order in $V(\mathbf{r})$. The only second order term in the expansion is

$$\psi^{(2)}(\mathbf{r}) = \int_{\text{target}} d^3 r_s\, G_k(\mathbf{r} - \mathbf{r}_s) V(\mathbf{r}_s) \psi^{(1)}(\mathbf{r}_s).$$

This term has no simple properties, in contrast to the first-order term's spatial Fourier transform. Ever higher orders can, in principle, be computed by iterating this procedure. In general, $\psi^{(n)}$ is nth-order in $V(\mathbf{r})$. Such an expansion of repeated Green's function integrals is called a Von Neumann series. For the approximation to be valid, if we keep terms up to $\psi^{(n)}$, then we must satisfy:

$$\left| \frac{\psi^{(n)}(r=0)}{\psi_{inc}(r=0)} \right| \ll 1 \qquad \text{(higher order Born validity).}$$

3.4 Partial Wave Method

In contrast to our discussion so far, the method of partial waves requires that the scattering target be spherically symmetric, so $V(\mathbf{r})=V(r)$, where $r \equiv |\mathbf{r}|$. In principle, the method of partial waves gives an exact series solution to the spherically symmetric scattering problem [2, p. 272b], though in general, neither the series, nor any of its terms, can be evaluated analytically. This section assumes you understand orbital (spatial) angular momentum (but not necessarily spin), and how and why the spherical harmonics are angular momentum eigenstates of spherically symmetric potentials. (We discuss orbital angular momentum in a later chapter.) This section is rather long. We proceed as follows:

- Angular momentum of a plane wave; inbound and outbound spherical waves.
- Conservation of angular momentum.
- Scattering of an angular momentum eigenstate.
- Calculating δ_l.
- Small k approximation.

Angular momentum of a plane wave What is the angular momentum of the incident plane wave? It might seem at first that it is zero, and indeed its *average* is zero: $\langle L_x \rangle = \langle L_y \rangle = \langle L_z \rangle = 0$. But we now show that it has nonzero squared magnitude, i.e., $\langle L^2 \rangle \neq 0$. This leads to an important method for computing scattering for *spherically symmetric* targets: partial waves. With partial waves, we decompose the incident plane wave into its L^2 eigenstate components (with quantum numbers l), and then treat each component separately. A single L^2 eigenstate can be easier to analyze than a plane wave, and we get a series solution in the quantum numbers l. Finally, we can estimate the magnitudes of the scattering as a function of l, and keep only the significant contributors. This allows an arbitrarily good approximation in a finite sequence of terms.

Consider the incident plane wave in Fig. 3.7, *left*. Recall that $\hat{L}_z = \frac{\hbar}{i}\frac{\partial}{\partial \phi}$. We easily show that $\langle L_z \rangle = 0$ using *cylindrical* coordinates:

$$\hat{L}_z \psi_{inc}(r,\phi,z) = \frac{\hbar}{i}\frac{\partial}{\partial \phi}e^{ikz} = 0 \qquad \Rightarrow \qquad \langle L_z \rangle \equiv \langle \psi_{inc} | L_z | \psi_{inc} \rangle = 0.$$

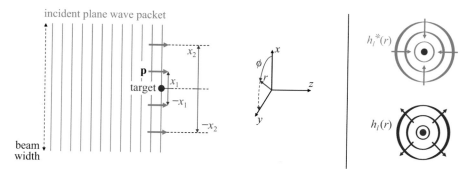

Fig. 3.7 (*Left*) An incident plane wave comprises angular momentum components up to arbitrary *l*. The momentum, **p**, is constant everywhere, but there are many component impact parameters, e.g., x_1, x_2. (*Right*) Spherical inbound and outbound waves

The left-hand equation says that the local $L_z(\mathbf{r})$ is everywhere 0. Physically, as we picture the particle traveling to the right, it has no momentum in the *x* or *y* directions, and therefore has no reason to spiral around the *z*-axis as it propagates. Hence $L_z = 0$.

Now consider L_y of the incident wave. Recall that $\mathbf{L} = \mathbf{r} \times \mathbf{p}$. Since $\mathbf{p} = p\hat{\mathbf{z}}$ is along *z*, $L_y = -xp$, i.e., the local angular momentum *about the target*, due to a point $\mathbf{r} = (x, y, z)$ on the wave-function $\psi(\mathbf{r})$, is $-xp$. We see from the diagram that each contribution to positive L_y from some $x < 0$ is canceled by an equal contribution to negative L_y from its mirror image with $x > 0$. Hence $\langle L_y \rangle = 0$. Similarly, $\langle L_x \rangle = 0$.

What about $L^2 = L_x^2 + L_y^2 + L_z^2$? Our prior result for L_z shows that $L_z^2(\mathbf{r}) = 0$ everywhere. However, L_y^2 is not zero. The two local components of L_y^2 now *add* rather than cancel. Similarly for L_x^2. Furthermore, for an infinitely wide plane wave, there are infinitely large values of *x* contributing, and therefore angular momentum components out to infinite magnitude.

Now recall that the spherical Hankel functions $h_l(\rho)$ are spherical radial functions propagating outward from the origin, and their complex conjugates $h_l^*(\rho)$ are spherical radial functions propagating inward. [h_l and h_l^* are sometimes written as $h_l^{(1)}$ and $h_l^{(2)}$.] The angular momentum eigenstates $h_l(kr)Y_{lm}(\theta, \phi)$ and $h_l^*(kr)Y_{lm}(\theta, \phi)$ are free-particle ($V = 0$) solutions to the Schrödinger equation ($l \equiv$ angular momentum quantum number), and so form an orthonormal basis. Therefore, we can write the incident plane wave as a sum of these angular momentum eigenstates. Since $L_z \propto \partial\psi/\partial\phi = 0$ everywhere, only the $m_l = 0$ functions contribute. Though it is traditional to write the expansion in terms of P_l rather than Y_{l0}, it is more consistent with QM notation to use the spherical harmonics Y_{l0}. Furthermore, we write it in terms of the Hankel functions, because we are interested in the outgoing and ingoing components separately [8, 23.34, p. 500b]:

$$\psi_{inc}(\mathbf{r}) = e^{ikz} = \sum_{l=0}^{\infty} i^l \sqrt{4\pi(2l+1)}\, \frac{1}{2}\left(\underbrace{h_l(kr)}_{out} + \underbrace{h_l^*(kr)}_{in}\right) Y_{l0}(\theta) \qquad (3.12)$$

This is a superposition of inbound and outbound *spherical* waves of *all* angular momenta *l*. They interfere in just the right way to add up to a plane wave. Note that while h_l and h_l^* are both irregular at the origin, their sum is the usual spherical Bessel function, and *is regular* at the origin:

$$\frac{1}{2}\left(h_l(kr) + h_l^*(kr)\right) = j_l(kr) \equiv \text{spherical Bessel function of order } l.$$

This explains why the expansion Eq. (3.12) is regular at the origin. However, we actually do *not* need regularity at the origin, since we will only use the Hankel functions in the vacuum outside the range of the scattering potential ($r>r_0$), and therefore excluding the origin.

Conservation of angular momentum Spherical symmetry of potentials is strongly tied to conservation of angular momentum, e.g., Noether's Theorem. We can see this physically because,

> A spherically symmetric potential exerts a force only radially, not tangentially. There is no torque about the origin. Therefore, scattering off such a target conserves the angular momentum about the origin of the incident particle state.

Alternatively, in QM, a spherically symmetric potential is a function only of *r*, making the Hamiltonian commute with the angular momentum operators, which comprise only $\partial/\partial\theta$ and $\partial/\partial\phi$. The angular momentum operators "pass right through" any $V(r)$ with no angular dependence: $\hat{L}_i V(r) = V(r)\hat{L}_i$. Recall the $\left[\hat{p}^2, \hat{L}_i\right] = 0$. Thus, $\left[\hat{H}, \hat{L}_i\right] = 0$, so we can have simultaneous eigenstates of $\hat{L}^2$ and $\hat{H}$.

To demonstrate this directly from the Schrödinger equation, consider a single angular momentum component of the incident wave, say $R_{kl}(r)Y_{l0}(\theta, \phi)$. Everywhere, including inside the target ($r<r_0$), the scattering stationary state, $\psi(\mathbf{r})$, satisfies Schrödinger's time-independent equation:

$$\left(-\frac{\hbar^2}{2m}\nabla^2 + V(r)\right)\psi(\mathbf{r}) = E_k\psi(\mathbf{r}) \quad where \quad V(r) \equiv \text{spherically symmetric potential.}$$
(3.13)

This form makes it easy to see that if ψ includes a spherical harmonic, then the Schrödinger equation reduces to an eigenfunction equation in *r* alone:

$$\text{If} \quad \psi(\mathbf{r}) = R_{kl}(r)Y_{l0}(\theta,\phi) \quad \text{then} \quad \nabla^2\psi = \left(\frac{1}{r^2}\frac{\partial}{\partial r}r^2\frac{\partial}{\partial r} - \frac{l(l+1)}{r^2}\right)\psi.$$

Substituting into Eq. (3.13) shows that when $V(r)$ is spherically symmetric, the Y_{l0} cancels:

$$\left[-\frac{\hbar^2}{2m}\left(\frac{1}{r^2}\frac{\partial}{\partial r}r^2\frac{\partial}{\partial r}-\frac{l(l+1)}{r^2}\right)+V(r)\right]R_{kl}(r)Y_{l0}(\theta,\phi)=E_kR_{kl}(r)Y_{l0}(\theta,\phi) \qquad \Rightarrow$$

$$\left[-\frac{\hbar^2}{2m}\left(\frac{1}{r^2}\frac{\partial}{\partial r}r^2\frac{\partial}{\partial r}-\frac{l(l+1)}{r^2}\right)+V(r)\right]R_{kl}(r)=E_kR_{kl}(r).$$

Thus, we are left with an eigenfunction equation in $R_{kl}(r)$. As expected, a spherically symmetric potential allows for the separation of eigenfunction solutions into radial and angular parts; the angular solutions are the spherical harmonics, with all $m_l=0$, and the scattered angular momentum equals the inbound angular momentum (angular momentum is conserved during scattering). Note that $R_{kl}(r)$ is a radial wave-function for one of an infinity of *unbound* states, where k is the *continuous* parameter defining the particle's momentum: $k = p/\hbar$. This is in contrast to *bound* states, labeled as $R_{nl}(r)$, where n is a *discrete* quantum number (e.g., an integer) labeling a subset of the bound states.

Scattering of an angular momentum eigenstate As with all scattering viewed far from the target, each scattered wave component approaches a spherical wave. However, even when the potential is spherical, for $l>0$, the scattered wave is *not* spherically symmetric, because the $Y_{l0}(\theta)$ are not spherically symmetric: they vary with θ (but not ϕ because $m_l=0$). Consequently, the *full* wave-function,

$$\psi(\mathbf{r})=\psi_{inc}(\mathbf{r})+\psi_{sc}(\mathbf{r})=\sum_{l=0}^{\infty}R_{kl}(r)Y_{l0}(\theta)\quad where\ k=p\,/\,\hbar\equiv particle\ spatial\ frequency,$$

is *also not* spherically symmetric. This is reasonable, since the entire system was never spherically symmetric: the incident wave-function violates that symmetry. Thus:

The method of partial waves applies only to spherically symmetric potentials, but even so, for $l>0$, the scattered spherical waves are *not* spherically symmetric.

However, the entire system is *axially* symmetric (about the z-axis), so ψ_{sc} and $f(\theta)$ are independent of ϕ.

As defined earlier, the "size" of the target is r_0. The spherical Hankel functions $h_l(kr)$ and $h_l*(kr)$ are the $V=0$ exact solutions to the Schrödinger equation, and therefore the approximate solutions for $r>r_0$, since $V(r>r_0)\approx0$. However, they are *not* the solutions within the range of the potential ($r<r_0$).

The inbound and outbound fluxes describe the particle flow. When we have both fluxes, we have a *superposition* of both inbound and outbound flows, which is *different* than zero flux. With or without a potential, the inbound particle flux must equal the outbound particle flux (for all r).

[Many references refer to probability current, however, we prefer particle flux over probability current, because fluxes can be broken into components. In contrast, individual j_{prob}s do *not*, in general, simply add because j_{prob} is *not* a linear operation on the wave-function [20, p. 528t]. See "Current Events: Probability Current and Electric Current," p. 223.]

In the absence of a target ($V(r)=0$), the free-particle solution of the time-independent Schrödinger equation for angular momentum eigenstates includes terms of this form at large r:

$$\psi_{inc,l}(r,\theta,\phi) \underset{\text{large } r}{\sim} \left(\underbrace{\frac{e^{ikr}}{r}}_{out} + \underbrace{\frac{e^{-ikr}}{r}}_{in} \right) Y_{l0}(\theta)$$

(angular momentum eigenstate for free particle).

The particle inbound and outbound flux densities must be exactly equal (the particle has nowhere else to go). Recall $\mathbf{j}=\mathbf{v}|\psi|^2=(\hbar\mathbf{k}/m)|\psi|^2$. By inspection, we see the previous outbound flux density exactly equals the inbound flux density, at all angles (θ, ϕ). In other words, ψ is a spherical standing wave.

Now we include the short-range target, $V(r)$. From far away, we send inward spherical waves, $h_l^*(kr)Y_{l0}(\theta,\phi)$. They "bounce" off the target, and come out. At large distances, the solution comprises an unmodified inbound component, and an outbound component different than the free-particle case. However, the flux densities must still be equal at all angles, since the inbound and outbound waves share the same Y_{l0}, and the particle still has nowhere else to go. So the outbound magnitude must equal the inbound magnitude at all angles. Therefore, the only modification the outbound wave can acquire from the target is a unit-magnitude phase-factor:

$$\psi_l(r,\theta,\phi) \equiv \psi_{inc,l} + \psi_{sc,l} \underset{\text{large } r}{\to} A\left(e^{i2\delta_l}\underbrace{\frac{e^{ikr}}{r}}_{out} + \underbrace{\frac{e^{-ikr}}{r}}_{in} \right) Y_{l0}(\theta)$$

where $e^{i2\delta_l} \equiv$ unit magnitude phase factor (δ_l real), $A \equiv$ complex amplitude.
(3.14)

We eliminate A shortly. The factor of 2 in the exponent is conventional. Since this is the only allowed form for each l, the scattering amplitude at all angles is defined by the infinite set of δ_l. We will see that at low energy (small k), only the lower l values are significant.

It might seem odd that the whole scattering processing is one which diverts some of the incident plane wave into outbound spherical waves, and yet we have just said that even after scattering, the inbound particle flux must equal the outbound particle flux. However, the outbound phase shift means the outbound wave is no longer exactly the right phase to reconstruct a plane wave. Thus:

> Partial wave scattering can be thought of as *changing the interference of angular momentum components.*

How much of the incident wave is scattered? The difference in complex amplitude between the plane wave's original *outbound* wave [from Eq. (3.12)], and the modified *outbound* wave [from Eq. (3.14)], is the part that was scattered:

$$\psi_{sc,l}(r \to \infty) = \psi_{out,l}(r \to \infty) - \psi_{inc,out,l}(r \to \infty)$$

$$= A \left[e^{2i\delta_l} \frac{e^{ikr}}{r} - \frac{e^{ikr}}{r} \right] = A \left(e^{2i\delta_l} - 1 \right) \frac{e^{ikr}}{r}. \qquad (3.15)$$

We find A from the expansion component of the incident wave with angular momentum l from Eq. (3.12). We require the large r limit, so we use the asymptotic form:

$$h_l(kr \to \infty) \to \frac{e^{i(kr - l\pi/2)}}{ikr} = \frac{e^{ikr} \left(e^{-i\pi/2} \right)^l}{ikr} = \frac{e^{ikr} (-i)^l}{ikr}.$$

The $(-i)^l$ here now cancels the i^l in the plane wave expansion coefficient:

$$\psi_{inc,out,l}(r \to \infty) = A \frac{e^{ikr}}{r} = i^l \sqrt{4\pi(2l+1)} \frac{1}{2} h_l(kr \to \infty) Y_{l0}(\theta) = \frac{\sqrt{4\pi(2l+1)}}{2ik} \frac{e^{ikr}}{r} Y_{l0}(\theta)$$

$$\Rightarrow A = \frac{\sqrt{4\pi(2l+1)}}{2ik}.$$

Then the scattered wave Eq. (3.15) becomes [cf 20, 19.5.13, p. 547]:

$$\psi_{sc,l}(r \to \infty, \theta) = \frac{\sqrt{4\pi(2l+1)}}{2ik} \left(e^{2i\delta_l} - 1 \right) \frac{e^{ikr}}{r} Y_{l0}(\theta). \qquad (3.16)$$

This is the scattering amplitude due to the l component, in terms of its phase shift δ_l. The phase-shift factor is often written in other forms:

$$e^{2i\delta_l} - 1 = e^{i\delta_l} \left(e^{i\delta_l} - e^{-i\delta_l} \right) = \boxed{2i e^{i\delta_l} \sin \delta_l}$$

$$= 2i \frac{\sin \delta_l}{e^{-i\delta_l}} = 2i \frac{\sin \delta_l}{\cos \delta_l - i \sin \delta_l} = \boxed{2i \frac{1}{\cot \delta_l - i}}.$$

Each angular momentum quantum number l has its own phase shift, δ_l, and as with all scattering results, the δ_l depend on k. Superposition applies, since we are finding solutions to the Schrödinger equation, which is linear. Therefore, the total scattering amplitude over all angular momentum components is simply the some of the amplitudes for each l:

$$\psi_{sc,l}(r \rightarrow \infty, \theta) \equiv f_l(\theta)\frac{e^{ikr}}{r} \quad \Rightarrow \quad f(\theta) = \sum_{l=0}^{\infty}\frac{\sqrt{4\pi(2l+1)}}{2ik}Y_{l0}(\theta)\left(e^{2i\delta_l}-1\right). \quad (3.17)$$

Since $e^{2i\delta_l}-1 = 2ie^{i\delta_l}\sin\delta_l$, we have:

$$f_l(\theta) \sim \frac{\delta_l}{k} \qquad \text{(in meters)}.$$

The infinite set of δ_l fully describes the scattering.

Calculating δ_l: Despite all this work, we have only shown a relationship between δ_l and $f(\theta)$, but we have not shown how to calculate either one. We now show how we might compute a δ_l. In general, δ_l is a function of the particle momentum $\hbar k$ (or equivalently, of its energy E), so some references write it as $\delta_l(k)$ or δ_{kl} (as in R_{kl}). We omit the argument (k) for brevity, but understand that our results depend on k.

We now describe a process for finding the cross sections, which can be summarized as follows (see Fig. 3.8):

- Find the stationary state solution (numerically if necessary) for R_{kl} *within* the target: $R_{kl}(r<r_0)$.
- Match $R_{kl}(r_0)$ and $R_{kl}'(r_0)$ to find the vacuum solution outside the target, $R_{kl}(r>r_0)$.
- Find δ_l from the matching conditions.

Then from δ_l, can compute the cross sections, as in Eq. (3.17). Recall that our total stationary-state wave-function for a given l component is given *everywhere* (all r) by:

$$\psi_l(r,\theta) = R_{kl}(r)Y_{l0}(\theta) \qquad where \qquad \psi_l \text{ and } Y_{l0} \text{ are independent of } \phi.$$

The potential has a range of r_0, beyond which it is taken to be zero: $V(r>r_0)=0$. We have already shown (from the arguments leading to (3.14)) that outside the range of the potential [1, 9.54, p. 204]:

$$R_{kl}(r>r_0) = \frac{1}{2}\left[e^{2i\delta_l}h_l(kr)+h_l^*(kr)\right] \qquad \text{(exact if } V(r>r_0)=0). \qquad (3.18)$$

Stationary state solution inside the target We could solve everything in terms of the spherical Hankel functions h_l and h_l^*, but they are complex. Within $r<r_0$, the stationary-state solution to the SE can be taken as real, so it is convenient to switch

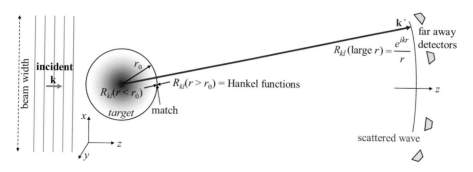

Fig. 3.8 There are three regions to the solution for $\psi = \psi_{inc} + \psi_{sc}$: (1) inside the target, we can take $\psi(r < r_0)$ to be real. (2) In vacuum just outside the target, we describe ψ with Hankel functions. (3) At large r, ψ_{sc} is closely approximated by e^{ikr}/r

to the real-valued spherical Bessel and Neumann functions. That makes it useful to also write $R_{kl}(r > r_0)$ in terms of them:

$$R_{kl}(r > r_0) = C j_l(kr) + D n_l(kr) \quad \text{(exact if } V(r > r_0) = 0\text{)}, \quad (3.19)$$

where the coefficients C and D need to be determined.

Inside the range of the potential, since we are given $V(r)$, we can in principle solve the Schrödinger equation, analytically if possible, or numerically to any desired accuracy. Then $R_{kl}(r)$ must be continuous, and have a continuous derivative, at r_0, the boundary between the target and the vacuum. Our solution inside r_0 provides the values of $R_{kl}(r_0)$ and $R_{kl}'(r_0)$. These two continuity conditions allow us to solve uniquely for the two coefficients C and D. We can compute δ_l from the (now known) values of C and D, as follows.

First, consider the large r limit of $R_{kl}(r)$, using [24, 6.2.23–24, p. 183]:

$$h_l(\rho \to \infty) \to -i \frac{e^{i(\rho - l\pi/2)}}{\rho}, \qquad h_l^*(\rho \to \infty) \to +i \frac{e^{-i(\rho - l\pi/2)}}{\rho}$$

$$\Rightarrow \quad R_{kl}(r \to \infty) = \frac{1}{2kr} \left[i\, e^{-i(kr - l\pi/2)} - e^{2i\delta_l} i e^{i(kr - l\pi/2)} \right].$$

We make the exponentials more symmetric by pulling out a factor of $(-i)\exp(i\delta_l)$, and writing the terms in reverse order:

$$R_{kl}(kr) = e^{i\delta_l} (-i) \frac{1}{2kr} \left[e^{i\delta_l} e^{i(kr - l\pi/2)} - e^{-i\delta_l} e^{-i(kr - l\pi/2)} \right]$$

$$= e^{i\delta_l} (-i) \frac{1}{2kr} \left[e^{i(kr - l\pi/2 + \delta_l)} - e^{-i(kr - l\pi/2 + \delta_l)} \right]$$

$$= e^{i\delta_l} (-i) \frac{1}{2kr} \left[2i \sin(kr - l\pi/2 + \delta_l) \right]$$

$$= e^{i\delta_l} \frac{1}{kr} \left[\sin(kr - l\pi/2 + \delta_l) \right].$$

Wave-functions are always arbitrary up to a complex phase, so we can discard the pure phase factor $\exp(i\delta_l)$, which is an artifact of our choosing the inbound spherical wave component to have the phase of Eq. (3.12), and has no physical significance.

Match $R_{kl}(r_0)$ and $R_{kl}{'}(r_0)$: In preparation for switching to the j_l and n_l forms of $R_{kl}(r)$, we separate out the δ_l from the previous equation using the identity:

$$\sin(\rho+\delta) = \cos\delta\sin\rho + \sin\delta\cos\rho \qquad \Rightarrow$$

$$R_{kl}(r\to\infty) = \frac{1}{kr}\left[\cos\delta_l\sin(kr-l\pi/2) + \sin\delta_l\cos(kr-l\pi/2)\right].$$

The first term looks like the asymptotic form of $j_l\,(r\to\infty)$ and the second term like $n_l\,(r\to\infty)$:

$$j_l(\rho\to\infty) = \frac{\sin(\rho-l\pi/2)}{\rho}, \qquad n_l(\rho\to\infty) = -\frac{\cos(\rho-l\pi/2)}{\rho} \qquad \Rightarrow \qquad (3.20)$$

$$R_{kl}(r\to\infty) = C\,j_l(kr) + D\,n_l(kr) = \left(\cos\delta_l\right)j_l(kr) - \left(\sin\delta_l\right)n_l(kr).$$

Thus, by inspection,

$$C = \cos\delta_l, \qquad D = -\sin\delta_l \qquad \Rightarrow \qquad \delta_l = -\tan^{-1}\frac{D}{C}.$$

You might think that knowing C and D separately provides more information than just their ratio D/C: having both C and D defines δ_l uniquely mod 2π radians; the ratio D/C defines δ_l only mod π radians. However, a difference of π in δ_l simply multiplies our wave-function by -1, which is of no physical significance. Therefore, the ratio D/C fully defines the physical situation.

Using the ratio D/C to find δ_l leads to using a "logarithmic derivative"$=d/dr$ ln $R_{kl}(r) = R_{kl}{'}(r)/R_{kl}(r)$ to establish the matching condition for R_{kl} at r_0. We use such a logarithmic derivative in the next section on the low-energy approximation. For other examples and further mathematical details, we defer to standard texts (e.g., [8], Sect. 23.3, p. 504).

Small k approximation This could also be called the "long wavelength," "low energy," or "low l" approximation. The Eq. (3.20) of $R_{kl}(r)$ leads to a very important approximation for the δ_l, and then the scattering amplitude. At low energy, where $kr_0 \ll 1$, we can approximate δ_l from the asymptotic expansions near the origin of $j_l\,(\rho\to0)$ and $n_l\,(\rho\to0)$, and their derivatives [1, 6.82, p. 166]:

$$j_l(\rho\to0) = \frac{\rho^l}{(2l+1)!!}, \qquad n_l(\rho\to0) = -\frac{(2l-1)!!}{\rho^{l+1}}. \qquad (3.21)$$

Simply taking the derivatives $d/d\rho$ gives the other identities we need:

$$j_l{'}(\rho\to0) = \frac{l\rho^{l-1}}{(2l+1)!!} \qquad\qquad n_l{'}(\rho\to0) = (l+1)\frac{(2l-1)!!}{\rho^{l+2}}.$$

To use this asymptotic limit, we must find D/C in terms of $j_l(r_0)$ and $n_l(r_0)$. Recall that our solution inside r_0 provides the values of $R_{kl}(r_0)$ and $R_{kl}'(r_0)$, from which we compute C and D, and then δ_l. We saw that δ_l depended only on the ratio D/C. In other words, if we scaled the $R_{kl}(r)$ by a constant factor, it would not affect the ratio D/C. We therefore use a common device to work directly in terms of D/C: we divide the derivative $R_{kl}'(r_0)$ by $R_{kl}(r_0)$, and any scale factor cancels. We can then solve for D/C, and thus δ_l, directly. This is just a shortcut to matching $R_{kl}(r_0)$ and $R_{kl}'(r_0)$ inside the target to $R_{kl}(r_0)$ and $R_{kl}'(r_0)$ outside [1, 9.56, p. 204]:

$$\frac{R_{kl}'(r_0)}{R_{kl}(r_0)} = \frac{Cj_l'(kr_0) + Dn_l'(kr_0)}{Cj_l(kr_0) + Dn_l(kr_0)} \equiv \alpha_l.$$

We define this logarithmic derivative as α_l for brevity, following [1], and emphasize that it depends on k and l (and of course, $V(r)$). Now solve for D/C [cf. 1, 9.58, p. 204]:

$$\alpha_l = \frac{j_l'(kr_0) + (D/C)n_l'(kr_0)}{j_l(kr_0) + (D/C)n_l(kr_0)}$$

$$\alpha_l j_l(kr_0) + \alpha_l(D/C)n_l(kr_0) = j_l'(kr_0) + (D/C)n_l'(kr_0)$$

$$\alpha_l j_l(kr_0) - j_l'(kr_0) = (D/C)n_l'(kr_0) - \alpha_l(D/C)n_l(kr_0)$$

$$\frac{D}{C} = \frac{\alpha_l j_l(kr_0) - j_l'(kr_0)}{n_l'(kr_0) - \alpha_l n_l(kr_0)} = -\tan\delta_l.$$

We now take the small k limit (long wavelength), $kr_0 \ll 1$, and use the small argument expansions, Eq. (3.21), and their derivatives. Note that $\frac{d}{dr}j_l(kr) = kj_l'(kr)$. The previous equation for D/C becomes:

$$\frac{D}{C} \rightarrow \frac{1}{(2l+1)!!(2l-1)!!} \cdot \frac{\alpha_l(kr_0)^l - lk(kr_0)^{l-1}}{(l+1)k(kr)^{-(l+2)} + \alpha_l(kr)^{-(l+1)}} \quad \left(\text{multiply by } \frac{r_0(kr_0)^{l+1}}{r_0(kr_0)^{l+1}}\right)$$

$$= \frac{1}{(2l+1)!!(2l-1)!!} \cdot \frac{\alpha_l r_0(kr_0)^{2l+1} - lkr_0(kr_0)^{2l}}{(l+1)kr_0(kr_0)^{-1} + \alpha_l r_0}$$

$$= \frac{1}{(2l+1)!!(2l-1)!!} \cdot \frac{\alpha_l r_0 - l}{l+1+\alpha_l r_0}(kr_0)^{2l+1},$$

Finally, we find δ_l from D/C:

$$\delta_l \approx \tan\delta_l = -\frac{D}{C} = \frac{1}{(2l+1)!!(2l-1)!!} \cdot \frac{l - \alpha_l r_0}{l+1+\alpha_l r_0}(kr_0)^{2l+1} \quad \text{(small } k \text{ approximation).}$$

This expression decreases rapidly with increasing l. (The worst case is when, for some l, $\alpha_l r_0$ is close to $-l-1$, making δ_l large.) Thus, for small k (low energy) particles, the δ_l are small and decrease rapidly:

$$\delta_l \sim \left(kr_0\right)^{2l+1} \quad \Rightarrow \quad \delta_{l+1} \sim \left(kr_0\right)^2 \delta_l.$$

Then the lth scattering amplitude, $f_l(\theta)$, and the l th partial cross section, σ_l, scale as:

$$f_l(\theta) \sim \frac{\sin \delta_l}{k} \sim \frac{\delta_l}{k} \sim r_0\left(kr_0\right)^{2l}, \quad \text{and} \quad \sigma_l \sim \left(f_l(\theta)\right)^2 \sim r_0^2\left(kr_0\right)^{4l}.$$

This rapid fall off of δ_l and σ_l with increasing l means that for low energy, only a few l values are needed, and often only $l=0$. The $l=0$ component of scattering is called the s-wave, just like the $l=0$ orbital of an atom is called the s-orbital. (It is important that we say $\delta_1 \sim (kr_0)^{2l}$, rather than just k^{2l}, because δ_l is dimensionless, and k has units. For example, k might be 0.01 cm^{-1}, in which case k^{2l} drops rapidly with l. But the same k is 1 m^{-1}, and k^{2l} does not drop at all. When comparing two numbers, they must both have the same units. kr_0 is dimensionless, as is δ_l, so the comparison between them is always valid in any unit system.)

The s-wave approximation and the low-l approximation in general, are examples of how the small structure of an object (the target) cannot be resolved by a long-wavelength probe. As we saw in the Born approximation, when only the s-wave is significant, the scattered wave is spherically symmetric, and the scattering resolves nothing of the target at all. Even when a few values of l are significant, such scattering has only large-scale angular structure. Details of the target much smaller than a wavelength produce no significant scattering, so only the coarsest features of the target can be "seen."

3.5   The Optical Theorem

The optical theorem is true for any coherent wave scattering, including EM waves, and a quantum particle. It holds with or without absorption, whenever the scattered wavelength is the same as the incident, even for asymmetric targets [1, p. 202b]. However, the optical theorem requires that waves, as they propagate through the medium, maintain their phase coherence over a long distance. Therefore, the medium must be uniform and the incident particles must have a very narrow bandwidth.

Many quantum references demonstrate the optical theorem only for one or two limited cases, but it is very general. The theorem illustrates a substantive application of interference, as well as some important mathematical methods. We prove the theorem here by direct calculation of the interference. The derivation is exact, as each step is a rigorous limit as $z \to \infty$ and then as $\Delta\theta \to 0$.

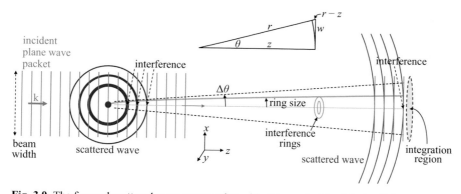

Fig. 3.9 The forward scattered wave gets weaker with distance, but the "shadow" rings compensate by getting bigger. (Many wave fronts are omitted here for clarity.) The beam width is actually much bigger than the rings

This section relies on the general principles of scattering described earlier, especially the idea of cross section. As always, we distinguish "flux" (particles/time) from "flux density" (particles/area/time).

Overview: Essentially, the optical theorem says that the flux scattered from a target exactly equals the flux depleted from the incident wave passing the target. For EM waves, the flux is energy/time or power, and the theorem is a statement of conservation of energy (power in = power out). For quantum mechanical scattering, the flux is particles/time, and it is a statement of conservation of particles (sometimes called conservation of probability): particles in = particles out. This simple idea has a surprisingly odd mathematical form:

$$\sigma_{total} = \frac{4\pi}{k} \operatorname{Im}\{f(\theta=0)\} \qquad \text{(the optical theorem).}$$

Figure 3.9 depicts the depletion of flux from the incident beam by the target. While this is sometimes called the "shadow" of the target, it bears little resemblance to an everyday macroscopic shadow. The "shadow" consists of circular interference rings across its face (Fig. 3.9, though the ring edges are actually gradual, not sharp). The rings are circular even for asymmetric targets, because (as required for our previous scattering analysis), any short-range target looks like a point source from large distances, and hence scatters spherical waves (but not spherically symmetric waves). We choose a small angular width $\Delta\theta$ so that the scattering amplitude $f(\theta, \phi)$ is essentially constant. The scattered wave has a wave front at $\theta=0$, called the **forward scattered wave**. As the propagating forward scattered wave amplitude weakens with distance ($\sim 1/r$), its destructive interference of the incident wave decreases, but the interference *area* grows with distance. The two effects exactly compensate, creating a decrease in flux that is independent of distance. This missing flux must go somewhere, and it goes into the total scattered flux.

We consider the intensity across the x-y plane at some large z. As shown in Fig. 3.9, the x and y distances covered by some small $\Delta\theta$ increase linearly with z. However, the radius of the interference rings increases more slowly, as $\sqrt{z}$. The two growth rates are shown in the figure. For a given $\Delta\theta$, there are more and more rings included as $z \to \infty$. This is important later, in choosing our limits of integration.

Calculation of depleted flux In this calculation, we first choose a small $\Delta\theta$, then take the limit as the observing distance $z \to \infty$, and finally take the limit as $\Delta\theta \to 0$.

Before calculating the depleted flux, we find the flux density near $\theta=0$. The variables and coordinates are illustrated in Fig. 3.9. Flux density $\mathbf{j} = \mathbf{v}|\psi|^2 = \mathbf{v}\psi^*\psi$, where $\mathbf{v}$ is the velocity of the wave. Since $\mathbf{v}$ is the same all along the z-axis, it cancels from all equations, and we can take the flux as:

$$j_z \propto |\psi|^2.$$

From our scattering results, we have the total wave-function, including the incident and scattered parts:

$$\psi(\mathbf{r}) = \psi_{inc}(\mathbf{r}) + \psi_{sc}(\mathbf{r}) = e^{ikz} + f\frac{e^{ikr}}{r} \qquad where \quad f \equiv f(\theta).$$

$$|\psi|^2 = \psi^*\psi = \left(e^{-ikz} + f^*\frac{e^{-ikr}}{r}\right)\left(e^{ikz} + f\frac{e^{ikr}}{r}\right)$$

$$= 1 + f^*\frac{e^{ik(z-r)}}{r} + f\frac{e^{ik(r-z)}}{r} + \frac{f^*f}{r^2}.$$

In the limit of large z, and therefore large r, the last term drops out compared to the others. Also, the two middle terms are conjugates of each other, and any $X + X^* = 2\text{Re}\{X\}$:

$$|\psi|^2 \xrightarrow[\text{large } z]{} 1 + 2\text{Re}\left\{f(\theta)\frac{e^{ik(r-z)}}{r}\right\}. \tag{3.22}$$

We now restrict ourselves to forward scattering (that near $\theta = 0$), still at large z and r, so we can evaluate the exponential in that limit. We choose a small $\Delta\theta$, and restrict θ so that $-\Delta\theta < \theta < +\Delta\theta$. Consider the x-y plane, at some fixed (large) z. The inset of Fig. 3.9 shows the geometry for a point in the x-y plane at a distance w from the z-axis, where $w \ll z$ and $\theta \ll 1$. The phase of the forward propagating incident wave ψ_{inc} is determined by z, but the phase of the spherically propagating scattered wave ψ_{sc} is determined by r. Thus the interference at the x-y plane varies with w (or θ) due to the path length difference $r - z$. To leading order:

$$r = \frac{z}{\cos\theta} = \frac{z}{1 - \theta^2/2} = z\left(1 + \frac{\theta^2}{2}\right), \quad and \quad r - z = z\frac{\theta^2}{2} = z\frac{(w/z)^2}{2} = \frac{w^2}{2z}.$$

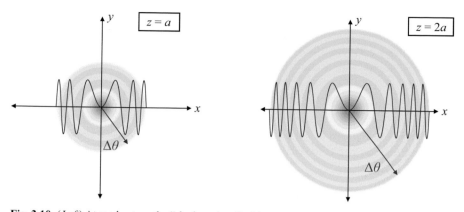

Fig. 3.10 (*Left*) At moderate z, the "shadow rings" of the target, overlaid with the imaginary component of $\exp(ix^2)$. (*Right*) At twice the z, more interference rings are included within $\Delta\theta$

(You can also see this from the binomial theorem and the Pythagoras relation:

$$r = \left(z^2 + w^2\right)^{1/2} = \left(z^2\right)^{1/2} + \frac{1}{2}\left(z^2\right)^{-1/2} w^2 = z + \frac{w^2}{2z}, \quad \text{and} \quad r - z = \frac{w^2}{2z}.)$$

This formula holds for any point (x, y) in the x-y plane, where $x^2 + y^2 = w^2$. Then, in x, y, z coordinates, and putting $r \to z$ (to leading order in small θ) in the denominator of Eq. (3.22):

$$\left|\psi(x, y, z)\right|^2 \underset{\text{large } z}{=} 1 + 2\,\mathrm{Re}\left\{ f(\theta \approx 0) \frac{e^{ik\left(x^2 + y^2\right)/2z}}{z}\right\}. \tag{3.23}$$

Now comes the delicate step: To find the total flux in an area around $\theta = 0$, we must integrate the previous $|\psi|^2$, over the area of the x-y plane near the z-axis. Our region of integration must be large enough to include the contributing region of the previous integrand, but small enough that $f(\theta \approx 0)$ is essentially constant. We can achieve both of these with small $\Delta\theta$ and large z, and thus take $f(\theta) = f(0)$, to first order. Surprisingly, in the large z limit, we can take our x-y limits of integration as $\pm\infty$, because as x or y get large, the exponential oscillates ever more rapidly, with adjacent lobes canceling. Figure 3.10 shows the overlay of the sine (imaginary) component of $\exp(ix^2)$. As z gets larger, the image of $f(\theta)$ on the x-y plane gets larger, and thus varies more slowly with distance w. At the same time, the number of oscillations within $\Delta\theta$ gets larger. Thus, the cancellation between lobes of the integral gets more and more exact with increasing z, and only the central region contributes to the integral (a region we can contain within $\Delta\theta$). A similar effect happens with the real component of $\exp(ix^2)$.

These plots show that as w increases, the complex phase of the spherical scattered wave rotates around the origin in the complex plane. Therefore, both the real and imaginary parts of the scattered wave alternately constructively and destructively interfere with the incident wave. The net flux loss depends on how much the destructive interference outweighs the constructive interference. We now calculate that effect, which shows that the real part's interference exactly cancels, and only the imaginary parts contribute to a net flux loss.

The z-directed flux in the "shadow rings" equals the area times the flux density:

$$j_z = v\rho_{particle}A = v|\psi|^2 A \qquad where \quad v = p/m \equiv \text{speed of the particle.}$$

We define A as the area of our region of integration in the x-y plane (near $\theta = 0$) and recalling that we can take our x and y limits of integration to ∞ for large z. Then from Eq. (3.23) we have:

$$j_z = v\int_A dA\,|\psi(x,y,z)|^2 \rightarrow v\left\{ A + 2\,\text{Re}\left[f(0)\frac{1}{z}\int_{-\infty}^{\infty} dx \int_{-\infty}^{\infty} dy\, e^{ik(x^2+y^2)/2z} \right]\right\} \quad (3.24)$$

$$where \quad A \equiv \text{area within } \Delta\theta.$$

The x and y integrals separate, and are standard Gaussian integrals, though with imaginary exponents. Thus,

$$\int_{-\infty}^{\infty} e^{-ax^2}\,dx = \sqrt{\frac{\pi}{a}} \qquad \Rightarrow \qquad \int_{-\infty}^{\infty} e^{ikx^2/2z}\,dx = \sqrt{\pi\frac{2z}{-ik}}, \quad and$$

$$\int_{-\infty}^{\infty} dx \int_{-\infty}^{\infty} dy\, e^{ik(x^2+y^2)/2z} = \left(\int_{-\infty}^{\infty} dx\, e^{ikx^2/2z}\right)\left(\int_{-\infty}^{\infty} dy\, e^{iky^2/2z}\right) = \pi\frac{2z}{-ik}.$$

The double integral is purely imaginary. Then Eq. (3.24) becomes:

$$j_z = v\left\{ A + 2\,\text{Re}\left[f(0)\frac{1}{z}\pi\frac{2z}{-ik} \right]\right\} = v\left\{ A + 2\,\text{Re}\left[f(0)\frac{2\pi}{-ik} \right]\right\}.$$

Only the imaginary component of $f(0)$ contributes to the Re{ } operator, and its factor of i cancels with the i in the denominator. In other words, only Im$[f(0)]$ contributes a net flux loss. The real part of the integral in Eq. (3.24) comes to zero, meaning the alternate constructive and destructive interferences cancel each other, and thus cancel any contribution from the real part of $f(0)$. Finally, then:

$$j_z = v\left\{ A - \frac{4\pi}{k}\text{Im}\left[f(0)\right]\right\} = v\{A - \sigma_{total}\} \qquad \Rightarrow \qquad \sigma_{total} = \frac{4\pi}{k}\text{Im}\left[f(0)\right].$$

In other words, the flux behind the target is reduced by an amount equal to the incident flux density over an area of σ_{total}. This is the flux scattered away from the incident wave by the target. Note that the flux lost, and therefore σ_{total} is independent of the observing distance z.

This result is exact, not an approximation. Every step was rigorous in the limit as $z \rightarrow \infty$, and then as $\Delta\theta \rightarrow 0$.

[Alternatively, we could have integrated over the area by using concentric rings of radius w, and differential area $2\pi w \, dw$:

$$\int_0^\infty 2\pi w \, dw \, \frac{e^{ikw^2/2z}}{z} = 2\pi \frac{2}{i^2 k}, \qquad \text{using} \qquad \int_0^\infty u \, du \, e^{iu^2} = \frac{1}{i}$$

This integral formally diverges, but can be regularized by letting $w \to w(1+i\varepsilon)$, evaluating the integral, and then taking the limit as $\varepsilon \to 0$. The Gaussian integrals used previously have already included similar regularization.]

In terms of the scattering operator $\hat{S}$ (sometimes called the scattering "matrix"):

$$\psi(\mathbf{r}) = \psi_{inc}(\mathbf{r}) + \psi_{sc}(\mathbf{r}) = \hat{S}\psi_{inc}(\mathbf{r}).$$

Conservation of particles implies that $\hat{S}$ is unitary. Hence the optical theorem is equivalent to the unitarity of the $\hat{S}$ operator.

Correspondence with macroscopic limit You might wonder why our everyday shadows do not seem to behave like the previous calculation. In fact, the central bright spot of a shadow is observed in careful experiments, and once served to definitively establish the wave nature of light. However, macroscopic shadows are usually observed in the near field (z small), where the previous results do not apply. Also, everyday light sources have short coherence lengths ($\sim\lambda$), and so small variations in the air easily destroy interference patterns.

Absorption With EM scattering, the target may absorb some power and scatter the rest. With particle scattering, a nucleus may absorb one electron (thus transmuting into a different element) while scattering others. Since the optical theorem computes the depletion of the incident wave, it counts the "losses" due to both scattering and absorption. Our conservation laws, and the optical theorem, then become:

$$\text{EM: } P_{in} = P_{out} + P_{abs}, \qquad\qquad \text{Particle: } N_{in} = N_{out} + N_{abs}.$$

$$\sigma_{total} = \sigma_{out} + \sigma_{abs} = \frac{4\pi}{k} \text{Im}\{f(0)\}.$$

3.6   Identical Particle Scattering

We mention briefly here some considerations in two-identical-particle scattering. Parts of this section rely on knowledge of complete spin-space quantum states.

The case of identical particles requires more care regarding reduction to one-body. As we will see in Multi-Particle Quantum Mechanics, identical particles always have a symmetry in the multiparticle wave-function. Identical fermions (matter particles such as electrons, muons, protons, quarks, ...) must have two-particle wave-functions that are antisymmetric under particle exchange. If the two spins are in the symmetric state of total $S=0$:

$$\psi_{12}(\mathbf{r}_1,\mathbf{r}_2) = -\psi(\mathbf{r}_2,\mathbf{r}_1) \qquad \text{(fermions with } S=0\text{)}.$$

For example, e^--e^- scattering must satisfy this antisymmetry. When reducing to one-body, this antisymmetry gets lost. For correct calculations, we must preserve the symmetry by imposing an equivalent one on the one-body scattering analysis. In particular, when the one body moves from a point to a diametrically opposite point (across the target), that is equivalent to exchanging the two actual electrons, so for the one-body wave-function, we must satisfy:

$$\psi(\mathbf{r}) = -\psi(-\mathbf{r}) \qquad \text{or} \qquad \psi(r,\theta,\phi) = -\psi(r,\pi-\theta,\phi+\pi).$$

This says the one-body wave-function must be odd parity.

As an example of the measurable consequence of this antisymmetry, consider e^--e^- scattering, where the two electrons collide "head on." For scattering into the x-y plane, $\theta=\pi/2$. In the one-body picture, we have the usual axial symmetry of the incident particle around the z-axis. This means $\psi(\pi/2, \phi)=\psi(\pi/2, \phi+\pi)$. But antisymmetry requires that $\psi(\theta, \phi)=-\psi(\theta, \phi+\pi)$. The only way to satisfy both requirements is for the outbound wave to be zero along $\theta=\pi/2$. In terms of $f(\theta, \phi)$:

$$f(\theta = \pi/2) = 0 \qquad \text{(identical fermions with } S=0\text{)}.$$

On the other hand, if the two fermions are in an antisymmetric spin state ($S=1$), then the two-particle wave-functions must be *symmetric* under particle exchange:

$$\psi(\mathbf{r}_1,\mathbf{r}_2) = +\psi(\mathbf{r}_2,\mathbf{r}_1) \qquad \text{(fermions with } S=1\text{, or bosons).}$$

This is also true of identical bosons with *symmetric* spin states. For example, alpha particles are bosons of zero spin, and so require symmetric two-particle wave-functions. This says the one-body wave-function must be even parity.

Further discussion is outside our scope.

3.7   Scattering Conclusion

Our goal has been to familiarize you with the conceptual background needed to delve into existing mathematical developments of quantum scattering. There are many different approaches to scattering, and every book has its own pedagogical methods. It is probably most beneficial to consult several different references, so as to obtain a broad view of various aspects of scattering theory.

Chapter 4
Matrix Mechanics

We have seen how quantum mechanical kets (states and other vectors) can be represented by wave-functions. Kets are the vectors of the vector space which encompasses a quantum system. Kets are either states or the result of operators on states. Most of the previous chapters deal with wave mechanics, where the kets are continuous functions of space, and therefore, the vector space has infinite dimension.

However, kets are often represented as finite-dimensional vectors, sometimes for convenience (e.g., orbital angular momentum), and sometimes by necessity (e.g., spin). Such a ket is called a **discrete ket** or "finite-dimensional ket." It may be written as a column of N complex numbers, where the vector space has finite dimension, N. Most notably:

> All of angular momentum, both orbital and spin, can be described by finite-dimensional vector spaces.

In addition, lots of other things can be represented or well-approximated by finite-dimensional states, including an ammonia atom's nitrogen position, electron configurations in atoms and molecules, or excitations of an oscillator where only a finite number of states are likely.

Column vectors, row vectors, and matrices are perfect for quantum mechanics (QM), since they are *defined* to be the elements of linear transformations, and so represent the fundamental axiom of QM: systems exist in a linear superposition of states. As such, no description of QM is complete without **matrix mechanics**: the QM of systems which can be represented by finite-dimensional vectors and matrices. Just as wave-functions are vectors in a vector space, finite-dimensional vectors are also vectors in a vector space. (Some quantum references call a finite-dimensional vector space a "Hilbert space," but mathematicians insist a Hilbert space must be infinite dimensional. Therefore, we use the generic term "vector space" for finite dimensional cases.)

E. L. Michelsen, *Quirky Quantum Concepts*, Undergraduate Lecture Notes in Physics, DOI 10.1007/978-1-4614-9305-1_4, © Springer Science+Business Media New York 2014

Because of the simple representation of discrete kets and operators as finite vectors and matrices, many QM concepts are easier to describe in finite-dimensional systems, even if they also apply to continuum systems. For example, density matrices are much easier to visualize and understand in finite-dimensional vector spaces.

Note that the dimension of the quantum state vector space describing a system has nothing to do with the dimension of the physical space of that system. Most of the systems we will consider exist in ordinary three-dimensional (3D) space, but are described by quantum state spaces of many different dimensions. For example, particles orbit in 3D space, but the state space of orbital angular momentum for $l=0$ is 1D, for $l=1$ is 3D, for $l=2$ is 5D, and for arbitrary l is $(2l+1)$D.

4.1   Finite-Dimensional Kets, Bras, and Inner Products

There is a strong analogy between continuous state space (wave-function space) and discrete state spaces. When written in Dirac notation, all the formulas of wave mechanics apply equally well to matrix mechanics, which illustrates again the utility of Dirac notation. Most of the wave mechanics algebra of kets, bras, and operators has simple analogs in finite dimensions. We describe those analogies as we go. Note that discrete space QM uses the standard mathematics of linear algebra, which is *not derived from* the continuous spaces, but is *analogous to* continuous spaces.

Finite-dimensional kets have N components and N basis kets, i.e., any ket can be written as a linear combination of N basis vectors. For example,

$$\text{For } N = 3, \qquad |\psi\rangle = \begin{bmatrix} a_1 \\ a_2 \\ a_3 \end{bmatrix} = a_1|\phi_1\rangle + a_2|\phi_2\rangle + a_3|\phi_3\rangle.$$

(We often use 3D quantum state spaces as examples, because they are nontrivial and illustrative. However, this has nothing to do with ordinary 3D space.) In wave mechanics, a ket $|\psi\rangle \leftrightarrow \psi(x)$, where given x, $\psi(x)$ is the complex value of the ket $|\psi\rangle$ at position x. In finite dimensions, $|\psi\rangle \leftrightarrow \psi_j$, where given an index j, ψ_j is the complex value of the jth component of $|\psi\rangle$. For general N:

$$|\psi\rangle = \begin{bmatrix} a_1 \\ a_2 \\ \vdots \\ a_N \end{bmatrix} = \sum_{k=1}^{N} a_k|\phi_k\rangle = a_1|\phi_1\rangle + a_2|\phi_2\rangle + ...,$$

where $|\phi_k\rangle \equiv$ basis kets and a_k are complex.

Inner products: Inner products and bras are analogous to wave mechanics:

Let

$$|\chi\rangle = \begin{bmatrix} c_1 \\ c_2 \\ c_3 \end{bmatrix}.$$

Then:

$$\langle\chi|\psi\rangle = \sum_{j=1}^{N} c_j^* a_j, \qquad \text{analogous to} \qquad \langle\alpha|\beta\rangle = \int_\infty \alpha^*(x)\beta(x)\, dx.$$

Therefore, bras are written as row vectors, with conjugated components, so that an inner product is given by ordinary matrix multiplication:

$$\langle\chi| = (|\chi\rangle)^\dagger = \begin{bmatrix} c_1^* & c_2^* & c_3^* \end{bmatrix} \quad\Rightarrow\quad \langle\chi|\psi\rangle = \begin{bmatrix} c_1^* & c_2^* & c_3^* \end{bmatrix} \begin{bmatrix} a_1 \\ a_2 \\ a_3 \end{bmatrix} = c_1^* a_1 + c_2^* a_2 + c_3^* a_3.$$

Recall that the dagger symbol acting on a ket produces the dual bras: $(|\chi\rangle)^\dagger \equiv \langle\chi|$.

Kets are written as column vectors, and bras are written as row vectors.

The squared magnitude of a vector is then:

$$\|\psi\|^2 = \langle\psi|\psi\rangle = \sum_{j=1}^{N} a_j^* a_j = \sum_{j=1}^{N} |a_j|^2.$$

All of these definitions comply with standard mathematical definitions.

4.2 Finite-Dimensional Linear Operators

Operators acting on kets (vectors): Matrix multiplication is *defined* to be the *most general* linear operation possible on a discrete vector. Therefore:

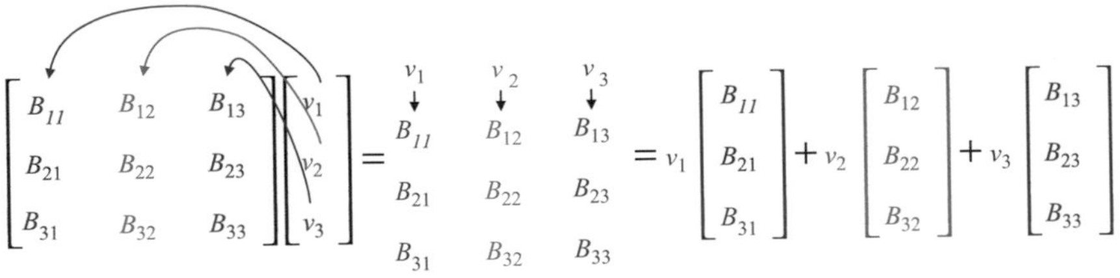

Fig. 4.1 Visualization of a matrix premultiplying a vector, yielding a weighted sum of the matrix columns

Any discrete linear operator can be written as a matrix, which operates on a vector by matrix multiplication.

The matrix elements are, in general, complex. For example, an operator in a 3D quantum state space (not physical 3D space) can be written:

$$\hat{B} = \begin{bmatrix} B_{11} & B_{12} & B_{13} \\ B_{21} & B_{22} & B_{23} \\ B_{31} & B_{32} & B_{33} \end{bmatrix}.$$

It is important to have a good mental image of a matrix multiplying a vector (Fig. 4.1).

Each component of the vector multiplies the corresponding column of the matrix. These "weighted" columns (vectors) are then added (horizontally) to produce the final result. Thus, when used as a linear operator, matrix multiplication of a vector converts each vector *component* into a *whole vector* with full N components, and sums those vectors. Matrix multiplication is linear, which means:

$$\hat{B}\big(a|v\rangle + |w\rangle\big) = a\hat{B}|v\rangle + \hat{B}|w\rangle$$

for all $a, |v\rangle, |w\rangle$.

Average values: Now let us look at the elements of the $\hat{B}$ matrix, B_{ij}, another way. First, consider the diagonal elements, B_{ii}. Recall that we compute the average value of B on a ket $|\psi\rangle$ as:

$$\langle B \rangle \text{ in state } |\psi\rangle = \langle \psi | \hat{B} | \psi \rangle.$$

If $|\psi\rangle$ is just a single basis vector, say $|\phi_1\rangle$, then

$$|\psi\rangle = \begin{bmatrix} 1 \\ 0 \\ 0 \end{bmatrix}, \qquad \langle \psi | \hat{B} | \psi \rangle = \begin{bmatrix} 1 & 0 & 0 \end{bmatrix} \begin{bmatrix} B_{11} & \cdot & \cdot \\ \cdot & \cdot & \cdot \\ \cdot & \cdot & \cdot \end{bmatrix} \begin{bmatrix} 1 \\ 0 \\ 0 \end{bmatrix} = B_{11}.$$

More generally, ψ is a superposition: $|\psi\rangle = a_1|\phi_1\rangle + a_2|\phi_2\rangle + a_3|\phi_3\rangle$. Looking at all the *diagonal* elements, $\langle\phi_i|\hat{B}|\phi_i\rangle = B_{ii}$, we see that each basis vector $|\phi_i\rangle$ in $|\psi\rangle$ contributes an amount B_{ii} to the average of B, weighted by $|a_i|^2$ (the squared magnitude of the amount of $|\phi_i\rangle$ in $|\psi\rangle$).

Now, what do the off-diagonal elements mean?

> When computing an average, off-diagonal elements of an operator matrix describe the *interaction* or "coupling" between two *different* basis vectors for the given operator.

That is, in addition to the diagonal contributions just described, if $|\psi\rangle$ contains two *different* components $|\phi_i\rangle$ and $|\phi_j\rangle$, then their interaction produces two *additional* contributions to the average of B. For example:

$$|\psi\rangle = \begin{bmatrix} a_1 \\ a_2 \\ 0 \end{bmatrix},$$

$$\langle\psi|\hat{B}|\psi\rangle = \begin{bmatrix} a_1^* & a_2^* & 0 \end{bmatrix} \begin{bmatrix} B_{11} & B_{12} & . \\ B_{21} & B_{22} & . \\ . & . & . \end{bmatrix} \begin{bmatrix} a_1 \\ a_2 \\ 0 \end{bmatrix}$$

$$= |a_1|^2 B_{11} + |a_2|^2 B_{22} + \underbrace{a_1^* a_2 B_{12}}_{\text{interaction}} + \underbrace{a_2^* a_1 B_{21}}_{\text{interaction}}.$$

The off-diagonal elements B_{12} and B_{21} produce the last two interaction terms on the right. When evaluating $\langle\psi|\hat{B}|\psi\rangle$ for a Hermitian operator, those terms are conjugates of each other, and their sum is real. This means that the ith component of $\hat{B}|\psi\rangle$ has half of the i-j interaction and the jth component of $\hat{B}|\psi\rangle$ has the other half (though they are complex conjugates of each other). This means the average value of a Hermitian operator in *any* state is real. In particular, the average value of an eigenstate is real, which means the eigenvalues are also real.

Thus, each off-diagonal element B_{ij} of $\hat{B}$ is the result of the interaction of two basis components $|\phi_i\rangle$ and $|\phi_j\rangle$, where i and j take on all values $1, 2, \ldots, N$. To find the numbers B_{ij}, the interactions are computed as if both a_i and a_j were 1. When we later multiply by some actual ket $|\psi\rangle$, the elements B_{ij} get properly weighted by the components a_i and a_j of $|\psi\rangle$. Thus, we can write the elements of $\hat{B}$ as the operation of $\hat{B}$ between all pairs of basis vectors:

$$B_{ij} = \begin{bmatrix} \langle\phi_1|\hat{B}|\phi_1\rangle & \langle\phi_1|\hat{B}|\phi_2\rangle & \langle\phi_1|\hat{B}|\phi_3\rangle \\ \langle\phi_2|\hat{B}|\phi_1\rangle & \langle\phi_2|\hat{B}|\phi_2\rangle & \langle\phi_2|\hat{B}|\phi_3\rangle \\ \langle\phi_3|\hat{B}|\phi_1\rangle & \langle\phi_3|\hat{B}|\phi_2\rangle & \langle\phi_3|\hat{B}|\phi_3\rangle \end{bmatrix}.$$

Again, each column i of the B matrix is a vector which will get weighted by the ket component a_i. The vector sum of these weighted vectors is the resultant vector $\hat{B}|\psi\rangle$.

Note that we sometimes take the inner product with some bra other than $\langle\psi|$. For example, in perturbation theory, we often take inner products $\langle\phi_k|\hat{H}|\phi_m\rangle$, where the bra and ket are different states. Our visualization works just as well for an inner product where the bra is different than the ket: $\langle\chi|\hat{B}|\psi\rangle$.

Recall our discussion of "local" and "nonlocal" operators on wave-functions. Local operators depend only on (at most) an infinitesimal neighborhood around a point. In finite dimensions, this is analogous to the diagonal elements of an operator. The off-diagonal elements are like "nonlocal" effects: the value of one component of the vector being acted on contributes to a *different* component of the result.

Visualization of an inner product with an operator: Besides the matrix multiplication shown previously, it is also sometimes helpful to visualize the inner product $\langle\psi|\hat{B}|\psi\rangle$ another way. This visualization will be used later for density matrices. An inner product with an operator is trilinear: it is linear in the row vector (bra), linear in the operator matrix, and linear in the column vector (ket). Therefore, the inner product must be able to be written as a sum of terms, each term containing exactly three factors: one from the bra, one from the operator, and one from the ket. We can see this by explicit multiplication of an example:

$$\langle\psi|\hat{B}|\psi\rangle = \underbrace{\begin{bmatrix} a_1^* & a_2^* & a_3^* \end{bmatrix}}_{\langle\psi|} \underbrace{\left[a_1\begin{bmatrix} b_{11} \\ b_{21} \\ b_{31} \end{bmatrix} + a_2\begin{bmatrix} b_{12} \\ b_{22} \\ b_{32} \end{bmatrix} + a_3\begin{bmatrix} b_{13} \\ b_{23} \\ b_{33} \end{bmatrix} \right]}_{\hat{B}|\psi\rangle}$$

$$= \begin{pmatrix} a_1^* a_1 b_{11} + a_1^* a_2 b_{12} + a_1^* a_3 b_{13} \\ + \\ a_2^* a_1 b_{21} + a_2^* a_2 b_{22} + a_2^* a_3 b_{23} \\ + \\ a_3^* a_1 b_{31} + a_3^* a_2 b_{32} + a_3^* a_3 b_{33} \end{pmatrix} \quad\begin{matrix} \} = P_{11} \\ \\ \} = P_{22} \\ \\ \} = P_{33}. \end{matrix} \qquad (4.1)$$

(We define P_{ij} shortly.) Indeed, the inner product is a sum of N^2 terms, of three factors each. Note that this is equivalent to:

$$\langle\psi|\hat{B}|\psi\rangle = \sum_{i=1}^{N}\sum_{j=1}^{N} a_i^* B_{ij} a_j. \qquad (4.2)$$

It will be useful (for density matrices) to separate the pieces of the inner product sum in Eq. (4.2) into the products of the bra and ket components, and then separately, the elements of the operator matrix. First, construct the $N\times N$ matrix of all combinations of the components of $|\psi\rangle$ and $\langle\psi|$:

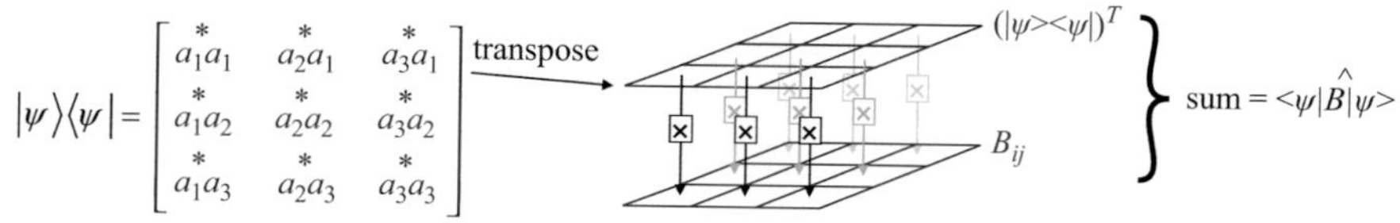

$$
|\psi\rangle\langle\psi| = \begin{bmatrix} a_1 a_1^* & a_2 a_1^* & a_3 a_1^* \\ a_1 a_2^* & a_2 a_2^* & a_3 a_2^* \\ a_1 a_3^* & a_2 a_3^* & a_3 a_3^* \end{bmatrix} \xrightarrow{\text{transpose}}
$$

Fig. 4.2 A visualization of the inner product, with an operator

$$
|\psi\rangle = \begin{bmatrix} a_1 \\ a_2 \\ a_3 \end{bmatrix}, \qquad |\psi\rangle \otimes \langle\psi| \equiv |\psi\rangle\langle\psi| = \begin{bmatrix} a_1 \\ a_2 \\ a_3 \end{bmatrix} \begin{bmatrix} a_1^* & a_2^* & a_3^* \end{bmatrix} = \begin{bmatrix} a_1^* a_1 & a_2^* a_1 & a_3^* a_1 \\ a_1^* a_2 & a_2^* a_2 & a_3^* a_2 \\ a_1^* a_3 & a_2^* a_3 & a_3^* a_3 \end{bmatrix},
$$

or
$$
\left[\, |\psi\rangle\langle\psi| \,\right]_{ij} = a_j^* a_i \,. \tag{4.3}
$$

This $N \times N$ matrix is the **outer product** (aka **tensor product**) of $|\psi\rangle$ with its dual bra. It is a 3×1 matrix times a 1×3 matrix, producing a 3×3 matrix, under the standard rules of matrix multiplication. (Also, the trace of the outer product of a bra and a ket is their inner product.) The $|\psi\rangle\langle\psi|$ matrix lists "how much" of each basis *pair* will contribute to the inner product $\langle\psi|\hat{B}|\psi\rangle$. Then from Eq. (4.2), a general inner product becomes:

$$
\langle\psi|\hat{B}|\psi\rangle = \sum_{i=1}^{N} \sum_{j=1}^{N} B_{ij} a_i^* a_j = \sum_{i=1}^{N} \sum_{j=1}^{N} B_{ij} \left[\, |\psi\rangle\langle\psi| \,\right]_{ji}. \tag{4.4}
$$

The last factor is the transpose of $|\psi\rangle\langle\psi|$. Figure 4.2 illustrates the computation of the inner product.

The total inner product, given in Eq. (4.1) and Fig. 4.2, is the sum of the inner products of the **B** matrix rows with the columns of $|\psi\rangle\langle\psi|$: (**B** row 1)· ($|\psi\rangle\langle\psi|$ column 1)+(**B** row 2)·($|\psi\rangle\langle\psi|$ column 2)+... . These N terms are the diagonal elements of the matrix product $\mathbf{B}\big[|\psi\rangle\langle\psi|\big]$, which we define for convenience as $\mathbf{P} \equiv P_{ij}$.

$$
\mathbf{B}\big[\, |\psi\rangle\langle\psi| \,\big] \equiv P_{ij} = \begin{bmatrix} P_{11} & \cdot & \cdot \\ \cdot & P_{22} & \cdot \\ \cdot & \cdot & P_{33} \end{bmatrix}.
$$

Then Eq. (4.4) becomes:

$$
\langle\psi|\hat{B}|\psi\rangle = \sum_{i=1}^{N} \left[\mathbf{B}\big[\, |\psi\rangle\langle\psi| \,\big]\right]_{ii} = P_{11} + P_{22} + P_{33} = \mathrm{Tr}\big(\mathbf{B}\big[\, |\psi\rangle\langle\psi| \,\big]\big). \tag{4.5}
$$

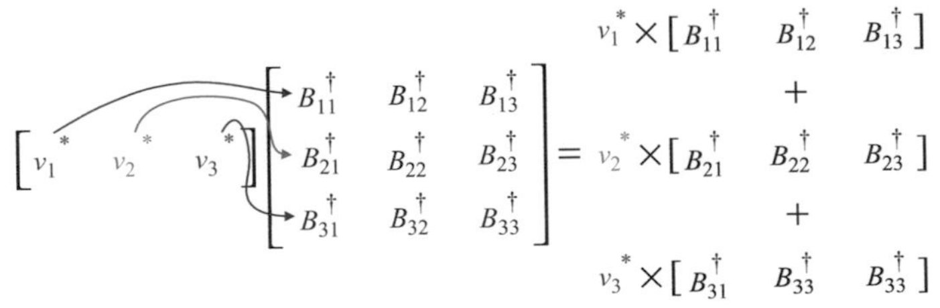

$$v_1^* \times \begin{bmatrix} B_{11}^\dagger & B_{12}^\dagger & B_{13}^\dagger \end{bmatrix}$$

$$+$$

$$\begin{bmatrix} v_1^* & v_2^* & v_3^* \end{bmatrix} \begin{bmatrix} B_{11}^\dagger & B_{12}^\dagger & B_{13}^\dagger \\ B_{21}^\dagger & B_{22}^\dagger & B_{23}^\dagger \\ B_{31}^\dagger & B_{32}^\dagger & B_{33}^\dagger \end{bmatrix} = v_2^* \times \begin{bmatrix} B_{21}^\dagger & B_{22}^\dagger & B_{23}^\dagger \end{bmatrix}$$

$$+$$

$$v_3^* \times \begin{bmatrix} B_{31}^\dagger & B_{33}^\dagger & B_{33}^\dagger \end{bmatrix}$$

Fig. 4.3 Visualization of a matrix postmultiplying a row vector, yielding a weighted sum of the matrix rows

Note, also:

$$\mathrm{Tr}(\mathbf{UV}) = \mathrm{Tr}(\mathbf{VU}) \qquad \Rightarrow \qquad \langle\psi|\hat{B}|\psi\rangle = \mathrm{Tr}\left(\mathbf{B}\left[\,|\psi\rangle\langle\psi|\,\right]\right) = \mathrm{Tr}\left(\left[\,|\psi\rangle\langle\psi|\,\right]\mathbf{B}\right).$$

Finite-dimensional adjoint operators: We now show that finite-dimensional adjoints are given by the conjugate transpose of the original operator matrix. Recall the definition of the adjoint $\hat{B}^\dagger$ of the operator $\hat{B}$:

$$\langle\psi|\hat{B}^\dagger \equiv \hat{B}|\psi\rangle^{DC} \equiv \left(\hat{B}|\psi\rangle\right)^\dagger.$$

In other words, the adjoint operator acts to the left the way the original operator acts to the right, i.e., the adjoint produces a bra from $\langle\psi|$ whose elements are the conjugates of the column vector $\hat{B}|\psi\rangle$. $\hat{B}^\dagger$ is pronounced "bee dagger."

To construct our adjoint matrix, first note the visualization of postmultiplying a bra (row vector) by a matrix (which is analogous to the visualization of premultiplying a vector by a matrix):

We see that postmultiplying a bra produces a weighted sum of the rows of the matrix (much like premultiplying a vector produces a weighted sum of the columns). By definition of the adjoint, the elements of the resulting bra (row vector) must be the conjugates of the corresponding ket (column vector). Comparing Fig. 4.3 to Eq. (4.1), we see that this happens if the rows of $\hat{B}^\dagger$ are the conjugates of the columns of $\hat{B}$. Therefore:

> The adjoint operator matrix is the conjugate-transpose of the original operator matrix.

E.g., $\qquad \hat{B} = \begin{bmatrix} 1 & 2i \\ -3i & -1-5i \end{bmatrix} \qquad \Rightarrow \qquad \hat{B}^\dagger = \begin{bmatrix} 1 & 3i \\ -2i & -1+5i \end{bmatrix}.$

Hermitian (self-adjoint) operators are very important in QM: all physical observable operators are Hermitian. The eigenvalues of a Hermitian matrix are real, and hence all observable properties, and their averages, are real. A Hermitian matrix equals its own conjugate-transpose, i.e., conjugate symmetric across the main diagonal. For example,

$$\hat{B} = \begin{bmatrix} 1 & 2 & 3+4i \\ 2 & -2 & 1-i \\ 3-4i & 1+i & 2 \end{bmatrix} \qquad \Rightarrow \qquad \hat{B} = \hat{B}^{\dagger}, \text{ and } \hat{B} \text{ is hermitian.}$$

It follows immediately that the diagonal elements of a Hermitian matrix are real.

A simple example of an operator and its adjoint is the spin-1/2 raising and lowering operators, which add (or subtract) 1 to the m_s quantum number of a spin ket (if possible):

$$\hat{s}_{+} \begin{pmatrix} 0 \\ 1 \end{pmatrix} = \begin{pmatrix} 1 \\ 0 \end{pmatrix}, \qquad \text{and} \qquad \hat{s}_{+} \begin{pmatrix} 1 \\ 0 \end{pmatrix} = \begin{pmatrix} 0 \\ 0 \end{pmatrix}$$

$$\Rightarrow \hat{s}_{+} = \begin{bmatrix} 0 & 1 \\ 0 & 0 \end{bmatrix}, \qquad \text{and} \qquad \hat{s}_{+}^{\dagger} = \begin{bmatrix} 0 & 0 \\ 1 & 0 \end{bmatrix}.$$

Then:

$$\begin{pmatrix} 0 & 1 \end{pmatrix} \hat{s}_{+}^{\dagger} = \begin{pmatrix} 1 & 0 \end{pmatrix}, \qquad \text{and} \qquad \begin{pmatrix} 1 & 0 \end{pmatrix} \hat{s}_{+}^{\dagger} = \begin{pmatrix} 0 & 0 \end{pmatrix}.$$

(We see explicitly that $\hat{s}_{+}{}^{\dagger} = \hat{s}_{-}$.)

Non-Hermitian operators: For non-Hermitian operators, the interactions between the ith and jth components are not conjugate, or necessarily related at all. Consider the $N=3$ space of angular momentum $j=1$, and $\hat{J}_{-}$ acting on vectors with components labeled (a_1, a_0, a_{-1}):

$$\hat{J}_{-} \begin{pmatrix} a_1 \\ 0 \\ 0 \end{pmatrix} = \begin{bmatrix} 0 & 0 & 0 \\ \sqrt{2} & 0 & 0 \\ 0 & \sqrt{2} & 0 \end{bmatrix} \begin{pmatrix} a_1 \\ 0 \\ 0 \end{pmatrix} = \begin{pmatrix} 0 \\ \sqrt{2}a_1 \\ 0 \end{pmatrix},$$

$$\hat{J}_{-} \begin{pmatrix} 0 \\ a_0 \\ 0 \end{pmatrix} = \begin{bmatrix} 0 & 0 & 0 \\ \sqrt{2} & 0 & 0 \\ 0 & \sqrt{2} & 0 \end{bmatrix} \begin{pmatrix} 0 \\ a_0 \\ 0 \end{pmatrix} = \begin{pmatrix} 0 \\ 0 \\ \sqrt{2}a_0 \end{pmatrix}.$$

We see that under $\hat{J}_-$, the a_1 component of the input vector contributes to the a_0 result, but the a_0 component of the input vector does *not* contribute to the a_1 result. There is no symmetry across the main diagonal of the $\hat{J}_-$ matrix, and therefore no symmetry between the effects of, say, the a_1 input on the a_0 output, and the a_0 input on the a_1 output.

These finite-dimensional operator matrix principles apply to any dimension N.

Representing wave-functions as finite-dimensional kets: Note that in some cases, wave-functions that we have thought to be infinite dimensional may actually be finite dimensional. For example, the $l = 1$ orbital angular momentum states are given by the three functions $Y_{1,1}(\theta, \phi)$, $Y_{1,0}(\theta, \phi)$, and $Y_{1,-1}(\theta, \phi)$. Although these can be written in the infinite-dimensional basis of θ and ϕ, the $l = 1$ space is only 3D.

4.3   Getting to Second Basis: Change of Bases

QM is a study of vectors, and vectors are often expressed in terms of components in some basis vectors:

$$\mathbf{r} = a\mathbf{e}_x + b\mathbf{e}_y + c\mathbf{e}_z \qquad where \quad \mathbf{e}_x, \mathbf{e}_y, \text{ and } \mathbf{e}_z \text{ are basis vectors,} \quad or$$

$$|\chi\rangle = a|z+\rangle + b|z-\rangle \qquad where \quad |z+\rangle, |z-\rangle \text{ are basis vectors.}$$

Our choice of basis vectors (i.e., our **basis**) is, in principle, arbitrary, since all observable calculations are independent of basis. However, most times one or two particular bases are significantly more convenient than others. Therefore, it is often helpful to *change our basis*: i.e., we transform our components from one basis to another. Note that such a transformation does *not* change any of our vectors; it only changes how we write the vectors, and how the internals of some calculations are performed, without changing any observable results.

> A basis change transforms the *components* of vectors and operators; it does *not* transform the vectors or operators themselves

Angular momentum provides many examples where changing bases is very helpful. The infamous Clebsch–Gordon coefficients are used to change bases.

Many references refer to the "transformation of basis vectors," but this is a misnomer. We do not transform our basis vectors; we choose new ones. We can write the new basis vectors as a superposition of old basis vectors, and we can even write these relations in matrix form, but this is fundamentally a different mathematical process than transforming the components of a vector.

In matrix mechanics, we can change the basis of a ket by multiplying it by a transformation matrix. Such a matrix is *not* a quantum operator in the usual sense. The usual quantum operators we have discussed so far change a ket into another ket; a transformation matrix changes the representation of a ket in one basis into a representation of the *same* ket in another basis. We will show that for orthonormal bases, the transformation matrix is unitary (preserves the magnitudes of kets), as it must be if it does not change the ket.

We describe here basis changes in the notation of QM, but the results apply to all vectors in any field of study. We consider mostly orthonormal bases, since more general bases are rarely used in QM. We rely heavily on our visualization of matrices and matrix multiplication, from Eq. (4.2). In matrix mechanics, there are two parts to changing bases: transforming vector components, and transforming the operator matrices.

We use a notation similar to [18, Chap. 6]. Note that some references use a notation in which the roles of U and $U^\dagger$ are reversed from our notation below. (And some references use both notations, in different parts of the book.)

4.3.1 Transforming Vector Components to a New Basis

We describe a general transformation matrix in two complementary ways. We start with an N-dimensional vector expressed in the orthonormal basis b_i:

$$|w\rangle = a_1|b_1\rangle + a_2|b_2\rangle + \ldots + a_N|b_N\rangle. \tag{4.6}$$

It is important to note that the previous is a *vector* equation and is true in *any* basis. The components of $|w\rangle$ in the b-basis are a_i, $i=1, \ldots, N$. Furthermore, the a_i are given by:

$$a_i = \langle b_i|w\rangle.$$

How would we convert the components a_i of $|w\rangle$ into a new basis, n_j? That is:

$$|w\rangle = c_1|n_1\rangle + c_2|n_2\rangle + \ldots c_N|n_N\rangle, \quad c_j = \langle n_j|u\rangle.$$

The inner product on the right, for each c_j, can be evaluated in any basis we choose. In particular, it can be evaluated in our old basis, the b-basis. In matrix notation, we have:

$$c_j = \langle n_j|w\rangle = \left(\mathbf{n}_j \; bra \rightarrow\right)_{old} \begin{pmatrix} a_1 \\ a_2 \\ \vdots \end{pmatrix}_{old}.$$

This is true for every j, so we can write the entire transformation as a single matrix multiplication:

$$|w\rangle = \begin{pmatrix} c_1 \\ c_2 \\ \vdots \end{pmatrix}_{new} = \underbrace{\begin{pmatrix} (\mathbf{n}_1\ bra \rightarrow)_{old} & \cdots \\ (\mathbf{n}_2\ bra \rightarrow)_{old} & \cdots \\ \vdots & \end{pmatrix}}_{N \times N\ \text{matrix} \equiv U} \begin{pmatrix} a_1 \\ a_2 \\ \vdots \end{pmatrix}_{old}.$$

(4.7)

(Recall that when multiplying a vector by a matrix, the jth element of the result is the dot-product (without conjugating the row) of the jth matrix row with the given vector.) In other words, the rows of the transformation matrix are the bras of the *new* basis vectors written in the *old* basis. We call the transformation matrix U.

We now consider a second view of the same transformation matrix. The vector Eq. (4.6) is true in any basis, so we can write it in the *new* basis. Recall that a matrix multiplied by a vector produces a weighted sum of the matrix columns, therefore:

$$|w\rangle = \begin{pmatrix} c_1 \\ c_2 \\ \vdots \end{pmatrix}_{new} = \underbrace{\begin{pmatrix} \begin{pmatrix} \mathbf{b}_1 \\ \downarrow \end{pmatrix}_{new} & \begin{pmatrix} \mathbf{b}_2 \\ \downarrow \end{pmatrix}_{new} & \cdots \end{pmatrix}}_{N \times N\ \text{matrix} \equiv U} \begin{pmatrix} a_1 \\ a_2 \\ \vdots \end{pmatrix}_{old} = a_1 \begin{pmatrix} \mathbf{b}_1 \\ \downarrow \end{pmatrix}_{new} + a_2 \begin{pmatrix} \mathbf{b}_2 \\ \downarrow \end{pmatrix}_{new} + \cdots$$

(4.8)

In other words, the columns of the transformation matrix are the *old* basis vectors written in the *new* basis. This view is equivalent to our first view of the transformation matrix (for orthonormal bases).

Transforming bra components: By the definition of adjoint, we can determine how to transform the components of a bra. We start with the transform of the ket components, and use the definition of adjoint:

$$|w\rangle = \begin{pmatrix} c_1 \\ c_2 \\ \vdots \end{pmatrix}_{new} = U \begin{pmatrix} a_1 \\ a_2 \\ \vdots \end{pmatrix}_{old} \quad \Rightarrow \quad \langle w| = \begin{pmatrix} c_1^* & c_2^* & \cdots \end{pmatrix} U^\dagger.$$

(4.9)

This follows directly from visualizing how premultiplying a column vector (ket) compares to postmultiplying a row vector (bra).

Summary of vector basis change: The transformation matrix can be viewed two ways: (1) as a set of row vectors, which are the bras of the new basis vectors written in the old basis; or (2) as a set of column vectors, which are the old basis vectors written in the new basis. These two views are equivalent.

In the rare case of a nonorthonormal basis, we note that the second view of the transformation matrix is still valid, since the vector Eq. (4.6) is always true, even in nonorthonormal bases. In contrast, the first view of the transformation matrix is *not* valid in nonorthonormal bases, because the components in such bases cannot be found from simple inner products. This means the transformation matrix for nonorthonormal bases is not unitary.

Basis change example: Again we use spin 1/2 as a simple example of a two-dimensional (2D) vector space, though we need not understand spin at all for this example. Suppose we want to transform the components of the 2D vector $|\chi\rangle$ from the z-basis to the x-basis, i.e., given a, b, we wish to find c and d such that:

$$|\chi\rangle = \begin{pmatrix} a \\ b \end{pmatrix}_z = \begin{pmatrix} c \\ d \end{pmatrix}_x \qquad \Rightarrow \qquad |\chi\rangle = a|z+\rangle + b|z-\rangle = c|x+\rangle + d|x-\rangle.$$

(This is analogous to transforming the components of a 2D spatial vector in the x-y plane to some other set of basis axes.) We are given the new x-basis vectors in the old z-basis:

$$|x+\rangle = \left(1/\sqrt{2}\right)|z+\rangle + \left(1/\sqrt{2}\right)|z-\rangle = \begin{pmatrix} 1/\sqrt{2} \\ 1/\sqrt{2} \end{pmatrix},$$

$$|x-\rangle = \left(1/\sqrt{2}\right)|z+\rangle - \left(1/\sqrt{2}\right)|z-\rangle = \begin{pmatrix} 1/\sqrt{2} \\ -1/\sqrt{2} \end{pmatrix}.$$

Then the rows of the transformation matrix U are the conjugates of the coefficients of $|x+\rangle$ and $|x-\rangle$ (since the coefficients are real, the conjugation is invisible):

$$c = \langle x+|\chi\rangle, \qquad d = \langle x-|\chi\rangle, \qquad \text{or}$$

$$\begin{pmatrix} c \\ d \end{pmatrix}_x = \underbrace{\begin{pmatrix} (x+ \, bra \to)_z \\ (x- \, bra \to)_z \end{pmatrix}}_{\text{2×2 matrix}} \begin{pmatrix} a \\ b \end{pmatrix}_z = \begin{pmatrix} 1/\sqrt{2} & 1/\sqrt{2} \\ 1/\sqrt{2} & -1/\sqrt{2} \end{pmatrix} \begin{pmatrix} a \\ b \end{pmatrix}_z.$$

Also, we can just as well interpret the very same transformation matrix as:

$$|\chi\rangle = \begin{pmatrix} c \\ d \end{pmatrix}_x = \underbrace{\begin{pmatrix} \begin{pmatrix} z+ \\ \downarrow \end{pmatrix}_x & \begin{pmatrix} z- \\ \downarrow \end{pmatrix}_x \end{pmatrix}}_{\text{2×2 matrix}} \begin{pmatrix} a \\ b \end{pmatrix}_z,$$

which tells us that the columns of the matrix are the z-basis vectors in the x-basis. Thus:

$$|z+\rangle = \left(1/\sqrt{2}\right)|x+\rangle + \left(1/\sqrt{2}\right)|x-\rangle, \qquad |z-\rangle = \left(1/\sqrt{2}\right)|x+\rangle - \left(1/\sqrt{2}\right)|x-\rangle.$$

4.3.2 The Transformation Matrix is Unitary

A basis-changing transformation matrix is often named U, because such a matrix is "unitary." This important property will help us in the next section on transforming operator matrices. Recall that a unitary matrix is one which preserves the magnitude of all vectors. We prove that any basis transformation matrix is unitary using two different methods. We also show that every unitary matrix satisfies $U^\dagger = U^{-1}$, which is an equivalent definition of "unitary."

The magnitude of a vector is a property of the vector itself, and is independent of the basis in which we represent the vector. Consider again our vector $|w\rangle$ in the two bases b_i and n_j. Then

$$\|w\|^2 = \langle w|w\rangle = \begin{pmatrix} a_1^* & a_2^* & \cdots \end{pmatrix} \begin{pmatrix} a_1 \\ a_2 \\ \vdots \end{pmatrix} = \begin{pmatrix} c_1^* & c_2^* & \cdots \end{pmatrix} \begin{pmatrix} c_1 \\ c_2 \\ \vdots \end{pmatrix}.$$

But the components a_i and c_j are related by:

$$|w\rangle = \begin{pmatrix} c_1 \\ c_2 \\ \vdots \end{pmatrix}_{new} = U \begin{pmatrix} a_1 \\ a_2 \\ \vdots \end{pmatrix}_{old} \quad \Rightarrow \quad \langle w| = \begin{pmatrix} c_1^* & c_2^* & \cdots \end{pmatrix}_{new} = \begin{pmatrix} a_1^* & a_2^* & \cdots \end{pmatrix}_{old} U^\dagger.$$

Then the inner product becomes:

$$\langle w|w\rangle = \begin{pmatrix} c_1^* & c_2^* & \cdots \end{pmatrix} \begin{pmatrix} c_1 \\ c_2 \\ \vdots \end{pmatrix} = \begin{pmatrix} a_1^* & a_2^* & \cdots \end{pmatrix} U^\dagger U \begin{pmatrix} a_1 \\ a_2 \\ \vdots \end{pmatrix}.$$

Since this is true for every possible vector $|w\rangle$, it must be that for every unitary matrix U, $U^\dagger U = \mathbf{1}_{N\times N}$ (the $N\times N$ identity matrix). This means:

$$U^\dagger = U^{-1}.$$

There is another way to see that the basis transformation matrix is unitary, using the descriptions of our transformation matrices from the previous section. The transformation matrix from the old b-basis to the new n-basis can be written [from Eq. (4.8)]:

$$U_{b\to n} = \underbrace{\left(\begin{pmatrix} \mathbf{b}_1 \\ \downarrow \end{pmatrix}_{new} \begin{pmatrix} \mathbf{b}_2 \\ \downarrow \end{pmatrix}_{new} \cdots \right)}_{N\times N \text{ matrix}}.$$

(4.10)

The reverse transformation, from n to b, can be written [from the form of Eq. (4.7), but going from n to b]:

$$U_{n \to b} = \overbrace{\begin{pmatrix} \left(\mathbf{b}_1 \; bra \to\right)_{old} & \cdots \\ \left(\mathbf{b}_2 \; bra \to\right)_{old} & \cdots \\ \vdots & \end{pmatrix}}^{N \times N \text{ matrix}}.$$

(4.11)

But $U_{b \to n}$ and $U_{n \to b}$ perform inverse transformations, and are therefore matrix inverses of each other. By inspection of the previous two equations, we see that they are adjoints (conjugate transposes) of each other. These statements are completely general, and true for any two orthonormal bases for any finite-dimensional vector space. Therefore, it is generally true that a basis transformation matrix is always unitary:

$$U^\dagger = U^{-1},$$

as before.

> A property of unitary matrices is that their rows (taken as vectors) are orthogonal to each other, and their columns are orthogonal to each other.

We can see this orthogonality from the preservation of inner products. Consider two basis vectors, $i \neq j$. Their inner product is zero in both the old and the new bases:

$$\left(\mathbf{b}_j \; bra \to\right)_{old} \begin{pmatrix} \mathbf{b}_i \\ \downarrow \end{pmatrix}_{old} = 0 = \left(\underbrace{\left(\mathbf{b}_j \; bra \to\right)_{old} U^\dagger}_{row \; j} \right)_{new} \left(\underbrace{U \begin{pmatrix} \mathbf{b}_i \\ \downarrow \end{pmatrix}_{old}}_{column \; i} \right)_{new} .$$

Since $U^\dagger$ is the conjugate-transpose of U, row j of $U^\dagger$ is the conjugate of column j of U, so the previous expression is the dot product of column j of U with column i. By the rules of matrix multiplication, it is also the ji element of the matrix product $U^\dagger U = \mathbf{1}_{N \times N}$, which is zero since $i \neq j$. Hence the columns of a unitary matrix are orthogonal. Furthermore, since $U U^\dagger = \mathbf{1}_{N \times N}$, we can also say that the columns of $U^\dagger$ are orthogonal. Since the columns of $U^\dagger$ are the rows of U, the rows of U are also orthogonal.

A unitary matrix which is also real is called an **orthogonal matrix**. Rotation matrices for ordinary 3D spatial vectors are orthogonal.

4.3.3   *Transforming Operators to a New Basis*

Any operator can be defined in terms of its action on vectors. This action is conceptually independent of basis, though the components of both vectors and operator matrices clearly depend on our choice of basis. For a given operator $\hat{A}$, we write its matrix elements $\hat{A}_{ij}$ in the old basis as:

$$\left[\hat{A}\right]_{old} = \begin{bmatrix} A_{11} & A_{12} & & \cdots & \\ A_{21} & A_{22} & & & \\ & & \ddots & & \\ \vdots & & & A_{NN} \end{bmatrix}_{old}.$$

Then, by definition, the elements of $\hat{A}$ in the new basis must satisfy:

$$\left[\hat{A}\right]_{new} \begin{pmatrix} c_1 \\ c_2 \\ \vdots \end{pmatrix}_{new} = U\left[\hat{A}\right]_{old} \begin{pmatrix} a_1 \\ a_2 \\ \vdots \end{pmatrix}_{old}.$$

A simple way to find the matrix elements of $\hat{A}$ in the new basis is to have $\left[\hat{A}\right]_{new}$ transform the components of $|w\rangle$, which are given in the new basis, back to the old basis, act on the vector in the old basis with $\left[\hat{A}\right]_{old}$, and then transform the result to the new basis again:

$$\left[\hat{A}\right]_{new} = U\left[\hat{A}\right]_{old} U^{-1} \qquad \text{(operator matrix transformation)}.$$

For orthonormal bases, $U^{-1} = U^{\dagger}$, so this is sometimes written as $\left[\hat{A}\right]_{new} = U\left[\hat{A}\right]_{old} U^{\dagger}$.

We now consider how a basis change looks for an inner product with an operator. Such inner products are basis-independent. For example, for a given operator $\hat{A}$, every inner product $\langle v|\hat{A}|w\rangle$ must be the same in any basis. In our (old) b-basis, we can write the inner product:

$$\langle v|\hat{A}|w\rangle = \begin{pmatrix} v_1^* & v_2^* & \cdots \end{pmatrix}_{old} \left[\hat{A}\right]_{old} \begin{pmatrix} a_1 \\ a_2 \\ \vdots \end{pmatrix}_{old}. \qquad (4.12)$$

In our new basis n_j, with transformation matrix U, we must have, for every possible $\langle v|$ and $|w\rangle$:

$$\langle v|\hat{A}|w\rangle = \underbrace{\left(v_1^* \quad v_2^* \quad ...\right)_{old} U^\dagger}_{\langle v| \text{ in new basis}} \left[\hat{A}\right]_{new} \underbrace{U \begin{pmatrix} a_1 \\ a_2 \\ \vdots \end{pmatrix}_{old}}_{|w\rangle \text{ in new basis}}.$$

Now use $\left[\hat{A}\right]_{new} = U\left[\hat{A}\right]_{old} U^{-1}$, and $U^\dagger U = U^{-1}U = \mathbf{1}_{N \times N}$:

$$\langle v|\hat{A}|w\rangle = \left(v_1^* \quad v_2^* \quad ...\right)_{old} U^\dagger \underbrace{U\left[\hat{A}\right]_{old} U^{-1}}_{\left[\hat{A}\right]_{new}} U \begin{pmatrix} a_1 \\ a_2 \\ \vdots \end{pmatrix}_{old}$$

$$= \left(v_1^* \quad v_2^* \quad ...\right)_{old} \left[\hat{A}\right]_{old} \begin{pmatrix} a_1 \\ a_2 \\ \vdots \end{pmatrix}_{old},$$

which is the same as before, Eq. (4.12). Thus, our transformations for kets, bras, and operator matrices satisfy the requirement that inner products are independent of basis.

4.4 Density Matrices

Density matrices and mixed states are important concepts required for many real-world situations. Experimentalists frequently require density matrices to describe their results. Density matrices are the quantum analog of the classical concept of ensembles of particles (or systems). Ensembles are used heavily in Statistical Mechanics.

Up until now, we have described systems of distinct "particles," where the particles are in definite quantum states. Even in a definite state, though, the measurable (i.e., observable) properties may be statistical (and thus not definite). This latter property, quite different from classical mechanics, gives rise to a striking new quantum result for the classical concept of ensembles. An **ensemble** is just a bunch of identical particles (or systems), possibly each in a different state, but we know the *statistics* of the distribution of states. For example, a particle drawn from a thermal bath is in an unknown quantum state, due to the randomness of thermal systems. However, if we draw many such particles (an ensemble) from the bath, we can predict the statistics of their properties from the bath temperature. While a known quantum state of a particle may be given by a ket, the state of an ensemble, or of a single particle drawn from it, is given by a density matrix.

We consider here ensembles only for finite-dimensional (and therefore discrete) quantum systems, though the concept extends to more general (continuous) systems. We use some examples from the QM of angular momentum, which is a topic discussed later in this book.

Instead of having a single particle in a definite state, suppose we have an ensemble of particles. If all the particles in the ensemble are in identical quantum states, then we have nothing new. All our QM so far applies to every particle, and extends to the ensemble as a whole. But suppose the ensemble is a *mixture* of particles in several *different* quantum states. What then? Can we compute average values of measurable quantities? If we know the fractions of all the constituent states in the ensemble, then of course we can compute the average value of any observable, and we do it in the straightforward, classical way. We will see that this idea of a classical mixture of quantum particles leads to a "density matrix:" a way of defining all the properties of such a mixture. However, we will also see that quantum ensembles have a highly *nonclassical* nature.

> The density matrix is essentially the quantum state of an ensemble.

For example, suppose we have an ensemble of electrons, 3/4 are spin $|z+\rangle$, and 1/4 are spin $|x+\rangle$. If we measure spin in the z-direction of many particles from the ensemble, we will get an average which is the simple weighted average of the two states:

For $|z+\rangle$:

$$\langle s_z \rangle_{|z+\rangle} = \frac{\hbar}{2},$$

for $|x+\rangle$:

$$\langle s_z \rangle_{|x+\rangle} = 0.$$

Then:

$$[s_z]_{ensemble} = \frac{3}{4}\langle s_z \rangle_{|z+\rangle} + \frac{1}{4}\langle S_z \rangle_{|x+\rangle} = \frac{3}{4} \cdot \frac{\hbar}{2}.$$

Following [16], we use square brackets $[B]$ to explicitly distinguish the *ensemble* average of an observable $\hat{B}$, from a *"pure" state* average $\langle B \rangle \equiv \langle \psi | \hat{B} | \psi \rangle$, the average of an observable for a particle in a known quantum state (which may be a superposition). This average $[B]$ is a number, distinct from the matrix for the operator $\hat{B} \equiv [\hat{B}]$. Expanding the previous average values into bra-operator-ket inner products, we get:

$$\left\langle s_z \right\rangle_{|z+\rangle} = \left\langle z + | \hat{s}_z | z + \right\rangle = \frac{\hbar}{2},$$

$$\left\langle s_z \right\rangle_{|x+\rangle} = \left\langle x + | \hat{s}_z | x + \right\rangle = 0.$$

Then:

$$\left[s_z \right]_{ensemble} = \frac{3}{4} \left\langle z + | \hat{s}_z | z + \right\rangle + \frac{1}{4} \left\langle x + | \hat{s}_z | x + \right\rangle = \frac{3}{4} \cdot \frac{\hbar}{2}.$$

So far, this is very simple.

Now let us consider a more general case: an ensemble consists of a mix of an arbitrary number of quantum states, each with an arbitrary fraction of occurrence (i.e., probability). Note that even in finite-dimensional systems, there are an *infinite* number of quantum states, because the N basis vectors can be combined with complex coefficients in infinitely many ways. Therefore, the number of states in the mixture is unrelated to the Hilbert space dimension, N. Say we have a mix of M states, $\left| \psi^{(k)} \right\rangle$, $k = 1, \ldots, M$, each with a fraction of occurrence in the ensemble (or weight) w_k. As in the spin example, we can simply compute the average value of many measurements of particles from the ensemble by taking a weighted average of the quantum averages [16, 3.4.6, p. 177]:

$$[B] = \sum_{k=1}^{M} w_k \left\langle \psi^{(k)} \left| \hat{B} \right| \psi^{(k)} \right\rangle \qquad where \qquad w_k \text{ are real, and } \sum_{k=1}^{M} w_k = 1.$$

A mixed state is quite different from a superposition. For one thing, a mixed state has no phase information relating the constituent states $\left| \psi_k \right\rangle$: the w_k are real.

Everything we have done so far is independent of basis: the $\left| \psi^{(k)} \right\rangle$ are arbitrary states, and will be superpositions in some bases, but not others. We use the term **constituent** to mean one of the states, $\left| \psi^{(k)} \right\rangle$, of the mixture. This is distinct from "component," which refers to the complex coefficient of a basis function in a superposition. The constituents in a mixture are quantum states, independent of basis.

An ensemble with only one constituent ($M = 1$) is called a **pure state**: each particle is in a definite quantum state.

4.4.1   Development of the Density Matrix

We now use the outer-product method of computing the average value of an operator on a definite state (Fig. 4.2): we overlay each matrix $\left|\psi^{(k)}\right\rangle\left\langle\psi^{(k)}\right|^{T} = a_i{}^*a_j$ with B_{ij}, multiply the overlapping element pairs, and add the resulting products, as in Eq. (4.2):

$$[B] = \sum_{k=1}^{M} w_k \left(\sum_{i=1}^{N} \sum_{j=1}^{N} a_i^{(k)*} a_j^{(k)} B_{ij} \right) \quad \text{where} \quad \left|\psi^{(k)}\right\rangle = \begin{pmatrix} a_1^{(k)} \\ a_2^{(k)} \\ \vdots \\ a_N^{(k)} \end{pmatrix}.$$

The $a_i^{(k)}$ and $a_j^{(k)}$ are the components of each constituent state $\left|\psi^{(k)}\right\rangle$, in any chosen basis [16, 3.4.7.8 p. 177]. This essentially computes $\left\langle\psi^{(k)}\right|\hat{B}\left|\psi^{(k)}\right\rangle$ for each k, and takes the weighted sum.

Now instead, we can move the summation over k inside the other two sums, i.e., we compute the weighted average of the $\left|\psi^{(k)}\right\rangle\left\langle\psi^{(k)}\right|$ *matrices*, and *then* sum that over the operator matrix B_{ij}.

$$[B] = \sum_{k=1}^{M} w_k \left(\sum_{i=1}^{N} \sum_{j=1}^{N} a_i^{(k)*} a_j^{(k)} B_{ij} \right) = \sum_{i=1}^{N} \sum_{j=1}^{N} B_{ij} \underbrace{\left(\sum_{k=1}^{M} w_k a_i^{(k)*} a_j^{(k)} \right)}_{N \times N \ matrix}.$$

The advantage is that we have separated the definition of the *ensemble* from the operator $\hat{B}$: the last factor in parentheses depends only on the ensemble, and *not* on the operator in question. This factor is an $N \times N$ matrix whose transpose is so useful that we give it its own name, the **density matrix:** [16, 3.4.8, p. 177]:

$$\rho \equiv \rho_{ij} \equiv \sum_{k=1}^{M} w_k a_j^{(k)*} a_i^{(k)} = \sum_{k=1}^{M} w_k \left|\psi^{(k)}\right\rangle\left\langle\psi^{(k)}\right|, \qquad i, j = 1, \dots N.$$

Then [16, 3.4.7, p. 177]:

$$[B] = \sum_{i=1}^{N} \sum_{j=1}^{N} B_{ij} \left[\rho^T\right]_{ij} = \sum_{i=1}^{N} [B\rho]_{ii} = \mathrm{Tr}(B\rho).$$

Again, the matrix ρ is *independent* of the operator $\hat{B}$, but just like quantum operators, the elements of ρ depend on the basis in which we choose to write it. Off diagonal elements in ρ indicate one or more superpositions (in our chosen basis) in the constituent states $\left|\psi^{(k)}\right\rangle$. Note, though, that a zero off-diagonal element ρ_{ij} does *not* mean there is no superposition of $\left|\phi_i\right\rangle$ and $\left|\phi_j\right\rangle$ in the mixture, because it is possible that two constituents of the mixture have off-diagonal elements that cancel. ρ is complex, even though the w_k are real, since the constituent states $\left|\psi^{(k)}\right\rangle$

can have components with complex coefficients. Furthermore, $|\psi\rangle\langle\psi|$ is Hermitian (Fig. 4.2 and related equations), and therefore ρ is Hermitian.

We need to compute ρ only once, from the weights and states of the ensemble, and we can then use it for *any* operator. This independence of density matrix from observable operator leads to a new quantum concept, which does not exist in classical mechanics: any two ensembles which have the same density matrix must be considered physically identical, regardless of how they were prepared. This identity is because every physical measurement (every operator acting on the ensemble) has the same statistics for both ensembles.

If no physical measurement in the universe can distinguish two things (ensembles), we must consider them physically identical.

For example, consider an ensemble of spin 1/2 particles, 50% in $|z+\rangle$ and 50% in $|z-\rangle$. The density matrix is

$$|z+\rangle = \begin{bmatrix} 1 \\ 0 \end{bmatrix}, \qquad |z-\rangle = \begin{bmatrix} 0 \\ 1 \end{bmatrix} \qquad \Rightarrow$$

$$\rho = 50\% \left(\begin{bmatrix} 1 & 0 \\ 0 & 0 \end{bmatrix} \right) + 50\% \left(\begin{bmatrix} 0 & 0 \\ 0 & 1 \end{bmatrix} \right) = \begin{bmatrix} 0.5 & 0 \\ 0 & 0.5 \end{bmatrix}.$$

The average spin in the z-direction is $\langle \hat{s}_z \rangle = 0$. In fact, the average spin in *any* direction is $\langle \hat{s}_n \rangle = 0$.

Now consider a second ensemble, 50% in $|x+\rangle$, and 50% in $|x-\rangle$. We write the states in the z basis as:

$$|x+\rangle = \begin{bmatrix} 1/\sqrt{2} \\ 1/\sqrt{2} \end{bmatrix}, \qquad |x-\rangle = \begin{bmatrix} 1/\sqrt{2} \\ -1/\sqrt{2} \end{bmatrix} \qquad \Rightarrow$$

$$\rho = 50\% \left(\begin{bmatrix} 0.5 & 0.5 \\ 0.5 & 0.5 \end{bmatrix} \right) + 50\% \left(\begin{bmatrix} 0.5 & -0.5 \\ -0.5 & 0.5 \end{bmatrix} \right)$$

$$= \begin{bmatrix} 0.25 & 0.25 \\ 0.25 & 0.25 \end{bmatrix} + \begin{bmatrix} 0.25 & -0.25 \\ -0.25 & 0.25 \end{bmatrix} = \begin{bmatrix} 0.5 & 0 \\ 0 & 0.5 \end{bmatrix}.$$

You might think this second ensemble is different from the first. It was formed very differently, from different states. However, the average spin in any direction is again 0.

Furthermore, since the density matrix is identical to the first ensemble, the average of *any* operator is identical to that of the first ensemble. No measurement in the universe can distinguish the two ensembles; therefore as scientists, we must admit that *they are physically the same!*

Even though the constituent states are quite distinct, the resulting ensembles are the same. (Note that we have not yet shown that *all* statistics are the same, but we do so in the next section.)

Classically, this collapsing of ensembles does not occur, because classically, we can make as many measurements as we want on each particle of the ensemble, and each without disturbing its state. For example, we could measure both the x and z components of angular momentum, and distinguish between particles with angular momentum in the x or z direction. This would allow us to distinguish the two previous ensembles. But in the real world of QM, we can only measure one component of a particle's state, and in the process of that measurement, destroy other information about its state. For example, if we measure a particle's spin in the z-direction to be $+\hbar/2$, we cannot tell whether the particle was actually a $|z+\rangle$ particle, or whether it was an $|x+\rangle$ or $|x-\rangle$ particle which happened to measure $+\hbar/2$ in the z-direction. Hence, we cannot distinguish between the two ensembles described previously. And furthermore, no other measurement can distinguish the ensembles, either.

4.4.2   Further Properties of Density Matrices

The phases of the constituent $|\psi^{(k)}\rangle$ **are arbitrary**: Adding an arbitrary phase to a constituent state $|\psi^{(k)}\rangle$ has no effect on the density matrix, because any phase cancels in the outer product:

$$\left[\,\left|\psi^{(k)}\right\rangle\left\langle\psi^{(k)}\right|\,\right]_{ij} = a_j^{(k)*}a_i^{(k)}.$$

Diagonal elements are probabilities: In a basis $|\phi_i\rangle$, the diagonal elements ρ_{dd}, $d = 1, \ldots, N$, are the probabilities of measuring the basis state $|\phi_d\rangle$ in a particle in the mixed state $\boldsymbol{\rho}$. We see this by combining probabilities for the constituent states:

$$\rho_{dd} = \sum_{k=1}^{M} w_k a_d^{(k)*}a_d^{(k)} = \sum_{k=1}^{M} w_k \,\mathrm{Pr}\left(\left|\phi_d\right\rangle in \left|\psi^{(k)}\right\rangle\right) = \mathrm{Pr}\left(\left|\phi_d\right\rangle\right).$$

Since the sum of the probabilities is 1, we see that the sum of the diagonal elements (i.e., the trace) of $\boldsymbol{\rho}=1$. Put another way, consider a state $|\psi\rangle$ that exists in a 3D Hilbert space:

$$|\psi\rangle = \begin{bmatrix} a_1 \\ a_2 \\ a_3 \end{bmatrix}, \quad |\psi\rangle \otimes \langle\psi| \equiv |\psi\rangle\langle\psi| = \begin{bmatrix} a_1 \\ a_2 \\ a_3 \end{bmatrix}\begin{bmatrix} a_1^* & a_2^* & a_3^* \end{bmatrix} = \begin{bmatrix} a_1 a_1^* & a_1 a_2^* & a_1 a_3^* \\ a_2 a_1^* & a_2 a_2^* & a_2 a_3^* \\ a_3 a_1^* & a_3 a_2^* & a_3 a_3^* \end{bmatrix}.$$

$$(4.13)$$

The diagonal elements of $|\psi\rangle\langle\psi|$ are just the terms in the dot product $\langle\psi|\psi\rangle = a_1 a_1{}^* + a_2 a_2{}^* + a_3 a_3{}^* = 1$. The **trace** of a matrix is the sum of the diagonal elements, so we have:

$$\mathrm{Tr}\left(|\psi\rangle\langle\psi|\right) = 1 \quad \text{for a normalized state } |\psi\rangle.$$

$\mathrm{Tr}(\ )$ is a linear operator *on matrices*. The density matrix $\boldsymbol{\rho}$ is a weighted sum of matrices, where each matrix has a trace of 1, and the sum of the weights equals 1. Thus:

$$\mathrm{Tr}(\boldsymbol{\rho}) = \mathrm{Tr}\left(\sum_{k=1}^{M} w_k \left|\psi^{(k)}\right\rangle\left\langle\psi^{(k)}\right|\right) = \sum_{k=1}^{M} w_k \underbrace{\mathrm{Tr}\left(\left|\psi^{(k)}\right\rangle\left\langle\psi^{(k)}\right|\right)}_{1} = \sum_{k=1}^{M} w_k = 1.$$

Sometimes ρ_{dd} is written as $\langle\phi_d|\boldsymbol{\rho}|\phi_d\rangle$, but that notation can be misleading, because it does not make physical sense for a density matrix to act on a ket $|\phi_d\rangle$. As noted earlier, $\boldsymbol{\rho}$ is an ensemble state, and by extension, $\boldsymbol{\rho}$ is a *particle* state (a mixed state). It has no business acting on *another kind* of particle state (a pure state).

A different kind of operator: The density matrix $\boldsymbol{\rho}$ is a Hermitian matrix expressed in a particular basis. You can change $\boldsymbol{\rho}$ to a different basis using the same similarity transform as for an operator matrix:

$$\rho' = U\rho U^{-1},$$

where $U \equiv$ the unitary transformation matrix.

However, $\boldsymbol{\rho}$ *is not a quantum operator*, in the usual sense. Some books use the unfortunate term "density operator," but just because a matrix is expressed in a basis, that does not make it an ordinary operator. In fact, the density matrix is a kind of quantum *state*. It is used *with* ordinary quantum operators, but in a different way than kets are. Recall that a quantum operator acts on a ket to produce another ket; density matrices do not usually do that. [We can consider a density matrix to be a rank-2 tensor acting on another rank-2 tensor (a quantum operator) to produce a scalar.]

Basis independent density matrix: Recalling the definition of outer product from Eq. (4.3), we have seen that we can write the density matrix in a basis-independent way, as a sum of outer products of the constituent states with themselves:

$$\boldsymbol{\rho} \equiv \sum_{k=1}^{M} w_k \left|\psi^{(k)}\right\rangle\left\langle\psi^{(k)}\right| \qquad \text{(basis independent).}$$

As with vectors, if we want to write $\boldsymbol{\rho}$ as a set of numbers, we must choose a basis.

For an ensemble in a **pure state**, where every particle has the exact same quantum state $|\psi\rangle$, $M = 1$, and the density matrix is:

$$\boldsymbol{\rho} = |\psi\rangle\langle\psi| \qquad \text{(pure state).}$$

Average from trace: We saw in Eq. 4.5 that the average in a pure quantum state $|\psi\rangle$ can be computed from $[B] = \mathrm{Tr}(\mathbf{B}\,|\psi\rangle\langle\psi|)$. The average of a mixed state is just the weighted average of the pure averages:

$$[B] = \sum_{k=1}^{M} w_k \, \mathrm{Tr}\left(\mathbf{B}\left[\left|\psi^{(k)}\right\rangle\left\langle\psi^{(k)}\right|\right]\right) = \mathrm{Tr}\left(\sum_{k=1}^{M} w_k \mathbf{B}\left[\left|\psi^{(k)}\right\rangle\left\langle\psi^{(k)}\right|\right]\right)$$

$$= \mathrm{Tr}\left(\mathbf{B}\sum_{k=1}^{M} w_k \left[\left|\psi^{(k)}\right\rangle\left\langle\psi^{(k)}\right|\right]\right) = \mathrm{Tr}(\mathbf{B}\boldsymbol{\rho}) = \mathrm{Tr}(\boldsymbol{\rho}\mathbf{B}).$$

Note that the trace of a matrix is unchanged by a basis change (unitary transformation), so $[B] = \mathrm{Tr}(\boldsymbol{\rho}\mathbf{B})$ in *any basis*, and we can choose to evaluate it in any convenient basis.

   Trace of pure and mixed states: $\boldsymbol{\rho}$ is a Hermitian matrix, and therefore can be diagonalized. Then there exists a basis in which:

$$\boldsymbol{\rho} = \begin{bmatrix} \rho_{11} & 0 & 0 \\ 0 & \rho_{22} & 0 \\ 0 & 0 & \ddots \end{bmatrix}, \qquad \mathrm{Tr}(\boldsymbol{\rho}) \equiv \sum_{d=1}^{N} \rho_{dd} = 1, \qquad \text{and} \qquad \rho_{dd} \leq 1, \ d = 1, \dots N.$$

In a pure state, there is only one constituent quantum state $|\psi\rangle$, so $M = 1$, and $w_1 = 1$. Then we can choose a diagonal basis of $\boldsymbol{\rho}$ in which $|\psi\rangle$ is one of the basis vectors. Then $\boldsymbol{\rho}$ has the simple form:

$$\boldsymbol{\rho} = \begin{bmatrix} 1 & 0 & 0 \\ 0 & 0 & 0 \\ 0 & 0 & \ddots \end{bmatrix}.$$

In this case of a pure state, and *only* in this case, $\boldsymbol{\rho}^2 = \boldsymbol{\rho}$ and $\mathrm{Tr}(\boldsymbol{\rho}^2) = 1$. Therefore, since these equations are independent of basis, we have in *any* basis:

$$\boldsymbol{\rho}^2 = \boldsymbol{\rho}, \quad \text{and} \quad \mathrm{Tr}(\boldsymbol{\rho}^2) = 1 \qquad \text{(pure state, any basis).}$$

In a mixed state, in a basis where $\boldsymbol{\rho}$ is diagonal, all its elements ρ_{dd} are strictly less than one. Then $\boldsymbol{\rho}^2 = \mathrm{diag}(\rho_{11}{}^2, \rho_{22}{}^2, \dots)$ has the squares of all the diagonal elements of $\boldsymbol{\rho}$, and thus each diagonal element of $\boldsymbol{\rho}^2$ is strictly less than the corresponding element of $\boldsymbol{\rho}$. This implies that, since $\mathrm{Tr}(\boldsymbol{\rho}) = 1$, $\mathrm{Tr}(\boldsymbol{\rho}^2) < 1$. Since $\mathrm{Tr}()$ is basis-independent, all of these statements are true in any basis. Thus $\mathrm{Tr}(\boldsymbol{\rho}^2)$, in any basis, is a test of whether an ensemble is pure or mixed:

$$\mathrm{Tr}(\boldsymbol{\rho}^2) < 1 \qquad \text{(mixed state, any basis).}$$

Statistics beyond averages: Recall that QM predicts the possible outcomes of an experiment (the "spectrum" of results), *and* their probabilities of occurrence. What are the spectra and probabilities of mixed ensembles? Any observable $\hat{B}$ has a PDF of its possible values. The moments of the PDF (probability distribution function) (averages of powers of the observable) are computed from $\hat{B}$ and $\boldsymbol{\rho}$:

$$\left[\hat{B}^n\right] = \sum_{i=1}^{N} \sum_{j=1}^{N} \rho_{ij} \left[\mathbf{B}^n\right]_{ji}$$

where $\mathbf{B} \equiv$ operator matrix of $\hat{B}$ in our basis.

It is well known that the moments of a random variable fully define its PDF. Therefore, since $\boldsymbol{\rho}$ fully defines all the moments, it also fully defines the PDF of measurements for any operator. Furthermore, any two ensembles with the same density matrix, regardless of how they were prepared or if their constituent states are the same, are physically indistinguishable by any measurement or statistics of measurements. Ultimately, then, we must conclude that the two ensembles are physically the same.

Time evolution of the density matrix: The state of an ensemble may change with time (e.g., it may be approaching equilibrium). Then its density matrix is a function of time, because its constituent states are evolving in time. We find the equation of motion (EOM) for $\boldsymbol{\rho}$ from the EOMs for its constituent kets and bras. From the Schrödinger equation:

$$i\hbar \frac{\partial}{\partial t}|\psi\rangle = \hat{H}|\psi\rangle, \qquad \text{and} \qquad -i\hbar \frac{\partial}{\partial t}\langle\psi| = \langle\psi|\hat{H} \qquad \Rightarrow$$

$$\frac{\partial}{\partial t}\left(|\psi\rangle\langle\psi|\right) = \left(\frac{\partial}{\partial t}|\psi\rangle\right)\langle\psi| + |\psi\rangle\frac{\partial}{\partial t}\langle\psi| - \frac{1}{i\hbar}\left[\hat{H}|\psi\rangle\langle\psi| - |\psi\rangle\langle\psi|\hat{H}\right].$$

Note that $\left(\hat{H}|\psi\rangle\right)\langle\psi| = \hat{H}\left[\,|\psi\rangle\langle\psi|\,\right]$ is just the matrix product of the matrix $\hat{H}$ with the matrix $|\psi\rangle\langle\psi|$, and also $|\psi\rangle\left(\langle\psi|\hat{H}\right) = \left[\,|\psi\rangle\langle\psi|\,\right]\hat{H}$ is a matrix product. We can see this by writing the matrix $|\psi\rangle\langle\psi|$ explicitly, and using linearity of $\hat{H}$:

$$\text{Let }\ |\psi\rangle = \begin{bmatrix} a_1 \\ a_2 \\ a_3 \end{bmatrix}. \quad \text{Then:}\ \ |\psi\rangle\langle\psi| = \begin{bmatrix} a_1 \\ a_2 \\ a_3 \end{bmatrix}\begin{bmatrix} a_1^* & a_2^* & a_3^* \end{bmatrix} = \begin{bmatrix} a_1^*\begin{bmatrix} a_1 \\ a_2 \\ a_3 \end{bmatrix} & a_2^*\begin{bmatrix} a_1 \\ a_2 \\ a_3 \end{bmatrix} & a_3^*\begin{bmatrix} a_1 \\ a_2 \\ a_3 \end{bmatrix} \end{bmatrix}$$

$$\Rightarrow \qquad \left(\hat{H}|\psi\rangle\right)\langle\psi| = \begin{bmatrix} a_1^*\hat{H}\begin{bmatrix} a_1 \\ a_2 \\ a_3 \end{bmatrix} & a_2^*\hat{H}\begin{bmatrix} a_1 \\ a_2 \\ a_3 \end{bmatrix} & a_3^*\hat{H}\begin{bmatrix} a_1 \\ a_2 \\ a_3 \end{bmatrix} \end{bmatrix}$$

$$= \hat{H}\begin{bmatrix} a_1^*\begin{bmatrix} a_1 \\ a_2 \\ a_3 \end{bmatrix} & a_2^*\begin{bmatrix} a_1 \\ a_2 \\ a_3 \end{bmatrix} & a_3^*\begin{bmatrix} a_1 \\ a_2 \\ a_3 \end{bmatrix} \end{bmatrix} = \hat{H}\left(|\psi\rangle\langle\psi|\right).$$

Then, using linearity over the constituents of $\boldsymbol{\rho}$:

$$\boldsymbol{\rho} \equiv \sum_{k=1}^{M} w_k \left| \psi^{(k)} \right\rangle \left\langle \psi^{(k)} \right| \quad \Rightarrow$$

$$\frac{\partial}{\partial t}\boldsymbol{\rho} = \sum_{k=1}^{M} w_k \frac{1}{i\hbar} \left(\hat{H} \left| \psi^{(k)} \right\rangle \left\langle \psi^{(k)} \right| - \left| \psi^{(k)} \right\rangle \left\langle \psi^{(k)} \right| \hat{H} \right) = \frac{1}{i\hbar} \left[\hat{H}\boldsymbol{\rho} - \boldsymbol{\rho}\hat{H} \right] = \frac{1}{i\hbar} \left[\hat{H}, \boldsymbol{\rho} \right].$$

Note that $\partial\boldsymbol{\rho}/\partial t$ is the time derivative of a matrix, which is itself a matrix. The previous form is reminiscent of the time evolution of the average value of an operator, but for the density matrix $\boldsymbol{\rho}$, the Hamiltonian comes first in the commutator, and there is no inner product (since $\boldsymbol{\rho}$ takes the place of the quantum state). In contrast, for the time evolution of the average of an operator, the Hamiltonian comes second, and we must take an inner product with the quantum state: $d\hat{A}/dt = \left\langle \psi \right| \left[\hat{A}, \hat{H} \right] \left| \psi \right\rangle / i\hbar$.

Note also that the time evolution depends only on $\boldsymbol{\rho}$, and not at all on its constituent states, once again confirming that two ensembles with the same density matrix are physically identical, regardless of how they were constructed or whether their constituent quantum states are the same.

Note that in some realistic systems, such as thermal ensembles, collisions and other multibody interactions make it essentially impossible to compute the long-term time evolution exactly. One generally turns, then, to statistical methods.

Continuous density functions: The concept of a density matrix extends to the continuum case, where the density matrices are replaced by density functions [16, p. 182]. We do not address that case.

4.4.3   Density Matrix Examples

We first consider two pure spin-1/2 states. In the z-basis:

$$\left| z+ \right\rangle = \begin{pmatrix} 1 \\ 0 \end{pmatrix} \quad \Rightarrow \quad \boldsymbol{\rho} = \left| z+ \right\rangle \left\langle z+ \right| = \begin{pmatrix} 1 \\ 0 \end{pmatrix} \begin{pmatrix} 1 & 0 \end{pmatrix} = \begin{bmatrix} 1 & 0 \\ 0 & 0 \end{bmatrix}, \quad \boldsymbol{\rho}^2 = \begin{bmatrix} 1 & 0 \\ 0 & 0 \end{bmatrix}.$$

$\boldsymbol{\rho}$ is Hermitian, and satisfies $\text{Tr}(\boldsymbol{\rho}) = 1$, as required for all density matrices, and also $\boldsymbol{\rho}^2 = \boldsymbol{\rho}$, as required for all pure states.

As another example:

$$\left| y+ \right\rangle = \begin{pmatrix} 1/\sqrt{2} \\ i/\sqrt{2} \end{pmatrix} \quad \Rightarrow \quad \boldsymbol{\rho} = \left| y+ \right\rangle \left\langle y+ \right| = \begin{pmatrix} 1/\sqrt{2} \\ i/\sqrt{2} \end{pmatrix} \begin{pmatrix} 1/\sqrt{2} & -i/\sqrt{2} \end{pmatrix} = \begin{bmatrix} 1/2 & -i/2 \\ i/2 & 1/2 \end{bmatrix},$$

$$\boldsymbol{\rho}^2 = \begin{bmatrix} 1/2 & -i/2 \\ i/2 & 1/2 \end{bmatrix} \begin{bmatrix} 1/2 & -i/2 \\ i/2 & 1/2 \end{bmatrix} = \begin{bmatrix} 1/2 & -i/2 \\ i/2 & 1/2 \end{bmatrix} = \boldsymbol{\rho}.$$

Again, $\boldsymbol{\rho}$ is Hermitian, satisfies $\mathrm{Tr}(\boldsymbol{\rho})=1$ (as required for all density matrices), and $\boldsymbol{\rho}^2=\boldsymbol{\rho}$ (as required for all pure states).

What is the density matrix for a thermal ensemble of spin-1/2 particles? We first solve this by brute force, and then describe some subtler reasoning. Since a thermal ensemble has no preferred direction, it must have zero average spin in *every* direction. We note the spin operator for a general direction, (θ, ϕ):

$$\hat{s}_{\theta,\phi} = \sin\theta\cos\phi\,\hat{s}_x + \sin\theta\sin\phi\,\hat{s}_y + \cos\theta\,\hat{s}_z$$

$$= \frac{\hbar}{2}\left(\sin\theta\cos\phi\begin{bmatrix}0 & 1\\ 1 & 0\end{bmatrix} + \sin\theta\sin\phi\begin{bmatrix}0 & -i\\ i & 0\end{bmatrix} + \cos\theta\begin{bmatrix}1 & 0\\ 0 & -1\end{bmatrix}\right)$$

$$= \frac{\hbar}{2}\begin{bmatrix}\cos\theta & \sin\theta e^{-i\phi}\\ \sin\theta e^{+i\phi} & -\cos\theta\end{bmatrix}.$$

We must find a constant density matrix such that for *every* (θ, ϕ) we have zero average:

$$[s_{\theta,\phi}] = \mathrm{Tr}(\boldsymbol{\rho}s_{\theta,\phi}) = \mathrm{Tr}\left(\begin{bmatrix}a & b\\ b^* & d\end{bmatrix}\frac{\hbar}{2}\begin{bmatrix}\cos\theta & \sin\theta e^{-i\phi}\\ \sin\theta e^{+i\phi} & -\cos\theta\end{bmatrix}\right) = 0,$$

where we must solve for the $\boldsymbol{\rho}$ matrix elements a, b, and d, and we have used that $\boldsymbol{\rho}$ is Hermitian. Then:

$$\mathrm{Tr}\left(\begin{bmatrix}a & b\\ b^* & d\end{bmatrix}\begin{bmatrix}\cos\theta & \sin\theta e^{-i\phi}\\ \sin\theta e^{+i\phi} & -\cos\theta\end{bmatrix}\right)$$

$$= a\cos\theta + b\sin\theta e^{+i\phi} + b^*\sin\theta e^{-i\phi} - d\cos\theta = 0 \quad\Rightarrow$$

$$(a-d)\cos\theta = 0, \qquad\qquad \sin\theta\,\mathrm{Re}\{be^{i\phi}\} = 0.$$

The constraint on a and d is required when $\theta=0$ (canceling any value of b). Since $\mathrm{Tr}(\boldsymbol{\rho})=a+d=1$, and $a=d$, we must have $a=d=1/2$. Then $\mathrm{Re}\{be^{i\phi}\}=0$ for *all* ϕ, which can only happen if $b=0$. So:

$$\boldsymbol{\rho} = \begin{bmatrix}1/2 & 0\\ 0 & 1/2\end{bmatrix} \qquad\text{(thermal ensemble)}.$$

We recognize the thermal ensemble as that of a 50-50 mix of $|z+\rangle$ and $|z-\rangle$. Since an ensemble is fully characterized by its density matrix, it must be that a thermal ensemble is identical to a 50-50 mix of $|z+\rangle$ and $|z-\rangle$. By rotational symmetry, any 50-50 mix of opposing spins is also a thermal ensemble.

We went to substantial effort to arrive at this simple result. Could we have deduced it more directly? Yes. The laws of physics are rotationally invariant: no matter what direction we face, the laws are the same. Therefore, any 50-50 mix of opposing spins has, for each probability to measure spin in some given direction, an equal probability to measure spin in the opposite direction. Therefore, the average spin in all directions is zero. This shows that any 50-50 mix of opposing spins is also a thermal ensemble.

4.4.4   Density Matrix Summary

A quantum state $|\psi\rangle$ characterizes a single particle. A density matrix ρ characterizes an *ensemble* of particles, and defines everything there is to know about it [16, p. 178t]. However, if we draw a single particle from the ensemble, we also say that the *particle* is in the "mixed state" ρ, because we cannot know exactly which quantum state the particle is in. The density matrix for such a particle then takes the place of a quantum state ket. The density matrix then defines everything that can be known about that particle.

> Particles in a definite quantum state have a state *vector*; particles drawn from an ensemble have a state *matrix*, the density matrix, ρ.

Two ensembles with the same density matrix are physically identical, even if their constituent quantum states are different. This is a purely quantum mechanical effect (i.e., it has no classical analog).

Chapter 5
Angular Momentum

Angular momentum is a critical part of quantum mechanics, with applications throughout spectroscopy, magnetism, and solid state physics. This chapter assumes you are somewhat familiar with quantum angular momentum, and therefore we focus on the more conceptually challenging aspects.

5.1 Rotation Notation

We introduce some notation here, but all of it should become clearer as we use it throughout the text. "$\hat{J}$" is a common symbol for an arbitrary angular momentum operator (orbital, spin, combination, …). To distinguish between operators and unit vectors, we use hats for operators, and bold **e** for unit vectors:

$\hat{L}_\theta$ is the operator for angular momentum in the θ direction.

$\mathbf{e}_\phi$ is the unit vector in the ϕ direction.

A quantum **vector operator** is a set of three operators, associated with the three basis directions in real space. For angular momentum vector operators we have:

$$\hat{\mathbf{J}} \equiv \hat{J}_x \mathbf{e}_x + \hat{J}_y \mathbf{e}_y + \hat{J}_z \mathbf{e}_z \quad \text{the angular momentum vector operator,}$$

where $\mathbf{e}_x \equiv$ unit vector in x-direction of 3-space ($\hat{\mathbf{x}}$ in non-quantum language), etc.

$$\hat{J}^2 \equiv \left|\hat{\mathbf{J}}\right|^2 = \left(\hat{J}_x\right)^2 + \left(\hat{J}_y\right)^2 + \left(\hat{J}_z\right)^2 \quad \text{the magnitude-squared operator}.$$

The eigenstates (orbital or spin or combination) of $\hat{J}^2$ and $\hat{J}_z$ are written:

$|j\ m\rangle$ or $|j, m\rangle$ where j is the quantum # for $\hat{J}^2$; m is the quantum # for $\hat{J}_z$.

For spin-1/2 particles in the z-basis, the following seven notations in common use are equivalent:

$$\chi_+ \equiv |\uparrow\rangle \equiv |+\rangle \equiv |z+\rangle \equiv |+z\rangle \equiv \left|\frac{1}{2}, \frac{1}{2}\right\rangle \equiv \begin{pmatrix} 1 \\ 0 \end{pmatrix}$$

$$\chi_- \equiv |\downarrow\rangle \equiv |-\rangle \equiv |z-\rangle \equiv |-z\rangle \equiv \left|\frac{1}{2}, -\frac{1}{2}\right\rangle \equiv \begin{pmatrix} 0 \\ 1 \end{pmatrix}.$$

E. L. Michelsen, *Quirky Quantum Concepts*, Undergraduate Lecture Notes in Physics, DOI 10.1007/978-1-4614-9305-1_5, © Springer Science+Business Media New York 2014

For combining two angular momenta (of arbitrary nature: spin, orbital, ...), the following notations are in common use, and use capital letters for total angular momentum, and lower-case letters for the constituent angular momenta:

$$\mathbf{J} = \mathbf{j}_1 + \mathbf{j}_2 \quad \text{(the vectors in 3-space)} \; where \; \mathbf{J} \equiv \text{total angular momentum,}$$

$$\hat{\mathbf{J}} = \hat{\mathbf{j}}_1 + \hat{\mathbf{j}}_2 \quad \text{(the quantum operators)} \, .$$

In the "uncoupled" basis, we write states of the combined system of angular momenta as combinations of the two original (uncoupled) angular momenta. There are several common, equivalent notations:

$$\left| j_1 \; m_1 \right\rangle \otimes \left| j_2 \; m_2 \right\rangle \equiv \left| j_1 \; m_1 \right\rangle \left| j_2 \; m_2 \right\rangle \equiv \left| j_1 \; m_1 ; j_2 \; m_2 \right\rangle \equiv \left| j_1 , j_2 ; m_1 , m_2 \right\rangle$$

$$\equiv \left| m_1 , m_2 \right\rangle \quad where \; j_1 , j_2 \; \text{are "understood"} \, .$$

The alternative basis is the "coupled" basis (combined total angular momentum), with quantum numbers J and M, and eigenstates:

$$\left| J \; M \right\rangle , \quad \text{eigenstates of} \quad \hat{J}^2 , \hat{J}_z \, .$$

Note the distinction between $\mathbf{J}$ and J: $\mathbf{J}$ is an angular momentum vector in three-space; J is an angular momentum quantum number of the combined total angular momentum. In particular, J is *not* the magnitude of $\mathbf{J}$:

$$J \neq \left| \mathbf{J} \right| , \quad \text{but} \quad \hat{J}^2 = \hat{\mathbf{J}} \cdot \hat{\mathbf{J}} = \left| \hat{\mathbf{J}} \right|^2 \, .$$

In all of the previous definitions, systems of purely orbital angular momentum might use $\mathbf{L}, L, l$, etc. in place of $\mathbf{J}, J, j$, etc. Systems of purely spin angular momentum might use $\mathbf{S}, S, s$, etc.

   Example: For spin 1/2:

$$\hat{\mathbf{s}} = \frac{\hbar}{2} \left(\sigma_x \mathbf{e}_x + \sigma_y \mathbf{e}_y + \sigma_z \mathbf{e}_z \right) \Rightarrow \hat{s}_x = \frac{\hbar}{2} \sigma_x , \quad \hat{s}_y = \frac{\hbar}{2} \sigma_y , \quad \hat{s}_z = \frac{\hbar}{2} \sigma_z \, .$$

   Eigenstates: $\left| \uparrow \right\rangle , \left| \downarrow \right\rangle$.

5.2   Dissection of Orbital Angular Momentum

5.2.1   Orbital Angular Momentum as Fluid

Imagine water circulating in a closed circular loop of frictionless pipe. The system is stationary (no property changes with time), however it has angular momentum. If we try to tilt the loop, we will experience the resistance typical of tilting a gyroscope (bicycle wheel, etc.). Thus, even though the system is stationary, we can measure its angular momentum. The system is only dynamic if we look on a microscopic scale at the individual water molecules.

An electron in orbit around a nucleus is similar: it is stationary, i.e., no property changes with time. However, it has angular momentum. The electron is like a compressible fluid, distributed in space, with a particle density and momentum density given by its wave-function. In contrast to the previous water example, so far as anyone knows, electrons have no smaller microscopic scale to look at: there are no "molecules" composing the electron. The electron appears to be a sort of "perfect fluid," no matter how closely we look.

5.2.2   Spherical Harmonics as Motion

We now look beyond the surface of spherical harmonics and orbital angular momentum, to provide a deeper description of their properties. This section assumes you understand that

$$\hat{\mathbf{L}} = \hat{\mathbf{r}} \times \hat{\mathbf{p}} = -\hat{\mathbf{p}} \times \hat{\mathbf{r}}.$$

because only commuting operators in $\hat{\mathbf{r}}$ and $\hat{\mathbf{p}}$ are mixed in $\hat{\mathbf{L}}$. We proceed along these lines:

1. Angular momentum depends only on angular functions, not on radial function.
2. An L_z eigenstate has unbounded linear momentum near the north pole.
3. Linear momentum p_ϕ^2 has $180°$ opposing phase to p_θ^2, and actually partially cancels it, yielding finite L^2 and energy.

Recall that all spherical harmonics separate into a product of θ and ϕ:

$$Y_{lm}(\theta, \phi) = NP_{lm}(\cos\theta)e^{im\phi} \quad where \quad P_{lm}(\cdot) \text{ is the associated Legendre function.}$$

N is a normalization constant. (See *Funky Electromagnetic Concepts* for more on spherical harmonics.) Consider a particle angular momentum eigenstate in spherical coordinates, with an arbitrary radial function (and ignoring normalization):

$$\psi(r, \theta, \phi) = R(r)P_{lm}(\cos\theta)e^{im\phi} \quad \text{(unnormalized)}.$$

Figure 5.1 shows that the local angular momentum $\mathbf{L}(\theta, \phi)$ depends only on the angular functions, and not on the distance from the origin (r). We can see this by comparing p_1 and p_2 in the figure. ψ at both points has the same angular derivative, $\partial\psi/\partial\phi$. However, p_1 is a smaller momentum than p_2 (wavelength is longer), but p_1 has a longer "lever arm" for angular momentum (larger r). The two effects exactly cancel and $\mathbf{L}$ has no radial dependence. Mathematically:

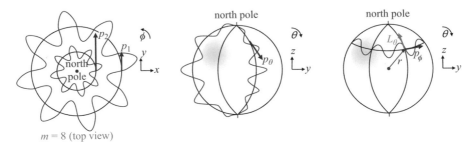

Fig. 5.1 (*Left*) Linear momenta in the ϕ-direction at two points for $m=8$: $e^{i8\phi}$. (*Middle*) Linear momentum in the θ-direction. (*Right*) Angular momentum in the $-\theta$ direction

$$\mathbf{L} \equiv \mathbf{r} \times \mathbf{p} \Rightarrow |\mathbf{L}| \propto rp. \quad \text{Also,} \ p \propto 1/r \Rightarrow |\mathbf{L}| = const.$$

Because $\mathbf{L}$ has no radial dependence, we limit ourselves from now on to the unit sphere, $r=1$, and $\psi(r, \theta, \phi) \rightarrow \psi(\theta, \phi)$. Also, there is no component of $\mathbf{L}$ in the r-direction, since any linear r-momentum goes through the origin, and has no "lever arm," i.e., $\mathbf{p}_r \times \mathbf{r} = 0$. In spherical coordinates, $\mathbf{L}$ has only θ and ϕ components:

$$\mathbf{L} = L_\theta \mathbf{e}_\theta + L_\phi \mathbf{e}_\phi = L_x \mathbf{e}_x + L_y \mathbf{e}_y + L_z \mathbf{e}_z.$$

> The local angular momentum $\mathbf{L}$ has no dependence on r, nor on the radial wave-function $R(r)$.

We now examine some of these components of angular momentum in more detail.

We consider L_z first, because it is simplest in our spherical coordinates. (Physically, there is nothing special or simple about any one component of $\mathbf{L}$ over another.) We must distinguish here between the *local value* of $L_z(\theta, \phi)$ evaluated at a point (θ, ϕ), and the *operator* $\hat{L}_z$. These are related by the standard definition for any operator:

$$\hat{L}_z \psi(\theta, \phi) = L_z(\theta, \phi) \psi(\theta, \phi).$$

Similarly, we distinguish between the local value $p_\phi(\theta, \phi)$ and the operator $\hat{P}_\phi$, related by:

$$\hat{p}_\phi \psi(\theta,\phi) = p_\phi(\theta,\phi)\psi(\theta,\phi).$$

Then (still on the unit sphere):

$$L_z(\theta,\phi) = r_\perp p_\phi(\theta,\phi) = \sin\theta\, p_\phi(\theta,\phi)$$

> *where* $r_\perp \equiv$ distance to z-axis, and $r = 1$,
>
> $p_\phi(\theta,\phi) \equiv$ linear momentum in the ϕ direction .

$$\hat{L}_z = (\sin\theta)\,\hat{p}_\phi = \sin\theta\frac{\hbar}{i}\frac{\partial}{\partial s}$$

> *where* $s \equiv$ linear distance along the unit sphere in the ϕ direction .

Using $ds = \sin\theta\,d\phi$, then at a given value of θ:

$$p_\phi(\theta,\phi) = \frac{\hbar}{i}\frac{\partial\psi}{\partial s} = \frac{\hbar}{i}\frac{1}{\sin\theta}\frac{\partial\psi}{\partial\phi} \Rightarrow L_z(\theta,\phi) = (\sin\theta)\,p_\phi(\theta,\phi), \text{ and } \boxed{\hat{L}_z = \frac{\hbar}{i}\frac{\partial}{\partial\phi}},$$

a well-known result. Now for our $Y_{lm}(\theta,\phi)$ eigenstate (still ignoring normalization):

$$\hat{L}_z\psi = \frac{\hbar}{i}\frac{\partial}{\partial\phi}P_{lm}(\cos\theta)e^{im\phi} = m\hbar P_{lm}(\cos\theta)e^{im\phi} \Rightarrow L_z(\theta,\phi) = m\hbar \quad \text{(a constant),}$$

as expected for an L_z eigenstate. However, even though L_z is constant (independent of θ, ϕ), linear momentum $p_\phi \sim L_z/\sin\theta$ is unbounded as $\theta \to 0$.

Note that $\partial/\partial\phi$ (off the equator) also contributes L_x and L_y components, which we will use later. At this point, you might be thinking, "Jeepers! Linear momentum in the ϕ direction, p_ϕ, goes to infinity at the north pole. That means p_ϕ^2 goes to infinity there, so L^2 and energy must also go to infinity there! Bummer." This is especially disturbing since you already know that this ψ is an eigenstate of $\hat{L}^2$ and also, with an appropriate $R(r)$, an eigenstate of energy. How can this be?

Here is the trick: *all* quantum functions of space, both states and operators acting on states, are complex valued, meaning they have both a magnitude and a phase at each point of space. A valid model is that these are the magnitudes and phases of oscillating sinusoids.

> Destructive interference is possible, even on results like p^2, so that one component of p^2 *can cancel* another component of p^2.

Classically, momentum in one direction cannot cancel momentum in another direction, but wave-interference is the essence of QM. Destructive interference can occur anywhere. (Note that in classical EM radiation, such destructive interference *does* occur, right in the same Y_{lm} as in QM, because EM waves *do* interfere.) Now, let us go through the mathematics, and physically interpret the steps.

We use Y_{11} as our test subject, but the concepts extend readily to all Y_{lm}:

$$Y_{11}(\theta, \phi) = \sqrt{3/8\pi} \sin \theta e^{i\phi}.$$

At this point, it is convenient to work with L_θ and L_ϕ components, rather than L_x, L_y, and L_z. First, p_ϕ contributes angular momentum in the $-\theta$ direction ($r=1$, and we ignore the normalization constant. See Fig. 5.1, *right*) [8, 11.18 p. 225]:

$$\hat{L}_\theta = -\hat{p}_\phi = -\frac{\hbar}{i} \frac{1}{\sin \theta} \frac{\partial}{\partial \phi}.$$

The local angular momentum component L_θ of Y_{11} is (canceling the normalization):

$$L_\theta(\theta, \phi)Y_{11} = -\hat{p}_\phi Y_{11} = -\frac{\hbar}{i} \frac{1}{\sin \theta} \frac{\partial Y_{11}}{\partial \phi} = -\frac{\hbar}{i} \frac{1}{\sin \theta} \frac{\partial}{\partial \phi} \sin \theta e^{i\phi}$$

$$= -\hbar \frac{1}{\sin \theta} \underbrace{\sin \theta e^{i\phi}}_{Y_{11}}$$

$$\Rightarrow L_\theta(\theta, \phi) = -\frac{\hbar}{\sin \theta} \left(\text{in general, } L_\theta(\theta, \phi) = -\frac{m\hbar}{\sin \theta} \right).$$

The other component is L_ϕ. p_θ contributes angular momentum in the $+\phi$ direction (again, $r=1$) [8, 11.18, p. 225]:

$$\hat{L}_\phi = \hat{p}_\theta = \frac{\hbar}{i} \frac{\partial}{\partial \theta}$$

$$L_\phi(\theta, \phi)Y_{11} = \hat{L}_\phi Y_{11} = \frac{\hbar}{i} \frac{\partial}{\partial \theta} \sin \theta e^{i\phi} = \frac{\hbar}{i} \cos \theta e^{i\phi} = \frac{\hbar}{i} \frac{\cos \theta}{\sin \theta} \underbrace{\sin \theta e^{i\phi}}_{Y_{11}}$$

$$\Rightarrow L_\phi(\theta, \phi) = \frac{\hbar}{i} \frac{\cos \theta}{\sin \theta}.$$

Combining L_θ and L_ϕ, we find the angular momentum vector:

$$\hat{\mathbf{L}} Y_{11} = \left(-\frac{\hbar}{\sin \theta} \mathbf{e}_\theta + \frac{\hbar}{i} \frac{\cos \theta}{\sin \theta} \mathbf{e}_\phi \right) Y_{11}.$$

We see that both L_θ and L_ϕ blow up at the north pole ($\theta=0$). But notice that L_θ is real, and L_ϕ is imaginary. This is crucial! It means that as oscillating sinusoids, they are 90° out of phase. And it means that when we square them, they will be 180° out of phase, i.e., they will destructively interfere. This is how the infinities at the North Pole are cancelled. To find their squares, though, we must act again with the operators $\hat{L}_\theta$ and $\hat{L}_\phi$, which involves tricky derivatives of unit vectors [8, p. 225]. The result for $\hat{L}^2$ is [8, 11.20, p. 226]:

$$\hat{L}^2 = -\hbar^2 \left[\frac{1}{\sin^2\theta} \frac{\partial^2}{\partial\phi^2} + \frac{1}{\sin\theta} \frac{\partial}{\partial\theta} \left(\sin\theta \frac{\partial}{\partial\theta} \right) \right].$$

The first term is from $\hat{L}_\theta$ (using $\partial/\partial\phi$); the second term is from $\hat{L}_\phi$ (using $\partial/\partial\theta$). Act with this $\hat{L}^2$ on Y_{11}:

$$\hat{L}^2 Y_{11} = -\hbar^2 \left[\frac{1}{\sin^2\theta} \frac{\partial^2}{\partial\phi^2} + \frac{1}{\sin\theta} \frac{\partial}{\partial\theta} \left(\sin\theta \frac{\partial}{\partial\theta} \right) \right] \sin\theta \, e^{i\phi}$$

$$= -\hbar^2 \left[\frac{1}{\sin\theta} i^2 + \frac{1}{\sin\theta} \frac{\partial}{\partial\theta} (\sin\theta\cos\theta) \right] e^{i\phi}$$

$$= -\hbar^2 \left[-\frac{\cos^2\theta + \sin^2\theta}{\sin\theta} + \frac{\cos^2\theta - \sin^2\theta}{\sin\theta} \right] e^{i\phi}$$

$$= -\hbar^2 \left[\frac{-2\sin^2\theta}{\sin\theta} \right] e^{i\phi} = 2\hbar^2 \underbrace{\sin\theta \, e^{i\phi}}_{Y_{11}} \Rightarrow L^2(\theta,\phi) = 2\hbar^2,$$

where $2 = l(l+1)$, as expected. Notice how the infinity as $\theta \to 0$ of ($\cos^2\theta/\sin\theta$) is cancelled. This is destructive interference between the L_θ^2 and L_ϕ^2 terms.

The infinite linear momenta similarly cancel, so $p^2(r, \theta, \phi)$ is finite and constant. Therefore, we can write the local kinetic energy in terms of the local L^2 or local p^2, which are both finite and constant over all angles (fixed r):

$$T(r,\theta,\phi) = \frac{L^2(r,\theta,\phi)}{2I} = \frac{(rp(r,\theta,\phi))^2}{2mr^2} = \frac{p^2(r,\theta,\phi)}{2m}.$$

Similarly, for any Y_{lm} at a fixed r, $T(r, \theta, \phi)$ is constant over all angles.

It is interesting to note that in a three-dimensional (3D) energy eigenstate of a spherical potential, $\psi(r, \theta, \phi) = R(r)Y_{lm}(\theta, \phi)$. As $r \to 0$, $T = L^2/2mr^2$ blows up, since the local L^2 is constant. Since ψ is an energy eigenstate, with constant total energy E, this infinity must be offset by a negatively infinite potential energy at the origin. Because of the spherical symmetry, there can be no out-of-phase canceling contribution to the kinetic energy from $R(r)$ near the origin.

5.3   Generating Station: A Different Spin on Angular Momentum

Particle "spin" is a kind of angular momentum with extremely important observable effects across a wide range of phenomena, but it has no classical analog. The magnitude of a particle's spin can never be changed, but its direction can. For charged particles, angular momentum comes with a magnetic dipole moment, which is a critical element of a chemical compound's radiation spectrum. (Even neutrons have a magnetic dipole moment, due to the charged quarks inside them.) Spin-related spectroscopy is an essential tool for understanding chemical structure, as well as many other microscopic phenomena. Also, spin magnetic moment is the essence of magnetic resonance imaging, an important medical diagnostic. And someday, spintronics, based on recording information in electron spins, may improve our electronic devices.

To facilitate understanding spin, and to illustrate its close association with rotations as well as with orbital angular momentum, we demonstrate here some of the mathematical tools that are universally used for analyzing spin. Motivated by our discussion of generators (Sect. 2.8.1), we follow this course:

- Orbital angular momentum and its commutation relations.
- Rotations in real space, and the commutators of rotation operators.
- Quantum rotations, which must have the same commutators as classical.
- Quantum generators of rotation.
- Commutation relations of quantum generators of rotation, derived from rotation commutators.

We follow a similar approach to [19, pp. 257–260].

Orbital angular momentum: We review here the commutation relations of orbital angular momentum, as derived from the commutation relations of position and momentum, $[x, p_x] = i\hbar$. Classically, angular momentum is $\mathbf{L} \equiv \mathbf{r} \times \mathbf{p}$. Quantum mechanically, $\hat{\mathbf{r}}$ and $\hat{\mathbf{p}}$ are observable operators, which means they are local (they depend only on a point, or an infinitesimal neighborhood around a point). New local operators can be derived from old local operators using the classical relationships. Thus,

$$\hat{L}_z = \hat{x}\hat{p}_y - \hat{y}\hat{p}_x.$$

Since there is nothing special about our choice of labeling the axes x, y, and z, the same relation must hold for any right-handed set of axes, such as a circular permutation of coordinate labels:

$$\hat{L}_z = \hat{x}\hat{p}_y - \hat{y}\hat{p}_x \Rightarrow \hat{L}_x = \hat{y}\hat{p}_z - \hat{z}\hat{p}_y, \quad \text{and} \quad \hat{L}_y = \hat{z}\hat{p}_x - \hat{x}\hat{p}_z.$$

From these definitions, and

$$[\hat{x}, \hat{p}_x] = [\hat{y}, \hat{p}_y] = [\hat{z}, \hat{p}_z] = i\hbar, \qquad \text{and}$$

$$[\hat{x}, \hat{p}_z] = [\hat{y}, \hat{p}_z] = [\hat{x}, \hat{p}_y] = [\hat{z}, \hat{p}_y] = [\hat{y}, \hat{p}_x] = [\hat{z}, \hat{p}_x] = 0,$$

we find the well-known commutators of angular momentum:

$$
\begin{aligned}
\left[\hat{L}_x, \hat{L}_y\right] &= \left(\hat{y}\hat{p}_z - \hat{z}\hat{p}_y\right)\left(\hat{z}\hat{p}_x - \hat{x}\hat{p}_z\right) - \left(\hat{z}\hat{p}_x - \hat{x}\hat{p}_z\right)\left(\hat{y}\hat{p}_z - \hat{z}\hat{p}_y\right) \\
&= \left(\hat{y}\hat{p}_z\hat{z}\hat{p}_x - \overline{\hat{y}\hat{p}_z\hat{x}\hat{p}_z} - \overline{\hat{z}\hat{p}_y\hat{z}\hat{p}_x} + \hat{z}\hat{p}_y\hat{x}\hat{p}_z\right) \\
&\quad - \left(\hat{z}\hat{p}_x\hat{y}\hat{p}_z - \overline{\hat{z}\hat{p}_x\hat{z}\hat{p}_y} - \overline{\hat{x}\hat{p}_z\hat{y}\hat{p}_z} + \hat{x}\hat{p}_z\hat{z}\hat{p}_y\right) \\
&= \left(\hat{y}\hat{p}_z\hat{z}\hat{p}_x - \hat{z}\hat{p}_x\hat{y}\hat{p}_z\right) + \left(\hat{z}\hat{p}_y\hat{x}\hat{p}_z - \hat{x}\hat{p}_z\hat{z}\hat{p}_y\right) \\
&= \left(\hat{p}_z\hat{z} - \hat{z}\hat{p}_z\right)\hat{y}\hat{p}_x + \left(\hat{z}\hat{p}_z - \hat{p}_z\hat{z}\right)\hat{x}\hat{p}_y \\
&= (-i\hbar)\hat{y}\hat{p}_x + (i\hbar)\hat{x}\hat{p}_y = i\hbar\left(\hat{x}\hat{p}_y - \hat{y}\hat{p}_x\right) = i\hbar\hat{L}_z.
\end{aligned}
$$

Again, the labeling of axes is arbitrary, so any circular permutation of indexes must also follow:

$$\left[\hat{L}_x, \hat{L}_y\right] = i\hbar\hat{L}_z \Rightarrow \left[\hat{L}_y, \hat{L}_z\right] = i\hbar\hat{L}_x, \text{ and } \left[\hat{L}_z, \hat{L}_x\right] = i\hbar\hat{L}_y. \qquad (5.1)$$

These angular momentum commutation relations turn out to be extremely important, and they describe more than just *orbital* angular momentum.

To expand on their significance, we now derive these relations in a more general way, from the mathematics of physical objects rotated in real space.

Rotations in real space: Before considering quantum rotations, we first consider classical rotations of a macroscopic body, such as a soccer ball. We imagine a point on the ball (perhaps where the inflation hole is), at coordinates (x, y, z). We then rotate the ball (about its center), and find the point has moved to a new position, (x', y', z'). More generally, we can replace the inflation hole by an arbitrary position vector, from the origin to the point (x, y, z). We consider "laboratory frame" rotations, where the x, y, z axes are fixed in the "laboratory," and do not rotate with the body.

Rotations are linear operators: (1) if I double the size of a vector, and then rotate it, that is the same as rotating first, then doubling its size; and (2) if I add two vectors, and rotate the sum, that is the same as rotating both vectors, and then adding them. All linear operators (in finite-dimensional vector spaces) can be written as matrices. From elementary trigonometry, we find for any angle α (the right-hand rule defines positive α) [19, 12.45, p. 258]:

$$R_x(\alpha) = \begin{pmatrix} 1 & 0 & 0 \\ 0 & \cos\alpha & -\sin\alpha \\ 0 & \sin\alpha & \cos\alpha \end{pmatrix}, \; R_y(\alpha) = \begin{pmatrix} \cos\alpha & 0 & \sin\alpha \\ 0 & 1 & 0 \\ -\sin\alpha & 0 & \cos\alpha \end{pmatrix},$$

$$R_z(\alpha) = \begin{pmatrix} \cos\alpha & -\sin\alpha & 0 \\ \sin\alpha & \cos\alpha & 0 \\ 0 & 0 & 1 \end{pmatrix}.$$

$$(5.2)$$

From our prior quantum experience, we know that commutators are important things (we also know it from Group Theory). Since even quantum systems can be rotated like classical systems (e.g., angular momentum can be made to have a definite value in any chosen direction), the commutators of quantum rotation operators must be the same as the commutators of classical rotation operators.

We first illustrate the commutator of rotations by an *infinitesimal* angle ε (Fig. 5.2, *left*). Consider the effect of rotating a unit vector by $R_y(\varepsilon)$ first, then $R_x(\varepsilon)$, compared to rotating by $R_x(\varepsilon)$ first, then $R_y(\varepsilon)$. The difference is a small vector of order ε^2.

We can directly compute that commutator from the rotation matrices, Eq. (5.2). Recall that in QM, operators acting to the right are evaluated from right to left, so the first rotation in the commutator later is about the y-axis, as illustrated. The exact result, even for large angles α, is:

$$\left[R_x(\alpha), R_y(\alpha)\right] = R_x(\alpha)R_y(\alpha) - R_y(\alpha)R_x(\alpha)$$

$$= \begin{pmatrix} 1 & 0 & 0 \\ 0 & \cos\alpha & -\sin\alpha \\ 0 & \sin\alpha & \cos\alpha \end{pmatrix} \begin{pmatrix} \cos\alpha & 0 & \sin\alpha \\ 0 & 1 & 0 \\ -\sin\alpha & 0 & \cos\alpha \end{pmatrix}$$

$$- \begin{pmatrix} \cos\alpha & 0 & \sin\alpha \\ 0 & 1 & 0 \\ -\sin\alpha & 0 & \cos\alpha \end{pmatrix} \begin{pmatrix} 1 & 0 & 0 \\ 0 & \cos\alpha & -\sin\alpha \\ 0 & \sin\alpha & \cos\alpha \end{pmatrix}$$

$$= \begin{pmatrix} \cos\alpha & 0 & \sin\alpha \\ \sin^2\alpha & \cos\alpha & -\cos\alpha\sin\alpha \\ -\cos\alpha\sin\alpha & \sin\alpha & \cos^2\alpha \end{pmatrix}$$

$$- \begin{pmatrix} \cos\alpha & \sin^2\alpha & \cos\alpha\sin\alpha \\ 0 & \cos\alpha & -\sin\alpha \\ -\sin\alpha & \cos\alpha\sin\alpha & \cos^2\alpha \end{pmatrix}$$

$$= \begin{pmatrix} 0 & -\sin^2\alpha & \sin\alpha(1-\cos\alpha) \\ \sin^2\alpha & 0 & \sin\alpha(1-\cos\alpha) \\ \sin\alpha(1-\cos\alpha) & \sin\alpha(1-\cos\alpha) & 0 \end{pmatrix} \quad \text{(classical)}.$$

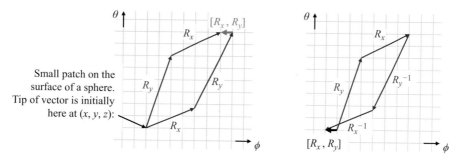

Fig. 5.2 (*Left*) $[R_x, R_y]$ acting on a vector produces a small, second order displacement (*green*). (*Right*) $1_{op} + [R_x, R_y]$ displaces the original vector by the amount of $[R_x, R_y]$ acting on it (*black*)

We will find that we need this only to second order in small rotations, so we expand the previous equation to that order:

$$\left[R_x(\varepsilon), R_y(\varepsilon)\right] = \begin{pmatrix} 0 & -\varepsilon^2 & 0 \\ \varepsilon^2 & 0 & 0 \\ 0 & 0 & 0 \end{pmatrix}, \quad \text{to } O\!\left(\varepsilon^2\right). \quad \text{NB: } \left(1 - \cos\varepsilon\right) \approx \varepsilon^2/2.$$

Note that:

> Rotations commute to first order in the angle ε, but not to second order.

Figure 5.2, right, shows that rotating a unit vector by $R_y(\varepsilon)$ first, then $R_x(\varepsilon)$, then $R_y^{-1}(\varepsilon)$, then $R_x^{-1}(\varepsilon)$, displaces the vector tip by the amount of the commutator $[R_x(\varepsilon), R_y(\varepsilon)]$, so is equivalent to operating on the unit vector with $(1_{op} + [R_x(\varepsilon), R_y(\varepsilon)])$, i.e., it adds the small vector to the given unit vector. Then:

$$1_{op} + \left[R_x(\varepsilon), R_y(\varepsilon)\right] = \begin{pmatrix} 1 & -\varepsilon^2 & 0 \\ \varepsilon^2 & 1 & 0 \\ 0 & 0 & 1 \end{pmatrix} = R_z\!\left(\varepsilon^2\right), \quad \text{to } O\!\left(\varepsilon^2\right).$$
(5.3)

The last equality follows because we recognize this matrix as $R_z(\varepsilon^2)$, to $O(\varepsilon^2)$, as shown here from Eq. (5.2):

$$R_z(\varepsilon) = \begin{pmatrix} 1 - \varepsilon^2/2 & -\varepsilon & 0 \\ \varepsilon & 1 - \varepsilon^2/2 & 0 \\ 0 & 0 & 1 \end{pmatrix} \Rightarrow R_z(\varepsilon^2) = \begin{pmatrix} 1 & -\varepsilon^2 & 0 \\ \varepsilon^2 & 1 & 0 \\ 0 & 0 & 1 \end{pmatrix}, \quad \text{to } O\!\left(\varepsilon^2\right).$$

Quantum rotations: Quantum rotations can be written as rotation operators, which act on a quantum state vector to produce a quantum state which is the rotated version of the given state. For a rotation by α radians around some given axis:

$$\left|\psi_{rotated}\right\rangle = \hat{R}(\alpha)\left|\psi\right\rangle.$$

As with classical rotations, $\hat{R}(\alpha)$ is linear because (1) the act of rotating a system commutes with scalar multiplication of the given vector and (2) the rotation of a sum of vectors equals the sum of the rotated vectors:

$$\hat{R}(\alpha)k\left|\psi\right\rangle = k\hat{R}(\alpha)\left|\psi\right\rangle, \quad \text{and}$$
$$\hat{R}(\alpha)\left(\left|\psi_1\right\rangle+\left|\psi_2\right\rangle\right) = \hat{R}(\alpha)\left|\psi_1\right\rangle + \hat{R}(\alpha)\left|\psi_2\right\rangle \Rightarrow \hat{R}(\alpha) \text{ is linear.}$$

$\hat{R}(\alpha)$ must be unitary, since it produces a normalized state from a normalized state.

Furthermore, these rotations are rotations in real space, which means they must obey the same commutation relations we derived in Eq. (5.3) for rotations in real space, to wit:

$$1_{op} + \left[R_x(\varepsilon), R_y(\varepsilon)\right] = R_z\left(\varepsilon^2\right).$$

Quantum Generators: In both classical and QM, angular momentum is the generator of rotations. We saw this in Sect. 2.8.1 for quantum orbital angular momentum from the definition of L_z, and a spherical coordinate wave-function:

$$\hat{L}_z = \frac{\hbar}{i}\frac{\partial}{\partial\phi}.$$

Rotation about z-axis:

$$\psi(r,\theta,\phi) \rightarrow \psi(r,\theta,\phi-d\phi) = \psi - \frac{\partial\psi}{\partial\phi}d\phi = \left(1-\frac{i}{\hbar}\hat{L}_z\,d\phi\right)\psi$$

$$\Rightarrow \hat{r}_z(d\phi) = 1 - \frac{i}{\hbar}\hat{L}_z\,d\phi \text{ (inifinitesimal rotation operator).}$$

Since there is nothing special about the z-axis, by relabeling axes (and $d\phi \rightarrow \varepsilon$) we must also have:

$$\hat{r}_x(\varepsilon) = 1 - \frac{i}{\hbar}\hat{L}_x\varepsilon, \quad \text{and} \quad \hat{r}_y(\varepsilon) = 1 - \frac{i}{\hbar}\hat{L}_y\varepsilon, \quad \varepsilon \text{ infinitesimal.}$$

These are the quantum infinitesimal rotation operators, to first order in infinitesimal rotation angles.

To compute rotation commutators, the first order effects cancel, so we need the rotation operators to second order. (We could do this by splitting the infinitesimal angle into two steps of $\varepsilon/2$ each, but instead we use a more general method.) We find the second order rotation operators by finding the exact operators for finite rotations, and then taking the second order expansion of the exact operators.

We construct a finite rotation operator (by α radians) from infinitesimal rotations by splitting α into N successive steps, then take $N \to \infty$. We demonstrate with $\hat{R}_z(\alpha)$, but the result applies to any axis:

$$\hat{R}_z(\alpha) = \lim_{N \to \infty} \hat{r}_z(\alpha/N)^N = \lim_{N \to \infty} \left(1 - \frac{i}{\hbar}\hat{L}_z \frac{\alpha}{N}\right)^N = \exp\left(-i\alpha \frac{\hat{L}_z}{\hbar}\right),$$

and similar for $\hat{R}_x$ and $\hat{R}_y$.

When a transformation operator with parameter, α, can be written as the exponential of $-i\alpha$ times another operator, the latter operator is called the **generator** of the transformation. In the previous case, the generator of rotations about the z-axis is $\hat{G}_z = \hat{L}_z / \hbar$. The generator can generate infinitesimal transformations without using an exponential.

We need only the second order expansion of this exact finite rotation, which we obtain from the standard Maclaurin series for the exponential (we abbreviate $1 \equiv 1_{op}$):

$$\hat{R}_z(\varepsilon) = \exp\left(-\frac{i}{\hbar}\varepsilon\hat{L}_z\right) = \left(1 - \frac{i}{\hbar}\varepsilon\hat{L}_z - \frac{1}{2}\frac{1}{\hbar^2}\varepsilon^2\hat{L}_z^2\right) \quad \text{to } O(\varepsilon^2)$$

$$= \left(1 - i\varepsilon\hat{G}_z - \frac{\varepsilon^2}{2}\hat{G}_z^2\right), \text{ where } \hat{G}_z \equiv \hat{L}_z / \hbar, \text{ and similar for } \hat{R}_x \text{ and } \hat{R}_y.$$

Finally, we obtain the commutation relations of quantum generators by combining our rotation commutation relations, Eq. (5.3), with the expansion of quantum rotations in terms of generators, all to second order. Thus, $\hat{G}_x$, $\hat{G}_y$, and $\hat{G}_z$ must satisfy:

$$1 + \left[\hat{R}_x(\varepsilon), \hat{R}_y(\varepsilon)\right] = \hat{R}_z(\varepsilon^2) \qquad \text{to } O(\varepsilon^2).$$

Expanding the commutator on the left, using generators for $\hat{R}_z(\varepsilon^2)$, and keeping terms only to second order:

$$\cancel{1} + \hat{R}_x(\varepsilon)\hat{R}_y(\varepsilon) - \hat{R}_y(\varepsilon)\hat{R}_x(\varepsilon) = \cancel{1} - i\varepsilon^2\hat{G}_z - \cancel{\frac{\varepsilon^4}{2}\hat{G}_z^2}.$$

Now write the rotation operators on the left in terms of generators, multiply, and truncate to second order:

$$\underbrace{\left(1 - i\varepsilon\hat{G}_x - \frac{\varepsilon^2}{2}\hat{G}_x^{\,2}\right)}_{\hat{R}_x(\varepsilon)}\underbrace{\left(1 - i\varepsilon\hat{G}_y - \frac{\varepsilon^2}{2}\hat{G}_y^{\,2}\right)}_{\hat{R}_y(\varepsilon)}$$

$$-\left(1 - i\varepsilon\hat{G}_y - \frac{\varepsilon^2}{2}\hat{G}_y^{\,2}\right)\left(1 - i\varepsilon\hat{G}_x - \frac{\varepsilon^2}{2}\hat{G}_x^{\,2}\right) = -i\varepsilon^2\hat{G}_z$$

$$\left(\cancel{1 - i\varepsilon\hat{G}_x - i\varepsilon\hat{G}_y} - \varepsilon^2\hat{G}_x\hat{G}_y - \cancel{\frac{\varepsilon^2}{2}\hat{G}_x^{\,2}} - \cancel{\frac{\varepsilon^2}{2}\hat{G}_y^{\,2}}\right)$$

$$-\left(\cancel{1 - i\varepsilon\hat{G}_y - i\varepsilon\hat{G}_x} - \varepsilon^2\hat{G}_y\hat{G}_x - \cancel{\frac{\varepsilon^2}{2}\hat{G}_y^{\,2}} - \cancel{\frac{\varepsilon^2}{2}\hat{G}_x^{\,2}}\right)$$

$$= -\varepsilon^2\left[G_x, G_y\right] = -i\varepsilon^2 G_z \Rightarrow \left[G_x, G_y\right] = iG_z \Rightarrow \left[L_x, L_y\right] = i\hbar L_z.$$

This is amazing! We derived these commutation relations purely from the properties of rotations in real space. We made no reference to $[x, p_x] = i\hbar$, yet we got exactly the same result as Eq. (5.1), derived from $[x, p_x] = i\hbar$ and $\mathbf{L} = \mathbf{r} \times \mathbf{p}$. We, therefore, suppose that (but have not proved):

Any quantum system which can be rotated has generators which satisfy these commutation relations.

We might further speculate that any such generators represent some form of angular momentum. Indeed, experiments show that there is another form of quantum angular momentum besides orbital; we call this angular momentum **spin**, though it is different than classical spin. Nonetheless, associated with spin is also a magnetic dipole moment, which evokes a similar rough analogy with a classical spinning blob of charge.

Finite rotations about arbitrary axes: Aside: It is common to generalize finite rotations about a coordinate axis to rotations about any axis by defining a rotation "angle" $\boldsymbol{\alpha}$, with magnitude and direction, but which is *not* a vector. (It is often incorrectly called a "vector.") We define the "components" of $\boldsymbol{\alpha}$ as those of a unit vector times the rotation angle in radians:

$$\boldsymbol{\alpha} \equiv \left(\alpha_x, \alpha_y, \alpha_z\right) \equiv \hat{\mathbf{n}}\alpha = \left(n_x\mathbf{e}_x + n_y\mathbf{e}_y + n_z\mathbf{e}_z\right)\alpha \quad \text{This is \textit{not} a vector!}$$

$$\textit{where } \alpha = |\boldsymbol{\alpha}| \equiv \text{rotation angle in rad, } \hat{\mathbf{n}} = \frac{\boldsymbol{\alpha}}{|\boldsymbol{\alpha}|} \equiv \text{rotation axis}.$$

$\boldsymbol{\alpha}$ is not a vector because we cannot add $\boldsymbol{\alpha}_1 + \boldsymbol{\alpha}_2$ to get the total "angle" of performing rotations $\boldsymbol{\alpha}_1$ and $\boldsymbol{\alpha}_2$ separately. Nonetheless, this vector-like notation allows us
to write a finite rotation as a dot product:

$$\hat{R}(\boldsymbol{\alpha}) = \exp\left(-\frac{i}{\hbar}\boldsymbol{\alpha}\boldsymbol{\cdot}\hat{\mathbf{L}}\right), \quad \hat{\mathbf{L}} \equiv \hat{L}_x\mathbf{e}_x + \hat{L}_y\mathbf{e}_y + \hat{L}_z\mathbf{e}_z.$$

We must be careful, though, because, unlike exponentials of numbers, the exponential of a sum of operators is *not* the composition ("product") of the exponentials of
each term:

$$\hat{R}(\boldsymbol{\alpha}) = \exp\left(-\frac{i}{\hbar}\left(\alpha_x\hat{L}_x + \alpha_y\hat{L}_y + \alpha_z\hat{L}_z\right)\right)$$

$$\neq \exp\left(-\frac{i}{\hbar}\alpha_x\hat{L}_x\right)\exp\left(-\frac{i}{\hbar}\alpha_y\hat{L}_y\right)\exp\left(-\frac{i}{\hbar}\alpha_z\hat{L}_z\right).$$

Exponentials of operators can only be factored in this way if the operators commute, which the $\hat{L}_j$ do not.

5.4 Spin 1/2

We reserve the term "wave-function" for spatial quantum states. We use the term
spin state to refer to the intrinsic spin of a particle (or particles). The combination
of a wave-function and a spin state is a complete quantum description of a particle,
a "quantum state."

5.4.1 Spin Kets, Bras, and Inner Products

Spatial kets and bras can be thought of as shorthand for functions of space (e.g.,
wave-functions), spin states cannot. Spin basis kets are abstract, formless things that
obey simple rules. They cannot be represented as functions of space, or anything
else. Spin inner products are defined using the presumed (or *defined*) orthonormality of spin basis kets.

   In particular, suppose we measure a fermion (spin-1/2 particle) spin along some
axis (perhaps using a Stern–Gerlach device). Experiment shows that only two values are possible; thus, we assume that any spin state along this axis is a general
superposition of exactly two spin eigenstates. Let us call them $|z+\rangle$ and $|z-\rangle$.
Subsequent measurements along the same axis always produce the same value of
spin, i.e., there is 100 % chance of measuring the same spin again and 0 % chance of

measuring the opposite spin. This behavior *defines* the inner product of spin states. If we believe that spin states behave similarly to spatial states, we must then say:

$$\langle z+|z+\rangle = 1, \quad \langle z-|z-\rangle = 1, \quad \text{and} \quad \langle z-|z+\rangle = 0.$$

Therefore, $|z+\rangle$ and $|z-\rangle$ form a complete, orthonormal basis for spin states along this axis. Furthermore, subsequent experiments performing measurements along different axes confirm that:

> $|z+\rangle$ and $|z-\rangle$ form a complete, orthonormal basis for aspin state that defines a particle's spin behavior along *all* axes.

Note that for spatial wave-functions, the kets and bras can be thought of as just "notation," in that we know that the ket $|\psi\rangle$ is "really" a complex-valued function of space. In other words, the ket notation is shorthand for a Hilbert space vector, which has a known internal structure: a complex function of space. This "internal structure" *does not exist* for spin kets.

> The spin kets are abstract vectors with *no further structure*.

All we know about $|z+\rangle$ is what we have described previously, and we know it from experiment. End of story.

5.4.2   Spinors for Spin-1/2 Particles

Spinors are kets (and bras) that represent the spin state of particles with intrinsic angular momentum. We discuss here spin-1/2 particles (**fermions**). When measured, spin-1/2 particles produce a component of angular momentum along the measurement axis that is always either $+\hbar/2$ ("up"), or $-\hbar/2$ ("down"). Since a particle's spin can be in a superposition of these two states, a spinor is a two-dimensional (2D) vector of complex numbers.

$$|\chi\rangle = \begin{pmatrix} a \\ b \end{pmatrix} \quad \text{(a spinor)}.$$

Every spin-1/2 spinor is an eigenspinor of *some* direction, so loosely, every spinor "points" in some direction. Be careful, though, that this is absolutely *not* the classical view of angular momentum having a definite direction in space. Spinors, like wave-functions, describe the two *probabilities* of measuring the two possible values of angular momentum along any axis. A spinor is a 2D vector of complex numbers, which represents a direction in 3D physical space. From it, you can compute the probability of measuring the spin being up or down along *any* axis.

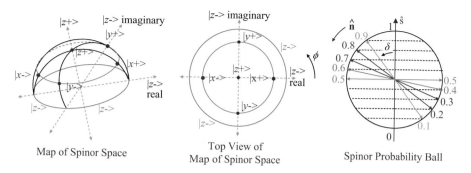

Map of Spinor Space   Top View of
                       Map of Spinor Space   Spinor Probability Ball

Fig. 5.3 (*Left* and *middle*) Spinor space mapped onto a hemisphere. (*Right*) Probabilities of measuring spin in various directions, given a particle in the state $|z+\rangle$

At first blush, a spinor might seem to have 4 degrees of freedom: two real numbers in each of two complex components. However, every quantum system of states has an arbitrary overall phase, i.e., you can multiply every ket in your system, at any time, by any unit-magnitude complex number, and you still have exactly the same quantum system. This accounts for one real-number degree of freedom out of the spinor's 4. Secondly, the spinor (like all quantum state kets) must be normalized: the sum of the squares of the component magnitudes must be one. This constraint removes another degree of freedom from the spinor, leaving just 2 degrees of freedom. It takes 2 degrees of freedom to represent any direction in 3D physical space (such as with two coordinates: θ, ϕ).

A common basis for spinors is the z-basis: i.e., we choose our z-axis to be along the measurement axis (sometimes called the "quantization axis"). The spinor can be written as the sum of $|z+\rangle$ and $|z-\rangle$ basis kets:

$$|\chi\rangle = a|z+\rangle + b|z-\rangle = \begin{pmatrix} a \\ b \end{pmatrix}, \quad where \quad \begin{array}{l} |z+\rangle \text{ component is always on top} \\ |z-\rangle \text{ component is always on bottom}. \end{array}$$

$$(5.4)$$

The $|z+\rangle$ ket is the eigenket for pointing in the $+z$ direction. A particle in the $|z+\rangle$ state, when measured along the z-axis, will always measure spin "up," i.e., positive. Thus, Pr($z+$)=1, and Pr($z-$)=0. A particle in the $|z-\rangle$ state, when measured along the z-axis, will always measure spin "down," i.e., negative. Thus, Pr($z+$)=0, and Pr($z-$)=1.

It might seem odd that a spinor can point in *any* direction in three-space with only the two "z-component" basis kets. But recall that each z-component is a complex number; they are interpreted as follows:

The magnitudes of the two spinor components define the z-axis spin properties of the particle. The relative phase angle between the two components defines the spin properties in the x-y directions.

We can illustrate how spinors map to 3D directions by drawing a spinor map on a hemisphere. The hemisphere is a surface (Fig. 5.3) in the ket space of spinor components a and b, Eq. (5.4), which has four real dimensions. We reduce the original four-dimensional spinor space to 3D by choosing a phase convention that makes a, the $|z+\rangle$ component, real and non-negative. We further reduce spinor space to 2D because we map only normalized spinors, so $|a|^2 + |b|^2 = 1$. Then the 2D surface of the "northern" hemisphere is a map of all normalized spinors (Fig. 5.3, *left* and *middle*).

[This is similar to a Bloch sphere; any two-component discrete quantum state can be similarly mapped.]

Every possible spinor maps to exactly one point on the hemisphere, except the $|z-\rangle$ eigenspinor.

The north pole is the $|z+\rangle$ eigenspinor $a=1$, $b=0$:

$$|z+\rangle = \begin{pmatrix} 1 \\ 0 \end{pmatrix}.$$

The whole equator is the $|z-\rangle$ eigenspinor $a=0$, $|b|=1$. Because the $|z+\rangle$ component is zero at the equator, it no longer fixes the phase of the spinor; suddenly the arbitrary complex phase for quantum states is allowed again for $|z-\rangle$. The line of 45° latitude is the set of spinors perpendicular to the z-axis in the x-y plane. Being midway between $|z+\rangle$ and $|z-\rangle$, their z-axis measurements have 50% probability of $|z+\rangle$, and thus also 50% probability of $|z-\rangle$.

Consider the complex phase of b (the $|z-\rangle$ component). As the phase of b rotates around the complex plane, the spinor direction in three-space rotates around the z-axis by the same angle (Fig. 5.3 *middle*). Recall that the x and y eigenspinors are [16, 1.4.17, p. 28]:

$$|x+\rangle = \begin{bmatrix} 1/\sqrt{2} \\ 1/\sqrt{2} \end{bmatrix}, \quad |x-\rangle = \begin{bmatrix} 1/\sqrt{2} \\ -1/\sqrt{2} \end{bmatrix}; \qquad |y+\rangle = \begin{bmatrix} 1/\sqrt{2} \\ i/\sqrt{2} \end{bmatrix}, \quad |y-\rangle = \begin{bmatrix} 1/\sqrt{2} \\ -i/\sqrt{2} \end{bmatrix}.$$

That is (for a real-valued $|z+\rangle$ component), the $|x+\rangle$ eigenspinor has a $|z-\rangle$ component phase of 0; $|y+\rangle$ has a $|z-\rangle$ phase of 90°, $|x-\rangle$ has a $|z-\rangle$ phase of 180°, $|y-\rangle$ has a $|z-\rangle$ phase of 270°. This is a choice of convenience [16, p. 27]. (Note that some references use a different convention, at least in some places.) This equality between the complex phase of the $|z-\rangle$ component and the real-space angle ϕ is true for all ϕ (not just the axes). In general, the eigenspinor for a direction in space given by the spherical angles (θ, ϕ), is [16, 1.9, p. 61]:

$$|\theta, \phi+\rangle = \begin{bmatrix} \cos(\theta/2) \\ \sin(\theta/2)e^{i\phi} \end{bmatrix},$$

where the equality of the real-space ϕ and the $|z-\rangle$ component complex angle ϕ is explicit. Be aware of other phases for the same spinor, such as [16, 3.2.52, p. 168]:

$$|\theta, \phi+\rangle = \begin{bmatrix} \cos(\theta/2)e^{-i\phi/2} \\ \sin(\theta/2)e^{i\phi/2} \end{bmatrix},$$

which is the same as previous, but multiplied by the unit-magnitude complex number $e^{-i\phi/2}$. In this latter spinor, the phase *difference* between the $|z+\rangle$ component and the $|z-\rangle$ component is still the real-space spherical angle ϕ.

Be careful to distinguish orthogonality (perpendicularity) in *real* space from orthogonality in *spinor* space. For example, in real space, the x-axis is perpendicular to the z-axis:

$$\mathbf{e}_x \perp \mathbf{e}_z, \text{ i.e. } \mathbf{e}_x \cdot \mathbf{e}_z = 0,$$

but in spinor space, the kets

$$|x+\rangle \not\perp |z+\rangle, \quad \text{i.e.} \quad \langle x+|z+\rangle = \begin{bmatrix} 1/\sqrt{2} & 1/\sqrt{2} \end{bmatrix} \begin{bmatrix} 1 \\ 0 \end{bmatrix} = \frac{1}{\sqrt{2}} \neq 0.$$

Notice that rotation by a polar angle θ in real space corresponds to rotation by a polar angle $\theta/2$ in spinor space. That means that rotating a $|z+\rangle$ eigenstate by π radians in real space takes the north pole to the equator in spinor space:

$$|z+\rangle = \begin{pmatrix} \cos 0 \\ \sin 0 \, e^{i0} \end{pmatrix} = \begin{pmatrix} 1 \\ 0 \end{pmatrix},$$

$$\hat{R}_y(\pi)|z+\rangle = \begin{pmatrix} \cos(\pi/2) \\ \sin(\pi/2) \, e^{i0} \end{pmatrix} = \begin{pmatrix} 0 \\ 1 \end{pmatrix} = |z-\rangle.$$

Spin-1/2 particles are really spin $\sqrt{3}/2$: The spin of any one component $(s_x, s_y, \text{ or } s_z)$ of a spin-1/2 particle is $1/2\ \hbar$, *but* $|\mathbf{s}| = \dfrac{\sqrt{3}}{2}\hbar$. Recall that

$$|\mathbf{s}| = \sqrt{s_x^2 + s_y^2 + s_z^2},$$

and if each component is $\pm 1/2\ \hbar$, the total magnitude is $\frac{\sqrt{3}}{2}\hbar$. This implies that $s^2 = (3/4)\ \hbar^2$. So they are called "spin-1/2" particles, but the magnitude of their spin is bigger than that. We can also see this from the well-known eigenvalue of $\hat{s}^2$:

$$\hat{s}^2 \binom{a}{b} = s(s+1)\hbar^2 \binom{a}{b} \Rightarrow |\mathbf{s}| = \sqrt{\frac{1}{2}\cdot\frac{3}{2}}\hbar = \frac{\sqrt{3}}{2}\hbar.$$

Spinor probabilities: Given a spinor, and a measurement direction along the unit vector $\mathbf{n}$, what is the probability that the particle measures in that direction? We now show it is:

$$\Pr(\mathbf{n}+) \quad = \cos^2\frac{\delta}{2} = \frac{1+\cos\delta}{2} = \frac{1+\mathbf{n}\cdot\mathbf{e_s}}{2},$$

$$\textit{where} \quad \delta = \text{angle between spinor direction, } \mathbf{e_s} \text{ (in 3-space),} \quad (5.5)$$
$$\text{and measurement direction, } \mathbf{n}.$$

The spinor probability ball shows this relationship (Fig. 5.3, *right*). Graphically, the probability equals the height of the tip of $\mathbf{n}$ projected onto the spin axis, as a fraction of the circle's *diameter*. When measuring "in-line" with the spinor, the probability of measuring "+" is 1 (certainty). When measured perpendicular to the spinor,

$$\Pr(\mathbf{n}+) = 0.5 \qquad \text{(perpendicular measurement)}.$$

We derive $\Pr(\mathbf{n}+)$ in two ways: conceptually and then from the eigenspinors.

Conceptual approach: We can actually deduce $\Pr(\mathbf{n}+)$, Eq. (5.5), conceptually, without solving for eigenvectors, or other quantum mathematics. We use the correspondence principle: taking a large number of quantum measurements must produce the classical result. This is a general result that often provides good insight and sanity checking for quantum problems.

The classical result of measuring angular momentum along an axis, of a body spinning about another axis, is just the projection of the body's angular momentum onto the measured axis:

$$L_{\mathbf{n}} = |\mathbf{L}|\cos\delta.$$

This equation must also hold for the average of quantum measurements. Hence,

$$\langle L_{\mathbf{n}} \rangle = \frac{\hbar}{2}\cos\delta = \Pr(\mathbf{n}+)\frac{\hbar}{2} - \Pr(\mathbf{n}-)\frac{\hbar}{2}.$$

Cancel the $\hbar/2$ and note that

$\Pr(\mathbf{n}-) = 1 - \Pr(\mathbf{n}+)$.   Then:

$$\cos\delta = \Pr(\mathbf{n}+) - \left[1 - \Pr(\mathbf{n}+)\right] = 2\Pr(\mathbf{n}+) - 1 \Rightarrow \Pr(\mathbf{n}+) = \frac{1+\cos\delta}{2}. \qquad (5.6)$$

More conceptually, as $\Pr(\mathbf{n}+)$ decreases with increasing δ, the average value decreases twice as fast as $\cos\delta$, because not only do we decrease the average from fewer (+) measurements, but we decrease it as much again with *more* (−) measurements. Hence, $\Pr(+)$ must decrease only half as fast as $\cos\delta$, which leads to the 2 in the denominator in the previous equation. For example, as δ goes from 0 to 90°, $\cos\delta$ goes from 1 to zero, but $\Pr(\mathbf{n}+)$ goes from 1 to 0.5 (half the change of $\cos\delta$).

   Eigenvalue approach: We can also derive $\Pr(\mathbf{n}+)$ from the expression for an eigenspinor in an arbitrary direction with spherical coordinate angles (θ, ϕ). As noted previously, in the z-basis, such an eigenspinor is:

$$|\mathbf{n}+\rangle = \begin{bmatrix} \cos(\theta/2) \\ \sin(\theta/2)e^{i\phi} \end{bmatrix} = \cos(\theta/2)|z+\rangle + \sin(\theta/2)e^{i\phi}|z-\rangle \quad where \ \mathbf{n} \equiv \mathbf{n}(\theta, \phi).$$

Then if we measure spin in the z-direction of a particle in this spin state,

$$\Pr(z+\,|\,\mathbf{n}) = \left|\langle z+|\mathbf{n}\rangle\right|^2 = \cos^2\frac{\theta}{2} = \frac{1+\cos\theta}{2}.$$

By symmetry, this same probability must apply to any eigenspinor and measurement direction separated by an angle θ (or δ), which is the same as before, Eq. (5.5).

5.4.3   Climbing the Ladder: Quantization and the Pauli Matrices

Starting with the commutation relations for all quantized angular momentum, and making a natural choice of complex phase, we can derive the Pauli matrices. We try here to motivate each step, so the development is more intuitive, and (we hope) less arbitrary. At this point, we take the experimentally observed values of J_z (and thus the eigenvalues of $\hat{j}_z$) to be $m\hbar$, with m a half-integer. We do not yet know the eigenvalues of $\hat{j}^2$, because it is hard to observe them directly, so for now we label $\hat{j}^2$ eigenstates with a quantum number j, whose meaning we do not yet know. We briefly review here the derivation of the eigenvalues of $\hat{j}^2$, and the raising and lower operators, since we must derive those along the way to the Pauli matrices.

   The fundamental relations, from which all of quantum angular momentum arises, are these [16, p. 158]:

$$\left[\hat{j}_x, \hat{j}_y\right] = i\hbar\,\hat{j}_z, \quad \left[\hat{j}_y, \hat{j}_z\right] = i\hbar\,\hat{j}_x, \quad \left[\hat{j}_z, \hat{j}_x\right] = i\hbar\,\hat{j}_y.$$

Writing the commutators of $\hat{j}_z$ with $\hat{j}_x$ and $\hat{j}_y$, we get:

$$\left[\hat{j}_z, \hat{j}_x\right] = i\hbar\,\hat{j}_y \qquad \left[\hat{j}_z, \hat{j}_y\right] = -i\hbar\,\hat{j}_x.$$

These are almost symmetric. By inspection, we see that putting an i in front of $\hat{j}_y$ in the second form makes them symmetric:

$$\left[\hat{j}_z, \hat{j}_x\right] = \hbar\,i\hat{j}_y \qquad \left[\hat{j}_z, i\hat{j}_y\right] = \hbar\,\hat{j}_x \;\Rightarrow\; \left[\hat{j}_z, \hat{j}_x + i\hat{j}_y\right] = \hbar\left(\hat{j}_x + i\hat{j}_y\right).$$

This resembles an eigenvalue equation, but it is an operator equation. For brevity, we define the standard operator $\hat{j}_+$:

$$\hat{j}_+ \equiv \hat{j}_x + i\hat{j}_y \qquad \left(\text{compare to } \hat{a} \text{ and } \hat{a}^\dagger \text{ for the simple harmonic oscillator (SHO)}\right).$$

$$\left[\hat{j}_z, \hat{j}_+\right] = \hbar\hat{j}_+ \;\Rightarrow\; \hat{j}_z\hat{j}_+ = \hat{j}_+\hat{j}_z + \hbar\hat{j}_+.$$

To eliminate the $\hat{j}_z$ on the right, we act with this operator (on both sides of the equation) on the $\hat{j}_z$ eigenstate $|j\,m\rangle$:

$$\hat{j}_z\hat{j}_+|j\,m\rangle = \left(\hat{j}_+\hat{j}_z + \hbar\hat{j}_+\right)|j\,m\rangle = \left(\hat{j}_+m\hbar + \hbar\hat{j}_+\right)|j\,m\rangle = (m+1)\hbar\hat{j}_+|j\,m\rangle.$$

In other words, $\hat{j}_+|j\,m\rangle$ is an eigenstate of $\hat{j}_z$, with eigenvalue $(m+1)\hbar$:

$$\hat{j}_z\left(\hat{j}_+|j\,m\rangle\right) = (m+1)\hbar\left(\hat{j}_+|j\,m\rangle\right) \;\Rightarrow\; \hat{j}_+|j\,m\rangle \propto |j, m+1\rangle. \qquad (5.7)$$

Similarly, we define $\hat{j}_-$, which by inspection, is the adjoint of $\hat{j}_+$:

$$\hat{j}_- \equiv \hat{j}_x - i\hat{j}_y \;\Rightarrow\; \hat{j}_- = \left(\hat{j}_+\right)^\dagger, \quad \text{and} \quad \hat{j}_-|j\,m\rangle \propto |j, m-1\rangle. \qquad (5.8)$$

Now, $\hat{j}_-|j\,m\rangle$ is an eigenstate of $\hat{j}_z$, with eigenvalue $(m-1)\hbar$. $\hat{j}_-$ and $\hat{j}_+$ are called **ladder operators**.

Note that, starting with a $\hat{j}_z$ eigenstate, $|j\,m\rangle$, repeated operation by $\hat{j}_+$ produces a sequence of kets with increasing m values. This sequence must terminate, because physically, the z-component of angular momentum, $m\hbar$, must be smaller than the magnitude of angular momentum. We have not yet determined what that magnitude is, in terms of the quantum number j, but physically, it must be finite.

We can now determine the constants of proportionality in $\hat{j}_+|j\,m\rangle$ and $\hat{j}_-|j\,m\rangle$, Eqs. (5.7) and (5.8). We use a similar method to that used in the SHO showing that $\hat{a}^\dagger\hat{a}$ has eigenvalues ≥ 0. The squared magnitude of any vector is ≥ 0, so:

$$\left|\hat{j}_+\,|\,j\,m\rangle\right|^2 = \langle j\,m\,|\,\hat{j}_+^{\,\dagger}\hat{j}_+\,|\,j\,m\rangle = \langle j\,m\,|\,\hat{j}_-\hat{j}_+\,|\,j\,m\rangle \geq 0. \tag{5.9}$$

We have explicit forms for $\hat{j}_+$ and $\hat{j}_-$, so we can evaluate:

$$\hat{j}_-\hat{j}_+ = \left(\hat{j}_x - i\hat{j}_y\right)\left(\hat{j}_x + i\hat{j}_y\right) = \underbrace{\hat{j}_x^{\,2} + \hat{j}_y^{\,2}}_{\hat{j}_x^{\,2}+\hat{j}_y^{\,2}} + i\underbrace{\left(\hat{j}_x\hat{j}_y - \hat{j}_y\hat{j}_x\right)}_{i\hbar\,\hat{j}_z} = \hat{j}^2 - \hat{j}_z^{\,2} - \hbar\hat{j}_z.$$

The right-hand side has only one unknown eigenvalue, that for $\hat{j}^2$. We call that eigenvalue $\lambda\hbar^2$, and substitute into Eq. (5.9):

$$\langle j\,m\,|\,\hat{j}_-\hat{j}_+\,|\,j\,m\rangle = \langle j\,m\,|\,\hat{j}^2 - \hat{j}_z^{\,2} - \hbar\hat{j}\,|\,j\,m\rangle$$
$$= \left(\lambda - m^2 - m\right)\hbar^2 \langle j\,m\,|\,j\,m\rangle \geq 0$$
$$\Rightarrow \lambda - m(m+1) \geq 0, \quad \text{or} \quad \lambda \geq m(m+1).$$

This relates the eigenvalue of $\hat{j}^2$, which is $\lambda\hbar^2$, to that of $\hat{j}_z$, which is $m\hbar$.

The same formula gives us the coefficient of the raising operator, to within a complex phase:

$$\left|\hat{j}_+\,|\,j\,m\rangle\right|^2 = \langle j\,m\,|\,\hat{j}_-\hat{j}_+\,|\,j\,m\rangle = \left(\lambda - m^2 - m\right)\hbar^2 \Rightarrow$$
$$\hat{j}_+\,|\,j\,m\rangle = e^{i\alpha}\hbar\sqrt{\lambda - m(m+1)}\,|\,j, m+1\rangle. \tag{5.10}$$

This allows us to define the meaning of the quantum number j. In addition, for convenience, we choose the phase so that the coefficient is real ($\alpha=0$), which allows us to fix the Pauli matrices.

First, to define j, we return to the fact that the sequence of kets formed from $\hat{j}_+$ must terminate, i.e., there is some maximum m, which we call j. This can only happen if the coefficient generated by $\hat{j}_+$ acting on the maximum m (called j) is zero:

$$\hat{j}\,|\,j, m = j\rangle = \mathbf{0_v} \Rightarrow \lambda - j(j+1) = 0 \Rightarrow \lambda = j(j+1).$$

Thus, the eigenvalue of $\hat{j}^2$ is $j(j+1)\hbar^2$, and the ladder of $\hat{j}_+$ kets terminates after the ket $|\,j, m = j\rangle$. We forevermore drop the use of λ, in favor of the quantum number j, now known to be a half-integer (just like m), and whose meaning is now fully specified: $j(j+1)\hbar^2$ are the eigenvalues of $\hat{j}^2$. We write our ladder operators in final form from Eq. (5.10):

$$\hat{j}_+\,|\,j\,m\rangle = \hbar\sqrt{j(j+1) - m(m+1)}\,|\,j, m+1\rangle,$$
$$\hat{j}_-\,|\,j\,m\rangle = \hbar\sqrt{j(j+1) - m(m-1)}\,|\,j, m-1\rangle.$$

[This method is an example of a more general group theory principle: all group characteristics derive from the commutators of the generators of the group. In our case, the group is the group of rotations in 3D space and the generators are the angular momentum operators. These calculations are the foundation of more advanced calculations in quantum field theory involving more advanced groups, such as SU(3).]

5.4.4 Deriving the Pauli Matrices

We can now derive the Pauli matrices by plugging into the previous formulas:

$$\hat{j}_+\begin{bmatrix}1\\0\end{bmatrix}=\begin{bmatrix}0\\0\end{bmatrix},\quad \hat{j}_+\begin{bmatrix}0\\1\end{bmatrix}=\hbar\sqrt{\frac{1}{2}\left(\frac{3}{2}\right)-\frac{-1}{2}\left(\frac{1}{2}\right)}\begin{bmatrix}1\\0\end{bmatrix}=\hbar\begin{bmatrix}1\\0\end{bmatrix}\Rightarrow \hat{j}_+=\hbar\begin{bmatrix}0&1\\0&0\end{bmatrix}.$$

Similarly for $\hat{j}_-$, and using $\hat{\mathbf{s}}\equiv(\hbar/2)\boldsymbol{\sigma}$,

$$\hat{j}_-=\hbar\begin{bmatrix}0&0\\1&0\end{bmatrix},\quad \hat{s}_x=\frac{\hat{j}_++\hat{j}_-}{2}=\frac{\hbar}{2}\sigma_x=\frac{\hbar}{2}\begin{bmatrix}0&1\\1&0\end{bmatrix},$$

$$\hat{s}_y=i\frac{-\hat{j}_++\hat{j}_-}{2}=\frac{\hbar}{2}\sigma_y=\frac{\hbar}{2}\begin{bmatrix}0&-i\\i&0\end{bmatrix}.$$

σ_z is defined diagonal, since we chose the z-axis to have definite angular momentum components, with experimentally determined eigenvalues $\pm\hbar/2$. So:

$$\hat{s}_z=\frac{\hbar}{2}\begin{bmatrix}1&0\\0&-1\end{bmatrix}.\Rightarrow \boldsymbol{\sigma}\equiv\left(\sigma_x,\sigma_y,\sigma_z\right)=\left(\begin{bmatrix}0&1\\1&0\end{bmatrix},\begin{bmatrix}0&-i\\i&0\end{bmatrix},\begin{bmatrix}1&0\\0&-1\end{bmatrix}\right).$$

These matrices have the properties we expect: the eigenvalues are all ±1, corresponding to spin angular momentum components of $\hbar/2$, since $\hat{\mathbf{s}}=(\hbar/2)\boldsymbol{\sigma}$. All the eigenvalues must be the same due to rotational invariance. Also, the operators are Hermitian, as required to be an observable property.

5.4.5 Matrices, Matrices, Matrices

There are several matrices used in describing angular momentum and rotations. We clarify the use and meaning of some common spin-1/2 ones here. In spin-1/2, besides the Pauli matrices, there are three important families of 2×2 matrices:

(1) the rotation matrix for spherical coordinate angles $\mathcal{D}^{1/2}(\theta,\phi)$;

(2) the rotation matrix $\mathcal{D}^{1/2}(\mathbf{n},\alpha)$ for an angle ϕ about an *arbitrary* axis $\mathbf{n}=(n_x, n_y, n_z)$; and

(3) the spin operator matrix for spin-1/2 along an arbitrary axis **n**, given by either **n**$=(n_x, n_y, n_z)$ or **n**(θ, ϕ).

Families 1 and 2 rotate any spin-1/2 spinor by the given parameters: either (1) standard spherical coordinate angles: first θ radians about the y-axis, then ϕ radians about the z-axis; or (2) α radians about the given **n**-axis. These are just two different ways to parameterize rotations in space.

Rotation operators take a quantum state into another quantum state. Therefore, rotation matrices are unitary (and not Hermitian). Given a rotation, and its rotation matrix, the matrix for the inverse-rotation is the Hermitian-conjugate (adjoint) of the given rotation matrix. We now list these three families of matrices.

(1) Rotation Matrix $\mathcal{D}^{1/2}(\theta,\phi)$:

$$\mathcal{D}^{1/2}(\theta,\phi) = \begin{bmatrix} \cos(\theta/2) & -\sin(\theta/2) \\ \sin(\theta/2)e^{+i\phi} & \cos(\theta/2)e^{+i\phi} \end{bmatrix}, \quad \text{or}$$

$$\mathcal{D}^{1/2}(\theta,\phi) = \begin{bmatrix} \cos(\theta/2)e^{-i\phi/2} & -\sin(\theta/2)e^{-i\phi/2} \\ \sin(\theta/2)e^{+i\phi/2} & \cos(\theta/2)e^{+i\phi/2} \end{bmatrix}. \tag{5.11}$$

Note that the second form is just $e^{-i\phi/2}$ times the first form, and any such overall complex phase is arbitrary, provided it is applied to every state in the system.

We derive this by noting that the first column (taken as a vector) must be the eigenspinor for (θ, ϕ), because rotating $|z+\rangle = (1\ 0)^T$ yields that eigenspinor. The second column must be orthogonal to the first (the matrix is unitary), which fixes the second column up to a complex phase. Furthermore, we wish $\mathcal{D}^{1/2}(0,0)$ (which is no rotation at all) to be the identity matrix, which then completely fixes the second column.

(2) Rotation Matrix $\mathcal{D}^{1/2}(\mathbf{n},\alpha)$: [16, 3.2.45, p. 166]:

$$\mathcal{D}^{1/2}(\mathbf{n},\alpha) = \begin{bmatrix} \cos(\alpha/2)-in_z\sin(\alpha/2) & (-in_x-n_y)\sin(\alpha/2) \\ (-in_x+n_y)\sin(\alpha/2) & \cos(\alpha/2)+in_z\sin(\alpha/2) \end{bmatrix}.$$

As a consistency check: for rotations about the z-axis, **n**$=(0, 0, 1)$, this reduces to the second form of Eq. (5.11) with $\theta=0$, $\phi=\alpha$.

(3) Spin operator matrix about a measurement axis **n**$=(n_x, n_y, n_z)$ or **n**(θ, ϕ):

Note that the spin operator is *not* a rotation operator; it is, however, the **generator** of rotations, which is quite different. The spin operator is a Hermitian matrix which produces a ket (*not a spinor!*) that encodes information about a spinor (see the section on operators, and how observable operators acting on state kets are *not* state kets). With this ket, you can directly find the average value of spin along the given measurement axis [1, 14.6, p. 303; 16, 3.2.38, p. 165].

$$\hat{\mathbf{s}} \cdot \mathbf{n} = \frac{\hbar}{2}\boldsymbol{\sigma} \cdot \mathbf{n} = \frac{\hbar}{2}\left[n_x \sigma_x + n_y \sigma_y + n_z \sigma_z \right] = \frac{\hbar}{2}\begin{bmatrix} n_z & n_x - in_y \\ n_x + in_y & -n_z \end{bmatrix}$$

$$= \frac{\hbar}{2}\begin{bmatrix} \cos\theta & \sin\theta e^{-i\phi} \\ \sin\theta e^{+i\phi} & -\cos\theta \end{bmatrix}.$$

The spin operator matrix is Hermitian, as all observable operators must be. Notice that the average value of spin component measured in some direction, given an eigenspinor $|z+\rangle$, follows the classical result of $\cos\theta$, as we expect from the correspondence principle for a large number of measurements:

$$\langle \hat{\mathbf{s}} \cdot \mathbf{n} \rangle = \langle z+|\frac{\hbar}{2}\boldsymbol{\sigma} \cdot \mathbf{n}|z+\rangle = \frac{\hbar}{2}\begin{bmatrix} 1 & 0 \end{bmatrix}\begin{bmatrix} \cos\theta & \sin\theta e^{-i\phi} \\ \sin\theta e^{+i\phi} & -\cos\theta \end{bmatrix}\begin{bmatrix} 1 \\ 0 \end{bmatrix}$$

$$= \frac{\hbar}{2}\begin{bmatrix} 1 & 0 \end{bmatrix}\begin{bmatrix} \cos\theta \\ \sin\theta e^{+i\phi} \end{bmatrix} = \frac{\hbar}{2}\cos\theta.$$

Consistency check: If we rotate the spin state $|z+\rangle$ by spherical angles (θ, ϕ), we should get an eigenvector of $\hat{\mathbf{s}} \cdot \mathbf{n}(\theta, \phi)$ with eigenvalue $\hbar/2$. Let us check by direct multiplication of the matrices given previously:

$$\mathcal{D}^{1/2}(\theta, \phi)|z+\rangle = \begin{bmatrix} \cos(\theta/2) & -\sin(\theta/2) \\ \sin(\theta/2)e^{+i\phi} & \cos(\theta/2)e^{+i\phi} \end{bmatrix}\begin{bmatrix} 1 \\ 0 \end{bmatrix} = \begin{bmatrix} \cos(\theta/2) \\ \sin(\theta/2)e^{+i\phi} \end{bmatrix}$$

We recognize this as the eigenspinor for direction (θ, ϕ). Then:

$$\hat{\mathbf{s}} \cdot \mathbf{n}\left(\mathcal{D}^{1/2}(\theta, \phi)|z+\rangle\right) = \frac{\hbar}{2}\begin{bmatrix} \cos\theta & \sin\theta e^{-i\phi} \\ \sin\theta e^{+i\phi} & -\cos\theta \end{bmatrix}\begin{bmatrix} \cos(\theta/2) \\ \sin(\theta/2)e^{+i\phi} \end{bmatrix}$$

$$= \frac{\hbar}{2}\begin{bmatrix} \cos\theta\cos(\theta/2) + \sin\theta\sin(\theta/2) \\ \sin\theta e^{+i\phi}\cos(\theta/2) - \cos\theta\sin(\theta/2)e^{+i\phi} \end{bmatrix}$$

$$= \frac{\hbar}{2}\begin{bmatrix} \cos(\theta/2) \\ \sin(\theta/2)e^{+i\phi} \end{bmatrix}.$$

Using: $\cos(a-b) = \cos a \cos b + \sin a \sin b$,

and $\sin(a-b) = \sin a \cos b - \cos a \sin b$.

This is $\hbar/2$ times the eigenspinor for (θ, ϕ), as required.

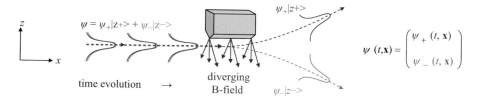

Fig. 5.4 Time evolution of a spin-1/2 particle moving through a Stern–Gerlach device. The up and down spatial wave-functions are the same for t before the device, but different for t after

5.5   Coupling Spin and Position

So far, we have considered wave-functions (spatial states), and separately, spin states. An electron is a spin-1/2 particle, and so has both kinds of states simultaneously. The complete quantum state of an electron, or any spin-1/2 particle, must specify both parts: space and spin. In the simplest case, the space and spin parts are independent of each other. The full quantum state is simply the concatenation of space and spin states. There are no consistent standards for notation. If the spin state has a definite value of "up" or "down," we may write it as:

$$|\psi\rangle|\uparrow\rangle, \quad \text{or} \quad |\psi\,\uparrow\rangle.$$

More general spin state can also be written in many ways, such as:

$$\psi(x)\begin{pmatrix} a \\ b \end{pmatrix}, \quad |\psi\rangle|\chi\rangle, \quad |\psi,\chi\rangle.$$

Note that when the spin is independent of the wave-function, we just "tack on" the spinor to the wave-function to get the complete quantum state. This is a simple "tensor product," or "product state."

However, in many cases, the spin and wave-function depend on each other. For example, in a Stern–Gerlach experiment (Fig. 5.4), the particle approaching the device might be a simple superposition (or even mixture) of up and down. Then after the device, the wave-function would be large for the upper path and spin-up, and also large for the lower path and spin-down. However, the probability of finding the particle spin down in the upper path is (ideally) zero, as is the chance of finding it spin up in the lower path.

In such a case, we have *different* wave-functions for spin-up (+) and spin-down (−). We thus have a two-component wave-function, with separate ψ s for each spin:

$$\chi(x) = \begin{bmatrix} \psi_+(x) \\ \psi_-(x) \end{bmatrix} \quad \text{(quantum state with coupled space and spin parts)}.$$

At any given point, measuring spin-up is mutually exclusive with measuring spin-down. Probabilities of mutually exclusive events add, so the probability density of finding the particle (with either spin) at position x is:

$$\text{pdf}_x(x) = |\psi_+(x)|^2 + |\psi_-(x)|^2 .$$

Such a wave-function is the most general possible for a particle with spin, allowing a full and arbitrary specification of joint space and spin properties at every point in space. This most general state can always be written as a superposition of tensor-product states:

$$\chi(x) = \begin{bmatrix} \psi_+(x) \\ \psi_-(x) \end{bmatrix} = \psi_+(x)|z+\rangle + \psi_-(x)|z-\rangle.$$

Normalization: The probability of measuring spin-up or down varies from point to point. However, the total probability of measuring all possible spins over *all* space must be 1. Therefore [1, 14.35, p. 309]:

$$\int_\infty dx \left(|\psi_+(x)|^2 + |\psi_-(x)|^2 \right) = 1.$$

In a given volume near a given point, $\left(|\psi_+|^2 + |\psi_-|^2 \right) dx$ is generally $\ll 1$. Therefore, the squared-magnitudes $|\psi_+|^2$ and $|\psi_-|^2$ give the *relative* probabilities of measuring the particle spin-up and spin-down *at the given point*. If we happen to measure the particle's location to be at x, we determine its spin probabilities by normalizing the relative spin probabilities to give the *absolute* spin probabilities:

$$\text{RelPr(spin-up at } x) = |\psi_+(x)|^2, \quad \text{RelPr(spin-down at } x) = |\psi_-(x)|^2 \Rightarrow$$

$$\text{Absolute Pr(spin-up at } x) = \frac{|\psi_+(x)|^2}{|\psi_+(x)|^2 + |\psi_-(x)|^2},$$

$$\text{Absolute Pr(spin-down at } x) = \frac{|\psi_-(x)|^2}{|\psi_+(x)|^2 + |\psi_-(x)|^2}.$$

Quantum properties and observables: Note that while the wave-function defines all the spatial and space-related properties of a particle (its motion, potential energy, charge density, etc.), the spin state defines the spin and spin-related properties of the particle. These two kinds of quantum states are quite distinct; they live in different vector spaces. The wave-function is vector in an infinite-dimensional complex space; the spin state is a vector in a 2D complex space.

However, there are some observables which get contributions from *both* space and spin parts of the quantum state. Angular momentum has an *orbital* contribution from the wave-function, as well as a *spin* contribution from the spin state:

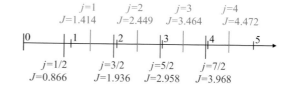

Fig. 5.5 The magnitude of angular momentum, J, vs. the quantum number, j, in units of $\hbar$.

$$\hat{\mathbf{L}}_{orbitel} = \hat{\mathbf{r}} \times \hat{\mathbf{p}}, \ \hat{\mathbf{s}}_{spin} = \frac{\hbar}{2}\boldsymbol{\sigma} \Rightarrow \hat{\mathbf{J}} = \hat{\mathbf{r}} \times \hat{\mathbf{p}} + \frac{\hbar}{2}\boldsymbol{\sigma} \quad where \ \boldsymbol{\sigma} \equiv \text{Pauli vector.}$$

Similarly, for a charged particle, magnetic dipole moment gets contributions from both states. For a spin-1/2 particle, we find:

$$\hat{\boldsymbol{\mu}}_{orbital} = \mu_B\hat{\mathbf{L}}, \quad \hat{\boldsymbol{\mu}}_{spin} \approx 2\mu_B\hat{\mathbf{s}} \quad where \quad \mu_B \equiv \text{Bohr magneton (a constant).}$$

Now, a magnetic dipole moment in a (classical) magnetic field has a potential energy associated with it, so energy must also get a contribution from both space and spin parts of the quantum state:

$$\hat{H}_{magnetic} = -\hat{\boldsymbol{\mu}}_{orbital}\cdot\mathbf{B} - \hat{\boldsymbol{\mu}}_{spin}\cdot\mathbf{B} = -\mathbf{B}\cdot\left(\hat{\boldsymbol{\mu}}_{orbital} + \hat{\boldsymbol{\mu}}_{spin}\right).$$

Entanglement: When two distinct states, such as space- and spin-, that are fundamentally separate properties, become interdependent due to outside influences (such as a Stern–Gerlach device), those two states are said to be **entangled**. We will see that even states of *separate* particles can become entangled, which leads to further unique and unexpected behaviors of QM.

5.6   Angular Momentum for Arbitrary *j*

Angular momentum, which may be a combination of orbit and spin, is quantized by a quantum number *j*, which is allowed values of 0, 1/2, 1, 3/2,…. For such a *j*, in an angular momentum eigenstate:

$$J \equiv |\mathbf{J}| = \hbar\sqrt{j(j+1)}.$$

As $j \to \infty$, $J \to \hbar\left(j + \frac{1}{2}\right)$,   because:

$$\sqrt{j(j+1)} = \sqrt{j^2 + j} = j\sqrt{1 + \frac{1}{j}} \approx j\left(1 + \frac{1}{2j}\right) = j + \frac{1}{2} \quad \text{(from the binomial theorem).}$$

We have seen the explicit $\hat{J}_x$ and $\hat{J}_y$ operator matrices (in the z-basis) for $s=1/2$. But there are general qualities of the $\hat{J}_x$ and $\hat{J}_y$ matrices that are not apparent in the overly simple $s=1/2$ case. Let us work out the $j=2$ case, from which one can imagine other cases. Recall that $j=2$ is five-dimensional, with z-basis kets $|22\rangle$, $|21\rangle$, $|20\rangle$, $|2-1\rangle$, $|2-2\rangle$. Since, for all j:

$$\hat{J}_x = \frac{\hat{J}_+ + \hat{J}_-}{2}, \quad \text{and} \quad \hat{J}_y = \frac{\hat{J}_+ - \hat{J}_-}{2i} = i\frac{-\hat{J}_+ + \hat{J}_-}{2} \quad \forall\, j,$$

we first find the $\hat{J}_+$ and $\hat{J}_-$ operator matrices (in the z-basis). $\hat{J}_+$ is the raising operator, which raises the m-value ($\hat{J}_z$ eigenvalue) by 1, and multiplies the resulting ket by an overall factor of $\hbar\sqrt{j(j+1)-m(m+1)}$. $\hat{J}_-$ is the lowering operator, which lowers the m-value ($\hat{J}_z$ eigenvalue) by 1, and multiplies the resulting ket by an overall factor of $\hbar\sqrt{j(j+1)-m(m-1)}$. Hence,

$$\hat{J}_+ = \hbar \begin{pmatrix} 0 & 2 & 0 & 0 & 0 \\ 0 & 0 & \sqrt{6} & 0 & 0 \\ 0 & 0 & 0 & \sqrt{6} & 0 \\ 0 & 0 & 0 & 0 & 2 \\ 0 & 0 & 0 & 0 & 0 \end{pmatrix} \quad \hat{J}_- = \hbar \begin{pmatrix} 0 & 0 & 0 & 0 & 0 \\ 2 & 0 & 0 & 0 & 0 \\ 0 & \sqrt{6} & 0 & 0 & 0 \\ 0 & 0 & \sqrt{6} & 0 & 0 \\ 0 & 0 & 0 & 2 & 0 \end{pmatrix} \quad (j=2).$$

Therefore,

$$\hat{J}_x = \hbar \begin{pmatrix} 0 & 1 & 0 & 0 & 0 \\ 1 & 0 & \sqrt{6}/2 & 0 & 0 \\ 0 & \sqrt{6}/2 & 0 & \sqrt{6}/2 & 0 \\ 0 & 0 & \sqrt{6}/2 & 0 & 1 \\ 0 & 0 & 0 & 1 & 0 \end{pmatrix},$$

$$\hat{J}_y = \hbar \begin{pmatrix} 0 & -i & 0 & 0 & 0 \\ i & 0 & -i\sqrt{6}/2 & 0 & 0 \\ 0 & i\sqrt{6}/2 & 0 & -i\sqrt{6}/2 & 0 \\ 0 & 0 & i\sqrt{6}/2 & 0 & -i \\ 0 & 0 & 0 & i & 0 \end{pmatrix}$$

In general, $\hat{J}_x|j\,m\rangle$ involves a (real-valued) weighted sum of $|j\,m+1\rangle$ and $|j\,m-1\rangle$. The $\hat{J}_y$ component for $|j\,m\rangle$ is also a weighted sum, using the same weight magnitudes as $\hat{J}_x$, but including a negative imaginary factor on the $|j\,m+1\rangle$ component, and a positive imaginary factor on the $|j\,m-1\rangle$ component.

For comparison, here are the $\hat{J}_x$ and $\hat{J}_y$ matrices for $j=1$ and $j=3/2$, along with some eigenvectors:

$$\hat{J}_x = \hbar \begin{pmatrix} 0 & 1/\sqrt{2} & 0 \\ 1/\sqrt{2} & 0 & 1/\sqrt{2} \\ 0 & 1/\sqrt{2} & 0 \end{pmatrix}, \qquad |1, m_x = 1\rangle = \begin{pmatrix} 1/2 \\ 1/\sqrt{2} \\ 1/2 \end{pmatrix},$$

$$|1, m_x = 0\rangle = \begin{pmatrix} 1/\sqrt{2} \\ 0 \\ -1/\sqrt{2} \end{pmatrix}, \qquad |1, m_x = -1\rangle = \begin{pmatrix} 1/2 \\ -1/\sqrt{2} \\ 1/2 \end{pmatrix}$$

$$\hat{J}_y = \hbar \begin{pmatrix} 0 & -i/\sqrt{2} & 0 \\ i/\sqrt{2} & 0 & -i/\sqrt{2} \\ 0 & i/\sqrt{2} & 0 \end{pmatrix}, \qquad |1\,m_y = 1\rangle = \begin{pmatrix} 1/2 \\ i/\sqrt{2} \\ -1/2 \end{pmatrix},$$

$$|1, m_y = 0\rangle = \begin{pmatrix} 1/\sqrt{2} \\ 0 \\ 1/\sqrt{2} \end{pmatrix}, \qquad |1, m_y = -1\rangle = \begin{pmatrix} 1/2 \\ -i/\sqrt{2} \\ -1/2 \end{pmatrix}$$

For $j=3/2$:

$$\hat{J}_x = \hbar \begin{pmatrix} 0 & \sqrt{3}/2 & 0 & 0 \\ \sqrt{3}/2 & 0 & 1 & 0 \\ 0 & 1 & 0 & \sqrt{3}/2 \\ 0 & 0 & \sqrt{3}/2 & 0 \end{pmatrix}$$

$$\hat{J}_y = \hbar \begin{pmatrix} 0 & -i\sqrt{3}/2 & 0 & 0 \\ i\sqrt{3}/2 & 0 & -i & 0 \\ 0 & i & 0 & -i\sqrt{3}/2 \\ 0 & 0 & i\sqrt{3}/2 & 0 \end{pmatrix}$$

5.7   Addition of Angular Momentum

Terminology: Given a particle of angular momentum quantum number j, its actual angular momentum magnitude is $J = \hbar\sqrt{j(j+1)}$. For brevity, we sometimes colloquially refer to the "angular momentum quantum number" as the "angular momen-

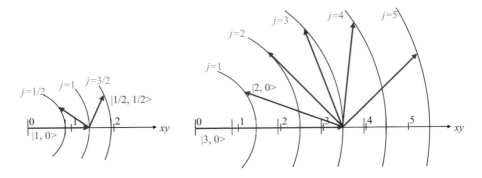

Fig. 5.6 (*Left*) Addition of $|1, 0\rangle$ and $|1/2, 1/2\rangle$ (*Right*) Addition of $|3, 0\rangle$ and $|2, 0\rangle$.

tum," i.e., we may say "a particle of angular momentum j," when we really mean "a particle with angular momentum quantum number j."

Consider two subsystems (or particles) of angular momentum quantum numbers j_1 and j_2. Together, their total angular momentum quantum number lies somewhere between the sum $j_1 + j_2$ and the difference $|j_1 - j_2|$, because angular momentum is a vector, and the two angular momenta can add constructively, destructively, or somewhere in between, depending on the angle between the two vector angular momenta.

Let us restate the previous in terms of actual angular momentum, instead of angular momentum quantum numbers. Consider two particles of angular momentum $\hbar\sqrt{j_1(j_1+1)}$ and $\hbar\sqrt{j_2(j_2+1)}$. Together, their total angular momentum lies somewhere between $\hbar\sqrt{(j_1+j_2)(j_1+j_2+1)}$ and $\hbar\sqrt{(j_1-j_2)(j_1-j_2+1)}$.

For example, suppose $j_1 = 1/2$ and $j_2 = 1$ (Fig. 5.6, *left*). Then $|J_1| = \sqrt{3/2}\ \hbar = 0.866$ $\hbar$, and $|J_2| = \sqrt{2}\ \hbar = 1.414\ \hbar$. At most, the total $j = j_1 + j_2 = 3/2$, and the total angular momentum is $\sqrt{15}/2\ \hbar = 1.936\ \hbar$, which is somewhat less than $\sqrt{3/2}\ \hbar + \sqrt{2}\ \hbar = 2.280\ \hbar$. At the least, $j_t = j_2 - j_1 = 1/2$, and the total angular momentum is $\sqrt{3/2}\ \hbar = 0.866\ \hbar$, which is more than $\sqrt{2}\ \hbar - \sqrt{3/2}\ \hbar = 0.548\ \hbar$. We see that with $j_1 \neq j_2$, the two angular momenta can never fully reinforce, nor fully oppose, each other. This is because total angular momentum J_t is quantized by the same rules as individual particles and that quantization does not allow for simply adding or subtracting the individual angular momenta. C'est la vie. Figure 5.6, *right*, shows a similar addition for $|3\ 0\rangle$ and $|2\ 0\rangle$.

In Fig. 5.6, notice that the addition of two angular momentum eigenstates does *not* (in general) add to a single total angular momentum eigenstate (however, they *do* add to a J_z eigenstate). Because the x-y components are *not* definite, the sum has some chance of being in several different total angular momentum states. In other words,

---

The sum of two *definite* angular momentum states may or may not be an eigenstate of total angular momentum, i.e., it may be a *superposition* of several *total* angular momentum states.

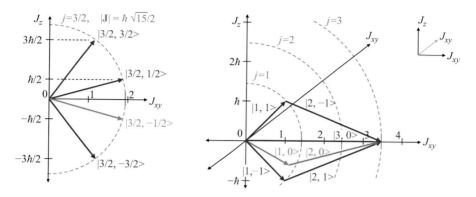

Fig. 5.7 (*Left*) "Side view" of angular momentum states, and (*right*) perspective view of addition of $|1\ m_1\rangle$ and $|2\ m_2\rangle$ to make $|3, 0\rangle$. These vectors can all be rotated around the z-axis.

For example,

$$|1, 0\rangle|1/2, 1/2\rangle = c_1|3/2, 1/2\rangle + c_2|1/2, 1/2\rangle,$$

$$|3, 0\rangle|2, 0\rangle = c_5|5, 0\rangle + c_4|4, 0\rangle + c_3|3, 0\rangle + c_2|2, 0\rangle + c_1|1, 0\rangle. \qquad (5.12)$$

where the c's could potentially be complex. In fact, the c's are the famous Clebsch–Gordon coefficients (CGs), and will turn out to be real (described later).

5.7.1 Two Indefinites can Make a Definite

Quantization of angular momentum leads to some curious consequences. We now show that two completely indefinite vectors can have a definite (real space) dot-product.

Recall that a $|j\ m\rangle$ eigenstate has definite angular momentum, J^2, or equivalently $|J|$, and definite z-component of angular momentum, J_z. Therefore the magnitude of its x-y components in the x-y plane is definite (Fig. 5.7):

$$J_x^2 + J_y^2 + J_z^2 = J^2 \Rightarrow |J_{xy}| = \sqrt{J_x^2 + J_y^2} = \sqrt{J^2 - J_z^2} = \hbar\sqrt{j(j+1) - m^2}.$$

The J_x and J_y component direction is uniformly distributed about the x-y plane. That is, we have no information about the *direction* of the x-y components in the x-y plane. Figure 5.7 labels the axis "xy," to indicate it is a superposition of all directions in the x-y plane. We cannot pin it down any more than that.

Figure 5.7, *right*, shows the three ways that real-space angular momentum vectors of $j_1 = 1$ and $j_2 = 2$ can add to produce an angular momentum vector corresponding to $|3\ 0\rangle$ (magnitude $\hbar\sqrt{12}$):

$|1\ 1\rangle|2 - 1\rangle$ these J vectors are almost in the plane of the paper, but slightly dimpled in (or out) of the paper, to make the vector sum magnitude exactly $\hbar\sqrt{12}$.

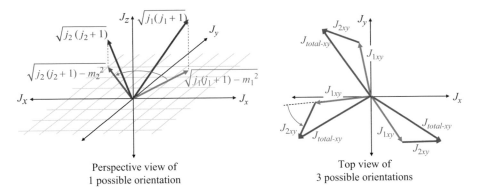

Fig. 5.8 (*Left*) Two angular momentum vectors, j_1 and j_2, with one possible orientation of their x-y components. (*Right*) Three possible orientations of the x-y components, with a fixed angle between J_{1xy} and J_{2xy}.

$|1\ 0\rangle|2-0\rangle$ these J vectors are entirely in the x-y plane. All the z-components are zero.

$|1\ 1\rangle|2-1\rangle$ these J vectors are almost in the plane of the paper, but slightly dimpled in (or out) of the paper, to make the vector sum magnitude exactly $\hbar\sqrt{12}$.

Shortly, we show that one specific superposition of $|1\ 1\rangle|2-1\rangle|1\ 0\rangle|2-0\rangle$, and $|1\ 1\rangle|2-1\rangle$ constructs the eigenstate $|3\ 0\rangle$. Much like Eq. (5.12), this superposition is also given by CGs.

To see how two indefinites can make a definite, consider two particles with angular momentum vectors $\mathbf{j}_1$ and $\mathbf{j}_2$ (quantum numbers j_1 and j_2), in the states $|j_1,m_1\rangle$ and $|j_2,m_2\rangle$. The magnitude in the x-y plane is definite. In units of $\hbar$,

$$j_1\left(j_1+1\right) = \mathbf{j}_{1xy}{}^2 + m_1{}^2 \Rightarrow \left|\mathbf{j}_{1xy}\right| = \sqrt{j_1\left(j_1+1\right)-m_1{}^2},$$

but the angle is indeterminate, i.e., the orientations in the x-y plane of the two angular momenta $\mathbf{j}_1$ and $\mathbf{j}_2$ are completely undetermined; they are uniformly distributed in all directions of the x-y plane. However, the $\mathbf{j}_1$ and $\mathbf{j}_2$ *may* sum to a definite magnitude of $\mathbf{j}_{total}$. If so, the angle between $\mathbf{j}_1$ and $\mathbf{j}_2$ must have a definite value; i.e., if $|\mathbf{j}_{total}|$ has a definite value, then $\mathbf{j}_1\cdot\mathbf{j}_2$ has a definite value (Fig. 5.8):

$$\mathbf{j}_{total}{}^2 = \left(\mathbf{j}_1+\mathbf{j}_2\right)^2 = \mathbf{j}_1{}^2 + \mathbf{j}_2{}^2 + 2\mathbf{j}_1\cdot\mathbf{j}_2 \Rightarrow \mathbf{j}_1\cdot\mathbf{j}_2 = \frac{1}{2}\left(\mathbf{j}_{total}{}^2 - \mathbf{j}_1{}^2 - \mathbf{j}_2{}^2\right).$$

Thus, the x-y orientations of $\mathbf{j}_1$ and $\mathbf{j}_2$ are individually completely uncertain, but the *relative* orientation is known.

The vector space of angular momentum states for the total (combined) system is different from the vector spaces of either system 1 or system 2. What, then, is this new, combined vector space? For that, we must understand tensor products.

5.7.2 Tensor Products

Oddly, you do not need to know anything about tensors to understand the tensor product of two vectors (kets). We will start by describing a tensor product informally, then get more precise. Each of the two vectors exists in a vector space. (Vector spaces are described in 2.4.6 p. 59.) We will assume the two original vector spaces are finite dimensional, though the results generalize straightforwardly to infinite dimensions. The two vector spaces are often structurally different, e.g., different dimensions.

We return to the earlier example of two subsystems of angular momentum, where $j_1 = 1/2$ and $j_2 = 1$. Considered separately, each subsystem has its own vector space of angular momentum:

System 1: dimension 2, $|j, m\rangle$ basis vectors: $\left|\frac{1}{2}, +\frac{1}{2}\right\rangle \left|\frac{1}{2}, -\frac{1}{2}\right\rangle$.

System 2: dimension 3, $|j, m\rangle$ basis vectors: $|1, +1\rangle \ |1, 0\rangle \ |1, -1\rangle$.

The tensor product *space* of these two vector spaces, $P = V_1 \otimes V_2$, has $2 \times 3 = 6$ dimensions, and its basis vectors are all six combinations of one basis vector from each of System 1 and System 2. In QM, we often write these pairs as just two kets, side by side:

$$\left|\frac{1}{2}, +\frac{1}{2}\right\rangle |1, +1\rangle \quad \left|\frac{1}{2}, +\frac{1}{2}\right\rangle |1, 0\rangle \quad \left|\frac{1}{2}, +\frac{1}{2}\right\rangle |1, -1\rangle$$

$$\left|\frac{1}{2}, -\frac{1}{2}\right\rangle |1, +1\rangle \quad \left|\frac{1}{2}, -\frac{1}{2}\right\rangle |1, 0\rangle \quad \left|\frac{1}{2}, -\frac{1}{2}\right\rangle |1, -1\rangle.$$

The scalar fields of V_1, V_2, and P are the complex numbers.

Generalizing: The **tensor product space** of two vector spaces, V_1 and V_2, is another vector space. The new vector space has dimension equal to the product of the dimensions of the two original vector spaces. The two original vector spaces must be over the same field of scalars, typically the complex numbers. The basis vectors of the new vector space are all the pairs of basis vectors you can form by taking one basis vector from V_1 and the other from V_2.

The **tensor product** of two *vectors* is the product of their scalar parts, and the juxtaposition of their vector parts. Sometimes, we write a tensor product explicitly with the $\otimes$ symbol. Some examples:

$$a|1/2, +1/2\rangle \otimes b|1, 0\rangle = ab|1/2, +1/2\rangle |1, 0\rangle$$

$$\left(a|1/2, +1/2\rangle + c|1/2, -1/2\rangle\right) \otimes b|1, 0\rangle$$
$$= ab|1/2, +1/2\rangle |1, 0\rangle + cb|1/2, -1/2\rangle |1, 0\rangle$$

$$\left(a|1/2, +1/2\rangle + c|1/2, -1/2\rangle\right) \otimes \left(b|1, 0\rangle + d|1, 1\rangle\right)$$
$$= ab|1/2, +1/2\rangle |1, 0\rangle + cb|1/2, -1/2\rangle |1, 0\rangle$$
$$+ ad|1/2, +1/2\rangle |1, 1\rangle + cd|1/2, -1/2\rangle |1, 1\rangle.$$

The tensor-product kets represent possible results of measuring the state of each constituent subsystem separately. Thus,

> The tensor product of two vectors gives a *single* complex-amplitude for each possible *set* of experimental results.

You can see that the scalar fields of the two vector spaces must be the same because we multiply the scalar coefficients together. However, the vector parts may be completely unrelated, coming from unrelated vector spaces. Unlike the scalar parts, they cannot be "multiplied" or otherwise combined in any way, so we simply write the basis vector parts of each factor next to each other: $|b_1\rangle|b_2\rangle$, or $|b_1, b_2\rangle$.

There are many notations for tensor products. Often, for example, the j_1 and j_2 are well known and understood. Then, it is cumbersome to write them in all the kets and bras (as you can see previously). So references frequently omit them, and keep only the m values. Thus, we define:

$$|1/2, +1/2\rangle \equiv |+1/2\rangle, \qquad |1/2, -1/2\rangle \equiv |-1/2\rangle$$
$$|1, 1\rangle \equiv |1\rangle, \qquad |1, 0\rangle \equiv |0\rangle, \qquad |1, -1\rangle \equiv |-1\rangle.$$

Then,

$$|+1/2\rangle \otimes |1\rangle = |+1/2\rangle|1\rangle \equiv |+1/2, 1\rangle, \quad \text{i.e.,} \quad |m_1\rangle \otimes |m_2\rangle = |m_1\rangle|m_2\rangle \equiv |m_1, m_2\rangle.$$

Space and spin: We have already seen another common example of a tensor product: the combination of the space and spin parts of a particle. For example, an electron has a spatial wave-function, ψ, and a spinor, χ. Their tensor product space spans the complete quantum state space which describes all properties of the electron. If the space part is independent of spin, then we can write the complete description as the tensor product of a spatial state and a spin state ($\psi(x)$ and $\chi = [a \ b]^T$). Then,

$$|electron\rangle = \psi(x) \otimes \chi = \psi(x) \otimes (a|z+\rangle + b|z-\rangle) \equiv \psi(x)\begin{bmatrix} a \\ b \end{bmatrix}.$$

If the space part depends on spin, as in a Stern–Gerlach experiment where the spin determines the path in space, then there are two wave-functions: one for spin up (ψ_+) and another for spin down (ψ_-). The resulting state *cannot* be written as a tensor product of two vectors. It can only be written as a sum of such tensor products:

$$|electron\rangle = \psi_+(x) \otimes |z+\rangle + \psi_-(x) \otimes |z-\rangle = \begin{bmatrix} \psi_+(x) \\ \psi_-(x) \end{bmatrix}.$$

5.7.3 Operators on Tensor Products

QM vector spaces also have linear operators. What are the linear operators of $P = V_1 \otimes V_2$? Recall that V_1 comes with a set of linear operators. V_2 comes with its own set, which are different from V_1, even though we may write them the same when discussing V_1 and V_2 separately. For example, when discussing V_1 alone, we will define operators $\hat{j}^2$ and $\hat{j}_z$. Similarly, when discussing V_2 alone, we will define operators $\hat{j}^2$ and $\hat{j}_z$. However, these two sets of operators are completely distinct. They operate on different kinds of vectors from different vector spaces of different dimension, so they cannot possibly be the same.

When discussing the combination of system 1 and system 2, we must write the operators distinctly, since they are distinct operators, so we write:

$\hat{j}_1^2$ and $\hat{j}_{1z}$ for the operators in V_1, and $\hat{j}_2^2$ and $\hat{j}_{2z}$ for the operators in V_2.

Note that these operators would be distinct even if system 1 and system 2 were both spin-1/2 particles, because $\hat{j}_1$ acts on particle 1 and $\hat{j}_2$ acts on particle 2, and we have to maintain this distinction.

So, what are the linear operators of P? All the operators of V_1, along with all the operators of V_2 and all the linear combinations of such operators. In our notation, the V_1 operators act on the vector factor from the V_1 vector space, and ignore the part from V_2. The V_2 operators act on the vector factor from V_2, and ignore the V_1 part. Some examples:

$$\hat{j}_{1z} \left| \frac{1}{2}, -\frac{1}{2} \right\rangle |1, 0\rangle = -\frac{1}{2}\hbar \left| \frac{1}{2}, -\frac{1}{2} \right\rangle |1, 0\rangle.$$

$$\hat{j}_{2z} \left| \frac{1}{2}, +\frac{1}{2} \right\rangle |1, 0\rangle = \left| \frac{1}{2}, +\frac{1}{2} \right\rangle \hat{j}_{2z} |1, 0\rangle = \mathbf{0}_v.$$

$$\left(\hat{j}_{1z} + \hat{j}_{2z} \right) \left| \frac{1}{2}, +\frac{1}{2} \right\rangle |1, 1\rangle = \left(\hat{j}_{1z} \left| \frac{1}{2}, +\frac{1}{2} \right\rangle \right) |1, 1\rangle + \left| \frac{1}{2}, +\frac{1}{2} \right\rangle \hat{j}_{2z} |1, 1\rangle$$

$$= \frac{3}{2}\hbar \left| \frac{1}{2}, +\frac{1}{2} \right\rangle |1, 1\rangle.$$

$$\hat{j}_1^2 \left| \frac{1}{2}, +\frac{1}{2} \right\rangle |1, 0\rangle = \frac{\sqrt{3}}{2}\hbar^2 \left| \frac{1}{2}, +\frac{1}{2} \right\rangle |1, 0\rangle.$$

Notice that the V_1 operators "ignore" the V_2 ket in the same way they "ignore" a scalar factor which multiplies a vector. This is one reason why the combination of V_1 and V_2 is called a tensor "product"; it is almost as if the V_2 vector part of P is a "multiplicative constant" from the viewpoint of V_1 operators.

Some notations would write operators on the V_1 part of the product space P as (say) $\hat{j}_1 \otimes I_2$, where I_2 is the identity operator on the V_2 kets of vectors in P. Similarly, they would write $\hat{j}_2$ as $I_1 \otimes \hat{j}_2$. For brevity, we do not use this notation.

5.7.4 Inner Products of Tensor Products

Of course, we can form inner products of vectors, and tensor products are vectors, so we must be able to form inner products of tensor products. The inner product of two tensor product vectors is simply the product of any scalars, times the inner products of all the bra–ket pairs of the constituent vectors:

$$\text{Define}\quad |v\rangle \otimes |w\rangle \equiv |v,w\rangle. \quad \text{Then}\quad \langle v,w|y,z\rangle = \langle v|y\rangle\langle w|z\rangle. \quad (5.13)$$

One way to think of this is that the inner product $|a|b\rangle$ tells what fraction of $|a\rangle$ is in $|b\rangle$, or how much "overlap" there is between $|a\rangle$ and $|b\rangle$. In the definition previously, if the fraction of overlap of $|v\rangle$ in $|y\rangle$ is 1/2, and the overlap of $|w\rangle$ in $|z\rangle$ is 1/3, then the overlap of the combination is just 1/6:

$$\text{If}\quad \langle v|y\rangle = \frac{1}{2}, \quad \text{and}\quad \langle w|z\rangle = \frac{1}{3}, \quad \text{then}\quad \langle v,w|y,z\rangle = \langle v|y\rangle\langle w|z\rangle = \frac{1}{6}.$$

This definition insures that the inner product is still a conjugate bilinear operation: it is linear in the ket and antilinear in the bra. (Recall this means it scales with any multiple of the ket, and with the complex conjugate of a multiple of the bra.)

Note that if *either* of the constituent vector pairs are orthogonal (i.e., either $\langle v|y\rangle = 0$ or $\langle w|z\rangle = 0$), then the full inner product is zero. Hence, two tensor product vectors are orthogonal if *any* pair of constituent vectors is orthogonal.

5.7.5 Clebsch-Gordon coefficients

If there is one subject that sends QM students screaming from the room, it seems to be Clebsch-Gordon coefficients (CGs) (second only, perhaps, to the related Wigner–Eckart theorem). CGs (as they are affectionately known) can be confusing, but in the end, it is all quite logical, and they can be understood by anyone who understands angular momentum addition (Sect. 5.7). Beyond angular momentum, CGs also apply to spherical tensor operators on angular momentum eigenstates, leading to the "one of the most important theorems in QM, the Wigner–Eckart theorem" [16, p. 239t].

CGs concern the combination of two subsystems (typically particles) of angular momentum, system 1 and system 2. We have seen that such a combination has a total angular momentum which is also quantized. However, eigenstates of the individual subsystem angular momenta are *not*, in general, eigenstates of the *total* angular momentum. Both the tensor products of individual eigenstates, and the total angular momentum bases, each compose a basis for the vector space of angular momentum of the combined system. The basis of eigenstate of the individual angular momenta is called the **uncoupled basis** because the basis states are those where the two angular momenta are separate and do not interact. In contrast, the **coupled basis** is the basis of quantized *total* angular momentum. CGs relate the basis vectors of the two bases. We now adopt some standard notation:

$$|j_1, m_1\rangle \equiv \text{eigenstates of system 1,} \quad |j_2, m_2\rangle \equiv \text{eigenstates of system 2,}$$
$$|J, M\rangle \equiv \text{eigenstates of } \textit{total} \text{ angular momentum .}$$

Note that "J" has a different meaning here than in previous sections; it is now a quantum number. Recall that $|J, M\rangle$ is an eigenstate of two operators: $\hat{J}^2$ (the square of total angular momentum) and $\hat{J}_z$ (z component of total angular momentum). In addition, if $|J, M\rangle$ is a result of combining two angular momenta of j_1 and j_2, then $|J, M\rangle$ is *also* an eigenstate of $\hat{j}_1^2$ and $\hat{j}_2^2$. That is, the j_1 and j_2 are fixed, and do not change, whether combined or not. In the uncoupled basis, $|j_1, m_1; j_2, m_2\rangle$ is defined as an eigenstate of four different operators: $\hat{j}_1^2, \hat{j}_{1z}, \hat{j}_2^2$, and $\hat{j}_{2z}$. This means it must also be an eigenstate of a fifth operator, the afore-mentioned total z-component, $\hat{J}_z$. Recall that $|j_1, m_1; j_2, m_2\rangle$ is a tensor product (combination of states of two subsystems in the system):

$$|j_1\ m_1; j_2\ m_2\rangle \equiv |j_1\ m_1\rangle |j_2\ m_2\rangle \equiv |j_1\ m_1\rangle \otimes |j_2\ m_2\rangle.$$

We saw previously that an uncoupled angular momentum eigenstate of two subsystems, $|j_1, m_1; j_2, m_2\rangle$, is an eigenstate of the *individual* $\hat{j}_1^2, \hat{j}_{1z}, \hat{j}_2^2$, and $\hat{j}_{2z}$ angular momentum operators, but it is generally a *superposition* of *total* angular momentum states, of the form $|J, M\rangle$. Similarly, a *total* angular momentum eigenstate of a combination of two systems, $|J, M\rangle$, is an eigenstate of $\hat{J}^2$ and $\hat{J}_z$, but it is generally a superposition of uncoupled eigenstates, of the form $|j_1, m_1; j_2, m_2\rangle$. CGs provide the coefficients of the superpositions in these relations.

> The key to Clebsch-Gordon coefficients is this: In all cases, j_1 and j_2 are known and fixed.

This means the magnitudes of angular momenta of the two subsystems are known and fixed. Then you can do two things with CGs:

1. You can write a state $|j_1, m_1; j_2, m_2\rangle$ as a linear combination (superposition) of several $|J, M\rangle$ states, where J varies, but (as always) j_1 and j_2 are known and fixed.
2. You can write a state $|J, M\rangle$ as a linear combination of several $|j_1, m_1; j_2, m_2\rangle$ states, where m_1 and m_2 vary, but (as always) j_1 and j_2 are known and fixed.

Let us consider each case in turn. We do not describe how to compute CGs, even though that is fun and interesting; for the most part, you look them up in a table.

$|j_1\,m_1;\,j_2\,m_2\rangle$ as a **Linear Combination of** $|J,\,M\rangle$

In this case for CGs, we are given the angular momenta and z-component quantum numbers for two subsystems of a system: $|j_1, m_1;\, j_2, m_2\rangle$. We want to write it as a sum of $|J, M\rangle$ states, that is, states of definite *total* angular momentum, and *total* z-component of angular momentum because the coefficients of such a superposition give us the probabilities of measuring all the possible *total* angular momenta. For example, suppose we want to write $|1, 0; 2, 1\rangle$ as a sum of $|J, M\rangle$ states:

$$|1,0;2,1\rangle = a|1,1\rangle + b|2,1\rangle + c|3,1\rangle. \qquad (5.14)$$

This is a straightforward decomposition of a ket into a basis (see Covering Your Bases: Decomposition, p. 54). a, b, and c are (potentially) complex coefficients, and are examples of CGs. The CGs tell you "how much" of each basis vector to include in the superposition to construct the given $|j_1, m_1;\, j_2, m_2\rangle$ vector. (It turns out that CGs are always real.) Note that the three basis kets on the right are, in a sense, not a "complete set," because you cannot create *all possible* $j_1 = 1, j_2 = 2$ states from them. But they are enough to create $|1, 0; 2, 1\rangle$.

How do we know there are only three basis vectors needed in this case? The j_1, m_1, j_2, and m_2 are all given constants. The m-values are proportional to the z-component of angular momentum, which have definite values. Therefore, the total angular momentum has a definite M value: the m-values of j_1 and j_2 add:

$$M = m_1 + m_2 = 0 + 1 = 1.$$

Hence, the only variable is J. But gosh, for any given M, there are an infinite number of J's that could have that M value (any $J \geq M$). How do we perform a finite decomposition? From simple addition of angular momentum, we know the possible J values are limited by j_1 and j_2. Assume that $j_1 < j_2$, then

$$j_2 - j_1 \leq J \leq j_2 + j_1 \qquad \Rightarrow \qquad 1 \leq J \leq 3,$$

so J has at most three possible values. In this example, the CG table tells us:

$$a = -\sqrt{\frac{3}{10}}, \quad b = \sqrt{\frac{1}{6}}, \quad c = \sqrt{\frac{8}{15}} \Rightarrow$$

$$|1,0;2,1\rangle = -\sqrt{\frac{3}{10}}|1,1\rangle + \sqrt{\frac{1}{6}}|2,1\rangle + \sqrt{\frac{8}{15}}|3,1\rangle.$$

More notation: There are many notations for CGs, which seems odd, because the existing Dirac notation of inner products works fine. Why invent anything new? Since the ket equation we started with, Eq. (5.14), is a straightforward decomposition, the coefficients are given by inner products. So we rewrite the ket equation as a direct decomposition:

$$\left|1,0;2,1\right\rangle = \underbrace{\left\langle 1,1\left|1,0;2,1\right\rangle\right.\left|1,1\right\rangle}_{a} + \underbrace{\left\langle 2,1\left|1,0;2,1\right\rangle\right.\left|2,1\right\rangle}_{b} + \underbrace{\left\langle 3,1\left|1,0;2,1\right\rangle\right.\left|3,1\right\rangle}_{c}.$$

Recall that in general, the inner product $\langle \phi \mid \psi \rangle$ is a complex number telling "how much" $|\phi\rangle$ is contained in $|\psi\rangle$ (and with what complex phase). Then in general:

$$\left|j_1\, m_1;j_2\, m_2\right\rangle = \sum_{J=\left|j_2-j_1\right|}^{j_2+j_1} \left\langle J,m_1+m_2\left|j_1\, m_1;j_2\, m_2\right\rangle\right.\left|J,m_1+m_2\right\rangle$$

where $\left\langle J,m_1+m_2\left|j_1\, m_1;j_2\, m_2\right\rangle\right.$ are Clebsch-Gordon coefficients .

Though this notation is somewhat clumsy, it is nothing more than ordinary Dirac notation and simple decomposition. Many references "simplify" the notation by omitting the j_1 and j_2. However, even though they may not write it down, j_1, m_1, j_2, and m_2 are always given and fixed within the sum; only J varies. Also, many references use a double sum (or worse). This is unnecessary. Given m_1 and m_2, M is fixed, so we need only a sum over J.

$|J, M\rangle$ as a Linear Combination of $|j_1\, m_1; j_2\, m_2\rangle$

In this case, you are given the state $|J, M\rangle$, which is to say you are given that the quantum numbers of total $\hat{J}^2$ and $\hat{J}_z$ are J and M, respectively. You want to express this state $|J, M\rangle$ in the "uncoupled" basis, that is, in the $|j_1\, m_1; j_2\, m_2\rangle$ basis, where j_1 and j_2 are given and fixed. For example, suppose $j_1=1$ and $j_2=2$, and we need to decompose $|2, 1\rangle$ into the $|1\, m_1; 2\, m_2\rangle$ basis. That is:

$$\left|J,M\right\rangle = \left|2,1\right\rangle = a\left|1,1;\,2,0\right\rangle + b\left|1,0;\,2,1\right\rangle + c\left|1,-1;\,2,2\right\rangle. \qquad (5.15)$$

Again, this is a straightforward decomposition of a ket into a basis. a, b, and c are (potentially) complex coefficients and are examples of CGs. The CGs tell you "how much" of each basis vector to include in the superposition to construct the given $|J, M\rangle$ vector. (It turns out that CGs are always real.) Note that the three basis kets on the right are, in a sense, not a "complete set," because you cannot create *all possible* $J=2$ states from them. But they are enough to create $|2, 1\rangle$.

How do we know there are only three basis vectors needed in this case? The m-values are proportional to the z-components of angular momentum, which have definite values. Therefore, the m-values of j_1 and j_2 add:

$$M = 1 = m_1 + m_2.$$

Therefore, we choose one of the j s (the smaller one is easier), call it j_1, and consider all its possible m-values. Of course, as always, $-j_1 \leq m_1 \leq +j_1$. Then for each m_1:

$$m_2 = M - m_1.$$

For each possible value of m_1, we see if the required m_2 is allowed by j_2. If it is, we must include that basis vector in the sum. In this example, since there are three values of m_1 ($-1, 0, +1$), and each has a valid m_2, where $-j_2 \leq m_2 \leq +j_2$, there are three basis vectors in the sum. In this example, the CG table tells us:

$$|2,1\rangle = \underbrace{\langle 1,1; 2,0|2,1\rangle}_{a}|1,1; 2,0\rangle + \underbrace{\langle 1,0; 2,1|2,1\rangle}_{b}|1,0; 2,1\rangle$$

$$+ \underbrace{\langle 1,-1; 2,2|2,1\rangle}_{c}|1,-1; 2,2\rangle$$

$$= \frac{-1}{\sqrt{2}}|1,1;2,0\rangle + \frac{1}{\sqrt{6}}|1,0;2,1\rangle + \frac{1}{\sqrt{3}}|1,-1;2,2\rangle.$$

In general,

$$|J,M\rangle = \sum_{m_1=-j_1}^{+j_1} \langle j_1,m_1; j_2, M-m_1|J,M\rangle |j_1,m_1; j_2, M-m_1\rangle$$

where $\langle j_1,m_1; j_2, M-m_1|J,M\rangle$ are the Clebsch-Gordon coefficients.

Note that J, M, j_1, and j_2 are always known and fixed within the sum. Many references use a double sum (or worse). This is unnecessary. Given m_1 and M, m_2 is determined, so only a sum over m_1 is needed. Note that any CG coefficient which has $M-m_1 > j_2$ is necessarily zero.

Now we have seen two forms of CGs:

$$\langle J,M|j_1\ m_1; j_2\ m_2\rangle, \quad and \quad \langle j_1\ m_1; j_2\ m_2|J,M\rangle.$$

But as with any inner product, $\langle a|b\rangle = \langle b|a\rangle^*$, and it happens that CGs are always real. Hence:

$$\langle J,M|j_1\ m_1; j_2\ m_2\rangle = \langle j_1\ m_1; j_2\ m_2|J,M\rangle.$$

Therefore, tables of CGs are usually arranged in some symmetric way, so you can look them up in either direction (coupled $\rightarrow$ uncoupled or uncoupled $\rightarrow$ coupled).

There is a general formula for an arbitrary CG due to Wigner. It is so complicated that it is probably only useful for computer programs. We give it in the list of formulas at the back of the book.

5.8   Just a Moment: The Landé g-Factor

The Landé g-factor is essential for understanding atomic spectra, and clarifies LSJ angular momentum coupling, which is essential for atomic electron structure. It is well known that both orbital and spin angular momentum of a charged particle contribute magnetic dipole moments to a system. (The spin-related magnetic moment is sometimes called the "intrinsic" magnetic moment.) However, there is a *quantitative* difference between orbit and spin in the ratio of magnetic dipole moment to angular momentum (sometimes called the **g-factor**). For a $|l\; m_l\rangle$ orbital state, and a $|s\; m_s\rangle$ spin state, experiment shows the magnetic moment is (in gaussian units):

$$\mu_{z,L} = g_L m_l \mu, \qquad \mu_{z,s} = g_s m_s \mu,$$

$$where \quad \mu \equiv \frac{|e|\hbar}{2(mass)c} \equiv magneton$$

$$e \equiv \text{particle charge}$$

$$g_L \equiv \text{orbital } g\text{-factor } (= -1 \text{ for electron})$$

$$g_s \equiv \text{spin } g\text{-factor } (\approx -2 \text{ for electron}).$$

These results are well understood theoretically, as well [17, p. 79]. The **magneton** is a function of the particle's mass and charge, and is therefore different for different particles. We follow the US National Institute of Standards and Technology terminology and sign conventions, which defines the **Bohr magneton** as the magneton for an electron and the **nuclear magneton** as the magneton for a proton. Both magnetons are defined positive. The magnetic moment vector for an electron is *opposite* its spin and orbital angular momentum because it is negatively charged, so its g-factors are negative. The intrinsic magnetic moment for the neutron is also opposite its spin (for no good reason), so its g-factor is also negative. Furthermore, the neutron g-factor is calculated using the proton mass and charge, i.e., the aforementioned nuclear magneton.

(Note that some references call the nuclear magneton a "Bohr" magneton. Some references define all g-factors as positive and put minus signs in "by hand," relying on the "understanding" of the correct sign, but that makes the neutron g-factor confusing.)

Because the particle mass appears in the denominator of the magneton formula, the nuclear magneton is about three orders of magnitude smaller than the Bohr (electron) magneton. This is why we usually ignore the magnetic moment of atomic nuclei: the mass is so large that μ is negligible.

Since the g-factor for spin is different than that for orbit, we must ask: what if our angular momentum eigenstate is a combination of spin and orbit? There are two important cases to consider: L-s eigenstates and LSJ eigenstates.

L-s eigenstates: Consider two angular momenta which add, say **L** and **s**. In the *uncoupled* basis, the system is in the state $|l, m_l; s, m_s\rangle$, which is an eigenstate of $\hat{L}^2, \hat{L}_z, \hat{s}^2, \hat{s}_z$, and $\hat{J}_z$, but not $\hat{J}^2$ (as we learned from our study of CGs: $j_1 \to l, j_2 \to s$).

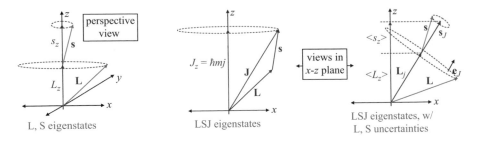

Fig. 5.9 (*Left*) L-s angular momenta eigenstates, with definite z-components. (*Middle*) LSJ angular momenta eigenstates, with definite components parallel to J. (*Right*) Uncertainties and averages for middle diagram

(Note that $m_j = m_l + m_s$, so is not an independent quantum number.) Figure 5.9 (*left*) shows these angular momenta and their uncertainties. The magnitudes of the L and s momenta are definite (no uncertainty), as are the z-components. The x-y components are (1) uncertain, (2) uniformly distributed in the x-y plane, and (3) average to zero.

Since the x-y components of angular momentum average to zero, they contribute nothing to the net magnetic moment. Therefore, the total magnetic moment is simply the sum of the orbital and spin z-components. For an electron:

$$\mu_z = \left(g_L m_L + g_s m_s\right)\mu_B \quad where \quad g_L = -1, \; g_s \approx -2$$

$$m_L = \text{orbital z-component quantum number}$$

$$m_s = \text{spin z-component quantum number.}$$

Note that even if $J_z = L_z + s_z = 0$, the total magnetic moment is *not* zero, because the spin g-factor is larger in magnitude than the orbital g-factor ($|g_s| > |g_L|$).

LSJ eigenstates: In the *LSJ* eigenstates (Fig. 5.9, *middle*), the system is in the state $|J\,M\rangle$, which is an eigenstate of $\hat{L}^2, \hat{L}_z, \hat{s}^2, \hat{s}_z$, and $\hat{J}_z$, but not $\hat{L}_z$ or $\hat{s}_z$ (as we learned from our study of CGs: $j_1 \rightarrow l, j_2 \rightarrow s$). For example, small atoms are in essentially this type of *coupled* state.

The key to this analysis is recognizing that while L_z and s_z do *not* have definite values (**L** and **s** states are sums of CG-weighted z-eigenstates), L_J and s_J (the components of **L** and **s** parallel to **J**) have *definite* values, just as the z-components are definite in Fig. 5.9, *left*. This is possible even though **J** is uncertain because the joint probability density of **J**, **L**, and **s** connects the three vectors in just the right way to satisfy this requirement.

> The key to LSJ analysis is recognizing while L_z and s_z do not have definite values, L_J and s_J (the components of L and s parallel to J) have definite values.

This is equivalent to saying that **L·J** and **s·J** have definite values, which we can show by writing them explicitly in terms of the defining quantum numbers l, s, and J. First, recall that,

$$\mathbf{J} = \mathbf{L}+\mathbf{s}, \quad |\mathbf{J}|^2 = (\mathbf{L}+\mathbf{s})^2 = |\mathbf{L}|^2 + 2\mathbf{L}\cdot\mathbf{s} + |\mathbf{s}|^2 \quad \Rightarrow \quad \mathbf{L}\cdot\mathbf{s} = \frac{1}{2}\left(|\mathbf{J}|^2 - |\mathbf{L}|^2 - |\mathbf{s}|^2\right).$$

Then,

$$\mathbf{L}\cdot\mathbf{J} = \mathbf{L}\cdot(\mathbf{L}+\mathbf{s}) = |\mathbf{L}|^2 + \mathbf{L}\cdot\mathbf{s} = |\mathbf{L}|^2 + \frac{1}{2}\left(|\mathbf{J}| - |\mathbf{L}|^2 - |\mathbf{s}|^2\right)$$

$$= \hbar^2\left[l(l+1) + \frac{1}{2}\left(J(J+1) - l(l+1) - s(s+1)\right)\right] \qquad (5.16)$$

$$= \frac{\hbar^2}{2}\left[J(J+1) + l(l+1) - s(s+1)\right].$$

Similarly:

$$\mathbf{s}\cdot\mathbf{J} = \frac{\hbar^2}{2}\left[J(J+1) - l(l+1) + s(s+1)\right]. \qquad (5.17)$$

Therefore, the **L** and **s** uncertainties are in the plane perpendicular to **J** (Fig. 5.9, *right*).

Is there an "effective" g_J for the system, applicable to the quantum number M? That is, can we express the magnetic moment of the system as:

$$\mu_z = g_J M \mu_B ?$$

Sure we can; we just need to find the *average* L_z and *average* s_z, and weight them with their associated g-factors. This produces an *effective* g-factor, different from both g_L and g_s. For a single electron, coupling its spin and orbit, we expect the "effective" g-factor is between g_L and g_s, and indeed it is. We now calculate this **Landé g-factor**. For an electron:

$$g_J M = g_L\langle m_L\rangle + g_e\langle m_s\rangle = g_L\frac{\langle L_z\rangle}{\hbar} + g_e\frac{\langle s_z\rangle}{\hbar}. \qquad (5.18)$$

Notice that $M = \langle m_L\rangle + \langle m_s\rangle$, where M has a definite value, but m_L and m_s do not.

The average of **L** is (averaging around the dotted circle in Fig. 5.9, *right*):

$$\langle \mathbf{L} \rangle = \mathbf{L}_J = \mathbf{L} \cdot \mathbf{e}_J = \frac{\mathbf{L} \cdot \mathbf{J}}{|\mathbf{J}|} \quad where \quad \mathbf{e}_J = \frac{\mathbf{J}}{|\mathbf{J}|} \equiv \text{unit vector in the } \mathbf{J} \text{ direction,}$$

$$and \quad \langle \mathbf{s} \rangle = \mathbf{s}_J = \frac{\mathbf{s} \cdot \mathbf{J}}{|\mathbf{J}|} .$$

Then $\langle L_z \rangle$ is just the z-component of $\langle L \rangle$, and $\langle S_z \rangle$ is the z-component of $\langle s \rangle$. Since $\langle L \rangle$ and $\langle s \rangle$ are parallel to $\mathbf{J}$, they project onto the z-axis the same way $\mathbf{J}$ does (similar triangles):

$$\langle L_z \rangle = \langle \mathbf{L} \rangle \frac{J_z}{|\mathbf{J}|} = \frac{\mathbf{L} \cdot \mathbf{J}}{|\mathbf{J}|} \frac{J_z}{|\mathbf{J}|} = \frac{\mathbf{L} \cdot \mathbf{J}}{|\mathbf{J}|^2} J_z, \quad and$$

$$\langle s_z \rangle = \frac{\mathbf{s} \cdot \mathbf{J}}{|\mathbf{J}|^2} J_z .$$

$$(5.19)$$

Note that we had to average to get these formulas, so they are *not* true with operators on the left-hand side, i.e., $\hat{s}_z \neq (\hat{\mathbf{S}} \cdot \hat{\mathbf{J}})(\hat{J}_z/\hat{J}^2)$.

We combine these results to find the **Landé g-factor**, the effective g-factor for a spin-orbit coupled angular momentum in the LSJ basis. The algebra is slightly tedious, but trivial. Plugging Eq. (5.19) into Eq. (5.18):

$$g_J M = g_L \frac{\langle L_z \rangle}{\hbar} + g_e \frac{\langle s_z \rangle}{\hbar} = \left(g_L \frac{\mathbf{L} \cdot \mathbf{J}}{|\mathbf{J}|^2} + g_e \frac{\mathbf{s} \cdot \mathbf{J}}{|\mathbf{J}|^2} \right) \frac{J_z}{\hbar} .$$

Use $M = J_z/\hbar$, divide M from both sides, and factor out $|\mathbf{J}|^2$:

$$g_J = \left(g_L \mathbf{L} \cdot \mathbf{J} + g_e \mathbf{S} \cdot \mathbf{J} \right) \frac{1}{|\mathbf{J}|^2} = \left(g_L \mathbf{L} \cdot \mathbf{J} + g_e \mathbf{S} \cdot \mathbf{J} \right) \frac{1}{\hbar^2 J(J+1)} . \qquad (5.20)$$

Finally, plugging Eqs. (5.16) and (5.17) into Eq. (5.20):

$$g_J = \frac{g_L \left[J(J+1) + l(l+1) - s(s+1) \right] + g_e \left[J(J+1) - l(l+1) + s(s+1) \right]}{2J(J+1)} . \qquad (5.21)$$

This is the Landé g-factor that applies to all M for given l, s, and J. Note that we did not have to use CGs to find it (though we could have).

For an electron, or other fundamental fermion (dropping the minus signs), $|g_L|=1$, and $|g_e| \approx 2$. Then

$$|g_J| \approx \frac{\left[J(J+1)+l(l+1)-s(s+1)\right]+2\left[J(J+1)-l(l+1)+s(s+1)\right]}{2J(J+1)}$$

$$= \frac{3J(J+1)-l(l+1)+s(s+1)}{2J(J+1)}$$

$$= \frac{3}{2}+\frac{s(s+1)-l(l+1)}{2J(J+1)}.$$

For a single electron, $j=1/2$ or $3/2$, and $|g_j|$ is between 1 and 2. As further checks, when $s=0$, $l=J$, the spin fraction is 0, and $g_J=1$. When $l=0$, $s=J$, the spin fraction is 1, and $g_J \approx 2$. The Landé g-factor is measurable in the energy splitting of a large atom in a magnetic field, called the "anomalous Zeeman effect."

Note that in Fig. 5.9 (*middle*), **L** and **s** need not be in the plane of the paper, i.e., the plane defined by **L**, **s**, and **J** need not include the z-axis. In fact, the uncertainties in **L** and **s** require that the head of **L** (and therefore the tail of **s**) are uniformly distributed in a circle around the **J** vector.

When considering the effect of many electrons in an atom with LSJ coupling, we can simply replace **s** by **S**, the total spin of all the electrons. Similarly, **L** becomes the total orbital angular momentum for all electrons. Then $|g_J|$ can be greater than 2. For example, for four electrons in a state with $S=2$ and $L=1$ and $J=1$, $|g_J|=5/2$. Also, if $l=s$ and $J=0$, then there is a dipole moment even thought $m_J=0$ (almost as if g_J is infinite).

The Landé g-factor generalizes to any two angular momenta, by simply inserting the proper g-factors into Eq. (5.21). Nothing in the derivation was particular to spin or orbit.

Chapter 6
Multiparticle Quantum Mechanics

6.1 Introduction

Quantum mechanics(QM) covers not just single particles, but systems of particles. Systems of particles exhibit further unexpected behaviors which explain atoms and molecules, enable new technologies (e.g., quantum cryptography), and require dedicated study. This chapter requires understanding single particle QM of waves and spins. We proceed along these topics:

1. Multiparticle wave-functions: distinct particles.
2. Multiparticle wave-functions: identical particles.
3. Multiparticle Schrödinger equation.
4. Multiparticle states with spin.
5. Symmetry of multiparticle wave-functions: fermions and bosons.
6. Atoms: Hund's rules.
7. Multiparticle entanglement

Many new phenomena arise in multiparticle systems. We will examine some of these, such as entanglement creating nonlocal behavior.

6.1.1 Multiparticle Wave-Functions: Distinct Particles

Suppose we have a system of one proton and one electron. It is described by a multiparticle wave-function, $\psi(\mathbf{r}_p, \mathbf{r}_e)$, which defines the joint probability density of finding a proton at $\mathbf{r}_p$, *and* an electron at $\mathbf{r}_e$:

$$\Pr(\text{proton around } \mathbf{r}_p \text{ and electron around } \mathbf{r}_e) = \left|\psi(\mathbf{r}_p, \mathbf{r}_e)\right|^2 d^3 r_p \, d^3 r_e.$$

Because the proton and electron may interact, their positions depend on each other. Therefore, ψ is a single wave-function which takes account of their interaction, and covers all possible combinations of positions of both the proton and the electron. If

E. L. Michelsen, *Quirky Quantum Concepts*, Undergraduate Lecture Notes in Physics, DOI 10.1007/978-1-4614-9305-1_6, © Springer Science+Business Media New York 2014

we take a measurement, we must find the two particles *somewhere*, so our normalization is:

$$\iint_\infty \left| \psi(\mathbf{r}_p, \mathbf{r}_e) \right|^2 d^3 r_p \, d^3 r_e = 1 \qquad \text{(a 6 dimensional integral)}$$

If we have two particles that do not interact, then their joint wave-function can be written simply as the product of the two individual wave-functions:

$$\psi(\mathbf{r}_1, \mathbf{r}_2) = \psi_1(\mathbf{r}_1) \psi_2(\mathbf{r}_2).$$

This is a kind of tensor product. The wave-functions are normalized separately according to:

$$\int_\infty \left| \psi_1(\mathbf{r}) \right|^2 d^3 r = 1, \qquad \text{and} \qquad \int_\infty \left| \psi_2(\mathbf{r}) \right|^2 d^3 r = 1,$$

so that:

$$\iint_\infty \left| \psi(\mathbf{r}_1, \mathbf{r}_2) \right|^2 d^3 r_1 \, d^3 r_2 = \int_\infty \left| \psi_1(\mathbf{r}) \right|^2 d^3 r \int_\infty \left| \psi_2(\mathbf{r}) \right|^2 d^3 r = 1.$$

In other words, when the two-particle wave-function separates, the individual normalizations insure the joint wave-function is normalized.

6.1.2   Multiparticle Wave-Functions: Identical Particles

We may have a system of two identical particles, such as two electrons. Like the proton–electron system, this system is also described by a two-particle wave-function, $\psi(\mathbf{r}_1, \mathbf{r}_2)$. This is the joint amplitude for finding *an* electron at $\mathbf{r}_1$ *and another* electron at $\mathbf{r}_2$. It is a crucial point that:

> Identical particles cannot be distinguished from each other. There is no concept of particle no. 1 and particle no. 2. There is only the concept of a particle at $\mathbf{r}_1$ and an identical one at $\mathbf{r}_2$.

Since the two particles are identical (indistinguishable), it must be the case that:

Pr(particle at $\mathbf{r}_1$ and another at $\mathbf{r}_2$)

$= $ Pr(particle at $\mathbf{r}_2$ and another at $\mathbf{r}_1$) $\Rightarrow$ $|\psi(\mathbf{r}_1,\mathbf{r}_2)| = |\psi(\mathbf{r}_2,\mathbf{r}_1)|$.

This is a symmetry requirement on the wave-function, and requires that the wave-function should differ by at most a phase under an $\mathbf{r}_1 \leftrightarrow \mathbf{r}_2$ interchange. Furthermore, if we swap particles 1 and 2 a second time, we must get the original two-particle wave-function back. We can write **particle exchange** as a linear operator, defined by:

$$\hat{P}_{12}\psi(\mathbf{r}_1,\mathbf{r}_2) \equiv \psi(\mathbf{r}_2,\mathbf{r}_1).$$

By definition, it follows that:

$$\left(\hat{P}_{12}\right)^2 \psi(\mathbf{r}_1,\mathbf{r}_2) = \psi(\mathbf{r}_1,\mathbf{r}_2).$$

This implies that any eigenvalue of $\hat{P}_{12}$, call it λ, must satisfy $\lambda^2 = 1$. Therefore, the only eigenvalues are $\lambda = \pm 1$. A function is **symmetric** if and only if $\psi(\mathbf{r}_1, \mathbf{r}_2) = \psi(\mathbf{r}_2, \mathbf{r}_1)$. It is an eigenstate of $\hat{P}_{12}$ with eigenvalue $+1$. A function is **antisymmetric** if and only if $\psi(\mathbf{r}_1, \mathbf{r}_2) = -\psi(\mathbf{r}_2, \mathbf{r}_1)$. It is an eigenstate of $\hat{P}_{12}$ with eigenvalue -1. Some functions are neither symmetric nor antisymmetric.

$\hat{P}_{12}$ also operates on spin states, in a similar way: it swaps the states for the particles:

$$\hat{P}_{12}\chi_{12} \equiv \chi_{21}, \quad \text{e.g.,} \quad \hat{P}_{12}|\uparrow\downarrow\rangle = |\downarrow\uparrow\rangle.$$

We define here the notation where χ_{21} is the spin state χ_{12} with the particle spins exchanged. For example,

$$\chi_{12} = |\uparrow_1 \downarrow_2\rangle \quad \Rightarrow \quad \chi_{21} = |\downarrow_1 \uparrow_2\rangle.$$

Putting space and spin together, we have:

$$\hat{P}_{12}\left[f(\mathbf{r}_1,\mathbf{r}_2)\chi_{12}\right] = f(\mathbf{r}_2,\mathbf{r}_1)\chi_{21}, \quad \text{e.g.,} \quad \hat{P}_{12}\psi(\mathbf{r}_1,\mathbf{r}_2)|\uparrow\downarrow\rangle = \psi(\mathbf{r}_2,\mathbf{r}_1)|\downarrow\uparrow\rangle.$$

Note that $\hat{P}_{12}$ operates on *both* the spatial state and the spin state.

6.1.3   Multiparticle Schrödinger Equation

We extend the Schrödinger equation to multiple-particles in a simple way: just add up the energies of all the particles:

two-particle
$$\frac{\hbar}{i}\frac{\partial}{\partial t}\psi(\mathbf{r}_1,\mathbf{r}_2,t)=\underbrace{\sum_{j=1}^{2}\frac{-\hbar^2}{2m_j}\nabla_j{}^2\psi(\mathbf{r}_1,\mathbf{r}_2,t)}_{\text{total kinetic energy}}+\underbrace{V(\mathbf{r}_1,\mathbf{r}_2)\psi(\mathbf{r}_1,\mathbf{r}_2,t)}_{\substack{\text{potential energy,}\\\text{including particle interactions}}},$$

where total energy is $\frac{\hbar}{i}\frac{\partial}{\partial t}\psi$.

n-particle
$$\underbrace{\frac{\hbar}{i}\frac{\partial}{\partial t}\psi(\mathbf{r}_1,\mathbf{r}_2,\dots\mathbf{r}_n,t)}_{\text{total energy}}=\underbrace{\sum_{j=1}^{n}\frac{-\hbar^2}{2m_j}\nabla_j{}^2\psi(\mathbf{r}_1,\mathbf{r}_2,\dots\mathbf{r}_n,t)}_{\text{total kinetic energy}}$$

$$+\underbrace{V(\mathbf{r}_1,\mathbf{r}_2,\dots\mathbf{r}_n)\psi(\mathbf{r}_1,\mathbf{r}_2,\dots\mathbf{r}_n,t)}_{\substack{\text{potential energy,}\\\text{including particle interactions}}}$$

where
$$\mathbf{r}_j\equiv(x_j,y_j,z_j),$$

$$\nabla_j{}^2\equiv\nabla_j\cdot\nabla_j=\left(\frac{\partial}{\partial x_j}\mathbf{e}_x+\frac{\partial}{\partial y_j}\mathbf{e}_y+\frac{\partial}{\partial z_j}\mathbf{e}_z\right)\cdot\left(\frac{\partial}{\partial x_j}\mathbf{e}_x+\frac{\partial}{\partial y_j}\mathbf{e}_y+\frac{\partial}{\partial z_j}\mathbf{e}_z\right)$$

$$=\frac{\partial^2}{\partial x_j{}^2}+\frac{\partial^2}{\partial y_j{}^2}+\frac{\partial^2}{\partial z_j{}^2}.$$

6.1.4   Multiparticle States with Spin

Recall that particles with spin have a state which comprises both a spatial wave-function *and* a spin state. Electrons are spin 1/2, and can therefore be either spin-up or spin-down (conventionally along the *z*-axis):

$$|z+\rangle=\begin{bmatrix}1\\0\end{bmatrix}\equiv|\uparrow\rangle \qquad\qquad |z-\rangle=\begin{bmatrix}0\\1\end{bmatrix}\equiv|\downarrow\rangle.$$

The total quantum state is a combination (tensor product) of wave-function and spin state, written as:

$$\psi(\mathbf{r})\chi, \qquad \text{where}\quad \chi \text{ is any superposition of } |\uparrow\rangle \text{ and } |\downarrow\rangle,$$

or a superposition of such states. For example, an excited hydrogen electron in a $2p$ state, with $n=2$, $l=1$, $m=1$, spin up, is in the state

$$\psi_{211}(\mathbf{r})|\uparrow\rangle=R_{21}(\mathbf{r})Y_{11}(\theta,\phi)|\uparrow\rangle.$$

Two-particle states with spin have two-particle wave-functions, and two-particle spin states, e.g.,

$$\psi(\mathbf{r}_1,\mathbf{r}_2)\chi_{12}, \qquad \text{where}\quad \chi_{12} \text{ is } |\uparrow\uparrow\rangle,|\uparrow\downarrow\rangle,|\downarrow\uparrow\rangle,|\downarrow\downarrow\rangle,$$

or a superposition of such states. It is easy to show that any quantum state can be written as a sum of a symmetric state and an antisymmetric state.

6.1.5   Symmetry of Multiparticle Wave-Functions: Fermions and Bosons

As mentioned previously, multiparticle wave-functions are subject to symmetry restrictions. Experimentally, there are two kinds of particles: fermions and bosons. Fermions are half-odd-integer spin and bosons are integer spin. It turns out that

> Multiparticle fermion states must be *anti*symmetric under interchange of any two identical fermions, and boson states must be *symmetric* under interchange of any two identical bosons.

[These symmetry requirements are proven by the spin-statistics theorem in quantum field theory (QFT).] Nonidentical fermions, such as an electron and a muon, are distinguishable and therefore have no symmetry requirement: the probability of finding the electron at $\mathbf{r}_1$ and the muon at $\mathbf{r}_2$ may be different than that of finding the muon at $\mathbf{r}_1$ and the electron at $\mathbf{r}_2$.

However, the symmetry requirement includes both the wave-function (i.e., spatial state) *and* the spin states: it is the *entire* state, wave-function with spin state, that must be antisymmetric for fermions. *If both particles have the same spin states* (say $|\uparrow\uparrow\rangle$), or if the spin states are otherwise symmetric (say $|\uparrow\downarrow\rangle + |\downarrow\uparrow\rangle$), the symmetry requirement falls only to the wave-function, as follows:

$$\left.\begin{array}{ll} \text{Fermions:} & \psi(\mathbf{r}_1,\mathbf{r}_2) = -\psi(\mathbf{r}_1,\mathbf{r}_2) \\ \text{Bosons:} & \psi(\mathbf{r}_1,\mathbf{r}_2) = \psi(\mathbf{r}_1,\mathbf{r}_2) \end{array}\right\} \text{(symmetric spin-states).}$$

Fermions: The antisymmetry of fermion states includes the **Pauli exclusion principle**. Two electrons cannot be in the same single-particle quantum state, because then (for symmetric spin):

$$\psi(\mathbf{r}_1,\mathbf{r}_2) = \psi(\mathbf{r}_1)\psi(\mathbf{r}_2) = \psi(\mathbf{r}_2,\mathbf{r}_1) \quad \text{in violation of} \quad \psi(\mathbf{r}_1,\mathbf{r}_2) = -\psi(\mathbf{r}_2,\mathbf{r}_1).$$

In fact, for $\mathbf{r}_1 = \mathbf{r}_2 = \mathbf{r}$ and symmetric spins, an antisymmetric state must have:

$$\psi(\mathbf{r},\mathbf{r}) = -\psi(\mathbf{r},\mathbf{r}) \quad \Rightarrow \quad \psi(\mathbf{r},\mathbf{r}) = 0, \quad \forall\, \mathbf{r}.$$

This means that there is zero probability of measuring two identical fermions at the same place. For n-particle systems, with fully symmetric spin states, there is zero probability of measuring *any* two identical fermions at the same place.

In contrast to the previous case of identical spins, if spin- and space states separate, this antisymmetry can be either:

$$\psi(\mathbf{r}_1,\mathbf{r}_2) = -\psi(\mathbf{r}_2,\mathbf{r}_1) \quad \text{and} \quad \chi_{12} = \chi_{21} \quad \text{or}$$

$$\psi(\mathbf{r}_1,\mathbf{r}_2) = \psi(\mathbf{r}_2,\mathbf{r}_1) \quad \text{and} \quad \chi_{12} = -\chi_{21}.$$

In words, to make the total fermion state antisymmetric (for separable spin/space states), we must have either the wave-function antisymmetric and the spin state symmetric, or the wave-function symmetric and the spin states antisymmetric.

But how can a spin state be antisymmetric under particle exchange, when it is not a function of position? There is no $\mathbf{r}_1$ and $\mathbf{r}_2$. A spin state can only be antisymmetric by being an antisymmetric superposition of two spin states e.g., (ignoring normalization),

$$\chi_{12} = \left|\uparrow\downarrow\right\rangle - \left|\downarrow\uparrow\right\rangle \qquad \Rightarrow \qquad \chi_{12} = -\chi_{21}, \text{ i.e. } \left|\uparrow\downarrow\right\rangle - \left|\downarrow\uparrow\right\rangle$$

$$= -\left(\left|\downarrow\uparrow\right\rangle - \left|\uparrow\downarrow\right\rangle\right) = \left|\uparrow\downarrow\right\rangle - \left|\downarrow\uparrow\right\rangle,$$

where on the right-hand side we have interchanged particles 1 and 2, which interchange their spins.

Symmetric spin states can be either z eigenstates, or symmetric superpositions, such as:

$$\chi_{12} = \left|\uparrow\uparrow\right\rangle, \qquad \chi_{12} = \left|\downarrow\downarrow\right\rangle, \qquad \chi_{12} = \left|\uparrow\uparrow\right\rangle - \left|\downarrow\downarrow\right\rangle, \qquad \chi_{12} = \left|\uparrow\downarrow\right\rangle + \left|\downarrow\uparrow\right\rangle.$$

> Note that *any* superposition of symmetric states is symmetric, even if some of the coefficients are negative or complex.

In the most general case, the spin states are entangled with the wave-function, and the antisymmetric requirement for fermions requires swapping both spatial positions ($\mathbf{r}_1$ and $\mathbf{r}_2$) and spin state labels. In other words, the total antisymmetric state may be any superposition of antisymmetric states.

The requirement for antisymmetric multiparticle fermion states can make calculating such states difficult. Sometimes, you may have a solution to the multiparticle Schrödinger equation, but which does not have the proper symmetry for the multiple particles. In such a case, you can "symmetrize" or "antisymmetrize" the solution to construct a valid solution with the proper symmetry. Using our particle exchange operator, and given a solution of arbitrary symmetry $f(\mathbf{r}_1, \mathbf{r}_2)\chi_{12}$, we can symmetrize and antisymmetrize that solution as follows:

$$\text{antisymmetric:} \quad \psi(\mathbf{r}_1,\mathbf{r}_2)\chi_{12} = f(\mathbf{r}_1,\mathbf{r}_2)\chi_{12} - \hat{P}_{12}\left[f(\mathbf{r}_1,\mathbf{r}_2)\chi_{12}\right]$$

$$= f(\mathbf{r}_1,\mathbf{r}_2)\chi_{12} - f(\mathbf{r}_2,\mathbf{r}_1)\chi_{21}$$

$$\text{symmetric:} \quad \psi(\mathbf{r}_1,\mathbf{r}_2)\chi_{12} = f(\mathbf{r}_1,\mathbf{r}_2)\chi_{12} + \hat{P}_{12}\left[f(\mathbf{r}_1,\mathbf{r}_2)\chi_{12}\right]$$

$$= f(\mathbf{r}_1,\mathbf{r}_2)\chi_{12} + f(\mathbf{r}_2,\mathbf{r}_1)\chi_{21}.$$

Note that $\hat{P}_{12}$ operates on *both* the spatial state and the spin state. For example, suppose $f(\mathbf{r}_1, \mathbf{r}_2)$ and χ_{12} are neither symmetric nor antisymmetric, e.g., $\chi_{12} = \left|\uparrow\uparrow\right\rangle + \left|\uparrow\downarrow\right\rangle$. Then antisymmetrizing the wave-function gives:

$$\psi(\mathbf{r}_1,\mathbf{r}_2)\chi_{12} = f(\mathbf{r}_1,\mathbf{r}_2)\big|\uparrow\uparrow\big\rangle + f(\mathbf{r}_1,\mathbf{r}_2)\big|\uparrow\downarrow\big\rangle - f(\mathbf{r}_2,\mathbf{r}_1)\big|\uparrow\uparrow\big\rangle - f(\mathbf{r}_2,\mathbf{r}_1)\big|\downarrow\uparrow\big\rangle$$

$$= \big[f(\mathbf{r}_1,\mathbf{r}_2) - f(\mathbf{r}_2,\mathbf{r}_1)\big]\big|\uparrow\uparrow\big\rangle + \big[f(\mathbf{r}_1,\mathbf{r}_2)\big|\uparrow\downarrow\big\rangle - f(\mathbf{r}_2,\mathbf{r}_1)\big|\downarrow\uparrow\big\rangle\big].$$

The first term is spatially antisymmetric and spin symmetric. The second term does not separate into space and spin, but it is still antisymmetric:

$$f(\mathbf{r}_1,\mathbf{r}_2)\big|\uparrow\downarrow\big\rangle - f(\mathbf{r}_2,\mathbf{r}_1)\big|\downarrow\uparrow\big\rangle \overset{?}{=} -\big[f(\mathbf{r}_2,\mathbf{r}_1)\big|\downarrow\uparrow\big\rangle - f(\mathbf{r}_1,\mathbf{r}_2)\big|\uparrow\downarrow\big\rangle\big]$$

$$\overset{!}{=} f(\mathbf{r}_1,\mathbf{r}_2)\big|\uparrow\downarrow\big\rangle - f(\mathbf{r}_2,\mathbf{r}_1)\big|\downarrow\uparrow\big\rangle. \qquad (6.1)$$

Thus, we see that superpositions of antisymmetric states are also antisymmetric, regardless of whether space and spin parts separate. Because antisymmetrization removes any symmetric part of a function, antisymmetrized wave-functions generally must be normalized after antisymmetrization.

Any number of fermions can have a fully symmetric spin state, but it is impossible to form a fully antisymmetric spin state for three or more fermions ([1], p. 457).

Bosons Bosons have multiparticle states that are *symmetric* under interchange of any two particles, and therefore, do not have Pauli exclusion restrictions. Two bosons *can* be in the same quantum state. In fact, n bosons can all be in the same quantum state. This is a critical property with many implications, from stimulated radiation (lasers) to superfluids.

We return to anti/symmetrization in the discussion of atoms later, and also extend it to three or more identical particles.

6.2   Atoms

Atoms are multiparticle systems comprising a nucleus, and one or more electrons. Understanding the electronic structure of atoms is both practical, and pedagogically important. Many of the concepts involved in atoms are readily applicable to other situations. Also, there is much confusion and many inaccurate statements about atoms in non-peer-reviewed sources; it is important to set the record straight.

This section requires that you understand spherical harmonics, the hydrogen atom, its electronic orbitals, and the n, l, and m quantum numbers. It also requires understanding that electrons can have two different spin states.

Note that the quantum number l is often called the "azimuthal" quantum number, even though m, not l, actually describes the electron's azimuthal motion. This term for l seems to be an anachronism from a time before atomic structure was understood ([5], p. 115). However, ([1], p. 156m) uses "azimuthal quantum number" for m, the z-component of angular momentum, which is consistent with the term "azimuthal," which means "along the horizon." Because of this confusion, we do not use the term "azimuthal quantum number."

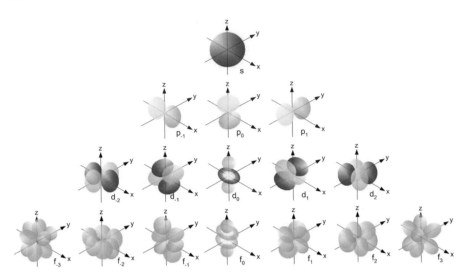

Fig. 6.1 The chemist's views of atomic orbitals are linear combinations of the spherical harmonics. "Single electron orbitals" by haade from http://commons.wikimedia.org/wiki/File:Single_electron_orbitals.jpg, used under a Creative Commons Attribution-Share Alike 3.0 Unported license: http://creativecommons.org/licenses/by-sa/3.0/deed.en

6.2.1   Forget the Bohr Model

Bohr was a great physicist, and did well with the information available to him. However, the Bohr model of the atom is completely wrong, and probably should not be taught any more. It does manage to compute the coarse electron energies for hydrogen, but for the wrong reasons. The idea of the electron wave-function of a 3D atom wrapping around in a flat 2D circle is unphysical (though for $l \geq 1$, this idea captures the ϕ dependence of the wave-function). The Bohr model predicts the wrong values for angular momentum, because it excludes $l=0$. It therefore calls the ground state $l=1$, instead of $l=0$. The fact that the coarse energies of hydrogen work out is just (bad) luck.

6.2.2   Why Spherical Harmonics are not Spherically Symmetric

As prelude to atoms, we consider the hydrogen-like orbitals. Why are spherical harmonics not spherically symmetric? Short answer: Because we are also in an $\hat{L}_z$ eigenstate, and choosing the z-axis breaks the spherical symmetry. But let us clarify.

Electron orbital shapes are frequently drawn in physics and chemistry books: the familiar sphere for s-orbitals, the infamous dumb-bell for p-orbitals, the mysterious dumb-bell through a donut for d-orbitals, etc (Fig. 6.1).

Of these, only the *s*-orbitals are spherically symmetric. But the orbital shapes derive from a central potential (the positive nucleus) which *is* spherically symmetric, so how can the orbitals be anything but spherically symmetric? What could break the symmetry and establish a preferred direction for the dumb-bells? Answer: a choice of angular momentum quantization axis.

Recall that only *energy* eigenstates are stationary. But for an atom (central potential), the energy eigenstates are degenerate, meaning there are multiple electron energy eigenstates with the same energy. To resolve the degeneracy into distinct quantum states, there are three other observables that have simultaneous eigenstates with energy: L^2 (the square of the magnitude of orbital angular momentum of the electron), L_z (the *z*-component of **L**, i.e., the electron orbital angular momentum about an axis we call the *z*-axis), and the electron spin (neglecting spin–orbit interaction). Therefore, to be in a definitive and stationary state, all four simultaneous eigenstates (particle observables) must be known.

By convention, all the spherical harmonic solutions are calculated in the "*z*-basis," which is similar to assuming that we have measured L_z, and are thus in an $\hat{L}_z$ eigenstate. This choice of axis makes the *z*-axis special. It breaks the spherical symmetry. Recall that in all cases except **L**=0, $|\mathbf{L}|^2 > L_z^2$, so L_x and L_y cannot both be zero. In fact, the generalized uncertainty principle says that neither L_x nor L_y can be known precisely when in an L_z eigenstate, and both have the same probability distribution function (PDF). Therefore, all the spherical harmonics, instead of being spherically symmetric, are rotationally symmetric *with respect to x and y* (i.e., axially symmetric about the *z*-axis).

6.2.3   Why are l-orbitals Called s, p, d, and f ?

$l=0$ is called *s*, $l=1$ is called *p*, $l=2$ is called *d*, and $l=3$ is called *f*. These letters derive from the early examinations of the spectral lines for electronic transitions involving these orbital. The brightest line was the *principal* line, corresponding to $l=1$. The others appeared sharp, diffuse, and otherwise unknown, called fundamental [26], corresponding to $l=0$, $l=2$, and $l=3$. (The "f" is sometimes called "fine.") For more, see [23].

6.3   Ground States of Small Atoms and Hund's Rules

Understanding the electronic structure of small-to-medium atoms requires understanding antisymmetry of electron (fermion) states and understanding Hund's rules. These concepts provide good insight into quantum mechanical principles, making atoms worth study for such insight alone. Hund's rules are frequently misunderstood and misapplied, so it is also helpful to review some common *invalid* reasoning, and why it is wrong. Antisymmetry and Hund's rules are sometimes presented as simple and glossed over. In fact, they are somewhat involved, but manageable. [1] and [5] are good references. Baym ([1] p. 452) has a complete periodic table with both electron configurations *and* their spectral term ground states. Goswami ([8], p. 426) and Gasiorowicz ([6], p. 307) have the same information in list form. We here discuss the ground states of atoms.

In addition to the prerequisites for the previous section (the hydrogen atom, spherical harmonics), this section requires you understand quantum addition of angular momentum, combined space/spin states, antisymmetry of multielectron states, and that you have been introduced to the fine structure and spin–orbit energy. We follow these topics:

1. Review of shell structure of atomic electrons, Hartree and Hartree–Fock approximations.
2. Angular momentum coupling of atomic electrons: S, L, and J.
3. Fermion antisymmetry.
4. Hund's rules, by rote.
5. Examples of Hund's rules for interesting atoms.
6. Why do Hund's rules work?

Our entire discussion is done within the Hartree–Fock approximation and perturbation theory. Sometimes, we speak loosely and say something like "The total angular momentum is J," by which we mean "The total angular momentum quantum number is J." Bold symbols indicate the actual angular momentum vector, such as $\mathbf{L}$, $\mathbf{S}$, or $\mathbf{J}$. Squared vectors are defined as magnitude squared, e.g., $\mathbf{L}^2 \equiv \mathbf{L} \cdot \mathbf{L} = |\mathbf{L}|^2$.

In mathematics and physics, the terms "symmetric" and "antisymmetric" have many meanings. In general, a symmetry is some invariant property under a given transformation. You are familiar with a function of one variable being symmetric under reflection about zero (aka an "even" function), or antisymmetric about zero (an "odd" function). In multiparticle wave-functions, we refer to anti/symmetry under particle interchange:

6.3.1 Introduction to Atomic Electron Structure

There are three major factors that determine the coarse and fine structure of atoms (energies and number of Zeeman splittings). They are, in order of significance:

1. The Coulomb energy binding the electron to the nucleus, giving a hydrogen-like shell structure. We take these states as our unperturbed basis.
2. The Coulomb energy between electrons in the outermost subshell, which drives toward antisymmetric spatial states of multielectron quantum states.
3. The electron spin–orbit coupling to the nuclear charge (responsible for the fine-structure of atoms).

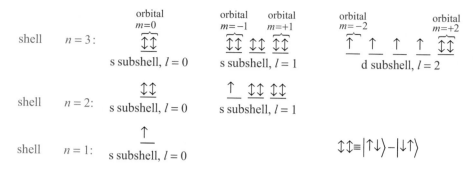

Fig. 6.2 Example shells, subshells, and orbitals. The notation "↕" is shorthand for "$|\uparrow\downarrow\rangle - |\downarrow\uparrow\rangle$"

The energy separation of subshells is typically a few electron volts; that of different multielectron states within a subshell is of order 0.1 eV, and the spin–orbit energies are roughly 1–10 meV ([1], p. 455). The observable properties are determined primarily by the electrons in unfilled subshells, and the nuclear charge. (We do not consider the hyperfine structure, which involves nuclear spin interactions.) The following shell/subshell/orbital structure of atoms is evident in the hydrogen atom, but applies to heavier atoms, as well:

Shell principal quantum number n, can hold $2n^2$ electrons.

Subshell a principal and orbital angular momentum quantum number pair, nl, can hold $2(2l+1)$ electrons.

Orbital spatial state with principal, orbital angular momentum, and orbital z-component angular momentum (aka magnetic) quantum numbers nlm; can hold two electrons.

(Note that [8] uses the term "orbital" where most use the term "subshell." [1] and [3] use the term "shell" where most use "subshell," and [1] uses "wave-function" where we use "state." We use **wave-function** to mean the spatial part of a state.)

> Shells, subshells, and orbitals appear in the hydrogen atom, but larger multi-electron atoms also have approximately the same structure.

This shell structure approximation is the **Hartree approximation**, and results from assuming that the electrons in small- to medium-sized atoms ($Z \lesssim \sim 23$) behave almost like independent electrons in a spherically symmetric net potential ([1], p. 451). This allows us to consider the electrons separately, and since the potential is approximately spherically symmetric, the Schrödinger solutions are products of radial functions and spherical harmonics (i.e., shells, subshells, and orbitals). In multielectron atoms, only the radial functions differ substantially from hydrogen ([5], p. 323). The Hartree approximation ignores electron antisymmetry, which makes it very inaccurate. The Hartree–Fock approximation adds the most important electron antisymmetry, which is an essential aspect of atomic structure.

The **electron configuration** of an atom specifies the n and l quantum numbers of all electrons, but not their m quantum numbers (L_z components), or spins. Neither does the configuration specify how the spins and orbital angular momenta combine. It is the purpose of Hund's rules to predict such properties of the atom that depend on these other quantum numbers. While Hund's rules were first determined empirically, we present here the quantum mechanical explanation of them.

Our analysis uses stationary-state degenerate perturbation theory. Our unperturbed basis for each electron is a set of eigenstates of the screened-nucleus Coulomb energy:

$$\hat{H} = \hat{H}_0 + \hat{H}_{so} \quad where \quad \hat{H}_0 = -\frac{Z_{eff}e^2}{r}; \; Z_{eff} \equiv \text{effective central charge;}$$

$$e \equiv \text{proton charge;}$$

$$\hat{H}_{so} \equiv \text{spin-orbit energy} \propto \mathbf{L \cdot S}.$$

The effective central charge is the nuclear charge, minus one unit for each electron in a full inner shell, because full electron shells are approximately spherically symmetric, and the negative electron charge cancels some of the true nuclear charge. This makes the effective central charge one positive unit for each valence electron, i.e., for each electron in the outermost shell.

The Hamiltonian previously allows our unperturbed basis (for each electron) to be either (l, m_l, s, m_s) or (l, s, j, m_j), since they all have the same energy. For our analysis, it is simpler to choose the (l, m_l, s, m_s) basis. In this basis, each electron has well-defined quantum numbers for n, l, m_l, and m_s. ($s = \frac{1}{2}$ always for an electron). Because these base states are degenerate in energy, we *cannot* use the simple (nondegenerate) perturbation energy $E_\psi^{(1)} = \langle \psi | H_{so} | \psi \rangle$.

Here is a summary of the symbols we use, which are explained in more detail as we go. In general, lowercase letters refer to a single electron and uppercase to the total over all subshell electrons:

k	the number of electrons in the subshell, $k \le 2(2l+1)$.
u	is a summation index.
l	the orbital angular momentum quantum number of the subshell we are "filling" with electrons.
m	the generic z-component *orbital* angular momentum quantum number.
$m_1, m_2, \ldots$	the z-component *orbital* angular momentum quantum number of electron 1, 2,...

S	the total spin angular momentum quantum number of all the electrons in the subshell.
L	the total orbital angular momentum quantum number of all the electrons in the subshell.
M_L	the total z-component *orbital L* quantum number of all the electrons in the subshell.
M_S	the total z-component *spin S* quantum number of all the electrons in the subshell.
J	the total (spin+orbital) angular momentum quantum number of the subshell.
M_J	the total z-component (spin+orbital) angular momentum quantum number of the subshell.
Z	the atomic number.

Bold letters refer to the actual vectors, still following the lower/uppercase notation for individual/total quantities, e.g., **l**, **s** for individual orbit and spin angular momentum vectors, and **L**, **S**, **J**, etc. for total subshell vectors.

Angular Momentum Coupling of Atomic Electrons: S, L, and J.

It can be shown by the Wigner–Eckart theorem that (within our approximations) the final electronic state has definite total orbital angular momentum L, total spin S, and total angular momentum J ([1] p. 461). This is equivalent to saying that **L·S** has a definite value. In fact, it is a minimum to minimize the spin–orbit energy. Thus L, S, and J are "good" (constant) quantum numbers ([8], p. 424), but not the z-components L_z or S_z ([5], p. 281m). Such an atom (with definite L and S) is said to have **LS coupling**, also known as Russell–Saunders coupling, or RS coupling. For atoms with a single unfilled subshell, the unfilled subshell electrons alone determine the atom's L, S, and J. We show that filled subshells have 0 spin, and half-full and completely full subshells always have $L=0$, because their spatial states are antisymmetric.

Recall, that the orientations in the x-y plane of orbital angular momentum **L** and spin **S** are completely undetermined, i.e., they are uniformly distributed in all directions of the x-y plane. Despite this, the angle between **L** and **S** may have a definite value, i.e., **L·S** may have a definite value, and therefore so does $|J|$:

$$\mathbf{J}^2 = (\mathbf{L}+\mathbf{S})^2 = \mathbf{L}^2 + \mathbf{S}^2 + 2\mathbf{L}\cdot\mathbf{S} \quad \Rightarrow \quad \mathbf{L}\cdot\mathbf{S} = \frac{1}{2}\left(\mathbf{J}^2 - \mathbf{L}^2 - \mathbf{S}^2\right).$$

In other words:

States of definite **L·S** are states of definite J, and vice-versa.

> Some references say that the spin or orbital angular momentum "precesses" around the total angular momentum, **J**. However, we are describing *stationary* states, whose properties do *not* change at all with time. Literally, there is no precession.

Instead, the spread of angular momentum around the **J** vector is due to uncertainty (Fig. 5.9): the x and y components are in a superposition of all angles, with uniform distribution around the x-y plane.

Most references use **spectroscopic notation**, which gives all three aggregate quantum numbers, S, L, and J in this form:

$$^{2S+1}L_J. \qquad \text{E.g.} \qquad S=1, L=0, J=1 \quad \rightarrow \quad {}^3S_1,$$

where the orbital angular momentum letter comes from the $S, P, D, F, G, H, I, K, ...$ sequence (which skips J), corresponding to $L=0, 1, 2,...$. The $2S+1$ superscript preceding the orbital angular momentum letter gives, for atoms where $S \leq L$, the number of nearby finely spaced spectral lines, which are due to spin–orbit coupling energy. This spacing can be widened by applying a magnetic field to the atom.

In addition, many references denote the *principal* quantum number (n) electron shell sequence by $K, L, M, N, O,...$. These are used primarily for x-ray spectra, involving inner-shell electrons, and we will not use them.

Fermion Antisymmetry

Electrons are spin-1/2 particles (fermions). A single-electron state comprises a space part (wave-function) and a spin part. For Hund's rules, we are concerned only with the angular parts of the wave-function, $|l,m\rangle$, but not the radial part, since it does not contribute to angular momentum, and therefore not to spin–orbit energy, either. We write the spin states as a series of arrows, up or down, one arrow for each electron, in order. For example, $|\uparrow\downarrow\rangle$ means the first electron is spin-up and the second is spin-down. Recall that all multielectron states must be antisymmetric under interchange of *any two* electrons; such a state is called **fully antisymmetric**. This required antisymmetry can come from the space part of the state, or the spin-part, or a combination of both (Eq. 6.1). When considering atoms, we approximate that the antisymmetry comes entirely from space or entirely from spin, but not both, since antisymmetric space states have lower Coulomb repulsion energy.

> Only *different* wave-functions can be antisymmetrized; e.g., two electrons in the same orbital (*nlm* spatial state) are necessarily symmetric in space, and therefore antisymmetric in spin.

Strictly speaking, in an atom, *all* of the electrons exist in a giant multielectron fully antisymmetric state. The **Hartree–Fock approximation** antisymmetrizes the outer subshell electron states (inner electrons do not make much difference, since they are close to spherically symmetric). The full antisymmetrization problem is "very involved" ([5], p. 363). We hit only the highlights of many particle antisymmetrization here. Recall from addition of angular momentum that,

- For two electrons, spin $S=0$ is antisymmetric under particle exchange, and $S=1$ is symmetric.
- It is impossible to antisymmetrize three or more spin-$\frac{1}{2}$ spin states. Therefore, for three or more electrons, the spin state is either fully symmetric or neither symmetric nor antisymmetric.

The Coulomb energy of antisymmetric electron space states is lower than that of symmetric space states, because antisymmetric states have no probability of the electrons being in the same place, but symmetric states can have the two electrons "on top of each other," with a consequently high Coulomb repulsion energy. Thus, to achieve lowest energy, antisymmetric spatial states are favored over symmetric spatial states.

Note that, despite common depictions, two electrons in the same orbital (spatial state) *do not* each have definite spin-orientation; we cannot say one is "up" and the other is "down." They are indistinguishable, so it is meaningless to think that way. All we can say is that if we measure one of them up, the other must measure down (they are "entangled"). They are together in a two-particle state, which (for fermions) must be antisymmetric. Since they both have the same single-particle spatial state $\psi_{nlm}(\mathbf{r})$, interchanging the particles does not change the space state, i.e., the space state is symmetric. Therefore, the spin state must be antisymmetric. For example,

$$\underbrace{|\psi_{12}\rangle}_{\substack{\text{2-particle space} \\ \text{and spin state}}} = \underbrace{\frac{1}{\sqrt{2}}\left(|\uparrow\downarrow\rangle - |\downarrow\uparrow\rangle\right)}_{\text{spin state}}\underbrace{\psi_{nlm}(\mathbf{r}_1)\psi_{nlm}(\mathbf{r}_2)}_{\text{space-state }\psi(\mathbf{r}_1,\mathbf{r}_2)}.$$

> Two electrons with opposite spin *can* occupy the same *space*, i.e., have the same single-particle wave-function. Then their spin state must be antisymmetric, yielding total $S=0$.

6.3.2   Hund's Rules: Overview

To find the ground state of an atom, we consider only states allowed by the antisymmetry of multielectron states, and then apply Hund's rules to those states to find the lowest energy state (i.e., the ground state). Hund's rules describe how electrons fill the outermost subshell (a set of states of fixed principal quantum number n and fixed orbital angular momentum quantum number l). Hund's rules give the total spin S, total orbital angular momentum L, and total angular momentum J for small-to-medium atoms. Hund's rules fail for heavier atoms, where the electrons are not in approximate L and S eigenstates or where multiple subshells are unfilled. Nonetheless, they work exactly for $Z \leq 23$ (vanadium), and often for even heavier atoms. However, $Z = 24$ (chromium) has two unfilled subshells ($1s^2\ 2s^2\ 2p^6\ 3s^2\ 3p^6\ 3d^5\ 4s$), for which Hund's rules do not work.

Hund's rules rely on:
- The shell/subshell/orbital approximation to the electron states of an atom;
- The antisymmetry requirement of electrons;
- LSJ coupling, which means the set of electrons is in an eigenstate of total $\hat{L}^2$, $\hat{S}^2$, and $\hat{J}^2$, and therefore $\mathbf{L} \cdot \mathbf{S}$ has a definite value;
- The spin–orbit interaction energy.

As noted earlier, the final perturbed states do not have definite values of the total z-components L_z and S_z, but *do* have definite values of total angular momentum J and M_J. Nonetheless, in our *un*perturbed basis (which excludes spin–orbit energy), each electron has definite quantum numbers m_l and m_s.

We imagine building up an atom, from the nucleus and adding one electron at a time. For a subshell of given n and l [which holds $2(2l+1)$ electrons], **Hund's rules** are these, in order of priority:

1. The first half-subshell, $(2l+1)$ electrons, align their spins in a direction we call "up." The last $(2l+1)$ electrons must then have spin "down." The spins combine to give total M_S, which is always up (positive) (or zero for a full subshell). (This rule is sometimes given as "electrons align for maximum spin," or "electrons align for maximum spin multiplicity.")

2a. These first half-subshell electrons (all spin up), fill distinct orbital ml values, starting with the most negative ($m_l = -1$) up to the most positive ($m_l = +1$). The last $(2l+1)$ electrons fill the ml values in the reverse order, from most positive to most negative. For example, the p-subshell fills in this order:

$$p \text{ subshell}: \frac{\overset{1}{\uparrow}}{m=-1}\ \frac{\overset{2}{\uparrow}}{m=0}\ \frac{\overset{3}{\uparrow}}{m=+1}\qquad \frac{\overset{1\ 6}{\uparrow\downarrow}}{m=-1}\ \frac{\overset{2\ 5}{\uparrow\downarrow}}{m=0}\ \frac{\overset{3\ 4}{\uparrow\downarrow}}{m=+1}$$

The m_l (z-components of orbital angular momenta of the electrons) adds to give M_L, which may be positive or negative. Each spin of the last $(2l+1)$ electrons must combine with the other electron in the orbital into an antisymmetric superposition of up and down, and therefore resulting in zero total spin: $|\uparrow\downarrow\rangle - |\downarrow\uparrow\rangle$.

2b. The total orbital quantum number $L = |M_L|$.

3. For less-than-half-full subshells, M_L and M_S oppose (which means $\mathbf{L}$ and $\mathbf{S}$ oppose), so that total angular momentum quantum number $J = |L - S|$. For more-than-half-full subshells, $\mathbf{L}$ and $\mathbf{S}$ align, so that $J = L + S$. (Half-full subshells have $M_L = L = 0$, so $J = S$.)

[The rules are often ambiguously worded. For example, 2a and 2b are sometimes given as "orbitals fill to maximize L," but they mean "maximum L" within a set of other constraints. In fact, considering only that each electron has orbital quantum number l, the atoms are *not* in states of maximum L. We show this explicitly later.]

Boron ($Z = 5$), carbon ($Z = 6$), nitrogen ($Z = 7$), and oxygen ($Z = 8$) illustrate almost all of the principles needed for understanding atomic ground states, so we discuss each atom in turn. Note that hydrogen ($Z = 1$) through beryllium ($Z = 4$) are all trivial, because they have only a single electron in any unfilled subshell, and all $l = L = 0$. Therefore, boron is the first interesting atom.

We now give several examples of applying Hund's rules, introducing some of *why* the rules work. Further explanation follows the examples.

6.3.3   Hund's Rules: Examples and First Principles

Boron $1s^2\, 2s^2\, 2p^1$. This is the first atom with nonzero L, and so the first with a spin–orbit interaction. With only one p-subshell electron, $l = L = 1$. Since the spin–orbit energy is proportional to $\mathbf{L \cdot S}$, it decreases the atom's energy (contributes negative energy) when $\mathbf{L}$ opposes $\mathbf{S}$ (and, therefore, $\mathbf{l}$ opposes $\mathbf{s}$):

$$\text{boron:} \qquad S = \frac{1}{2}: \qquad \frac{\uparrow}{m=-1}\ \frac{}{m=0}\ \frac{}{m=+1}.$$

$$L = |M_L| = |-1| = 1, \qquad J = |L - S| = \frac{1}{2} \quad \Rightarrow \quad {}^2P_{1/2}.$$

The J value comes from the "less than half-full" rule: $J = |L - S|$.

Carbon $1s^2\, 2s^2\, 2p^2$: The $2p$ subshell has three orbitals, and therefore six electron states: $|l,m\rangle = |1,1\rangle;\, |1,0\rangle;$ *and* $|1,-1\rangle$, with two spin states for each orbital:

$$\text{carbon:} \qquad S = \frac{1}{2} + \frac{1}{2} = 1: \qquad \frac{\uparrow}{m=-1}\ \frac{\uparrow}{m=0}\ \frac{}{m=+1}.$$

$$L = |M_L| = |-1 + 0| = 1, \qquad J = |L - S| = 0 \quad \Rightarrow \quad {}^3P_0.$$

Applying Hund's rules: (1) assign both electrons spin up, $S = M_S = 1$, a symmetric spin state. (2) These spin-up electrons go into $m = -1$ and $m = 0$ orbital states, making negative and zero spin–orbit energies (whereas, $m = 1$ has positive energy). $M_L = m_1 + m_2 = -1 + 0 = -1$. $L = |M_L| = 1$. (3) $J = L - S = 0$, because the subshell is less than half full. (Note that there is no need to think about "maximum L.")

Nitrogen ($1s^2\, 2s^2\, 2p^3$) has three electrons in the $2p$ subshell.

$$\text{Nitrogen:} \qquad S = \frac{1}{2} + \frac{1}{2} + \frac{1}{2} = \frac{3}{2}: \qquad \frac{\uparrow}{m=-1}\ \frac{\uparrow}{m=0}\ \frac{\uparrow}{m=+1}.$$

$$L = |M_L| = |(-1) + 0 + 1| = 0, \qquad J = |L - S| = \frac{3}{2} \quad \Rightarrow \quad {}^4S_{3/2}.$$

First, all electrons align spin-up, $S = M_S = 3/2$, a fully symmetric spin state. Second, orbits fill from negative to positive, so $M_L = m_1 + m_2 + m_3 = (-1) + 0 + 1 = 0$; $L = |M_L| = 0$. Third, the subshell is half full, so $J = S = 3/2$. Note that the spin–orbit energy of the $m=1$ electron is positive (unfavorable), but that is still lower than the Coulomb energy of putting the third electron in either of the two already occupied orbitals.

You might ask, why do not the electrons' orbital angular momenta add to produce, say, the $L=3$, $M_L=0$ state, $|3\ 0\rangle$? Since the three-electron *spin* state is symmetric, the three-electron spatial wave-function must be fully antisymmetric. We show in the next section that any consecutively filled antisymmetric subshell electron wave-function has $L=|M_L|$.

Oxygen $1s^2\ 2s^2\ 2p^4$: Oxygen is very interesting, because it is the first more-than-half full subshell. The spins are two up and two canceling, with the fourth electron going into orbital $m=+1$:

$$\text{Oxygen:} \qquad S=\frac{1}{2}+\frac{1}{2}+0=1: \qquad \underset{m=-1}{\uparrow}\quad \underset{m=0}{\uparrow}\quad \underset{m=+1}{\updownarrow\updownarrow},$$

$$\textit{where}\quad \updownarrow\updownarrow\equiv|\uparrow\downarrow\rangle-|\downarrow\uparrow\rangle.$$

$$L=|M_L|=|(-1)+0+1+1|=1, \qquad J=L+S=2 \qquad \Rightarrow \qquad {}^3P_2.$$

Note that for any full *orbital* (the previous two electrons with $n=2$, $l=1$, $m=+1$), the space state is the same for both electrons, and therefore the two-particle space state is symmetric. This means the *spin* state must be antisymmetric, and recall that the antisymmetric combination of spin-up and spin-down has total $S=0$:

$$\frac{1}{\sqrt{2}}\left(|\uparrow\downarrow\rangle-|\downarrow\uparrow\rangle\right)=|S=0, M_S=0\rangle.$$

So for oxygen, the sum of all four spins is $S=M_S=1$. Furthermore, the superposition of spin states for the $m=1$ electron pair has zero spin–orbit energy, eliminating the penalty of the positive single-electron spin–orbit energy in $m=+1$. (If the fourth electron went into $m=-1$, it would eliminate the *benefit* of the first electron's negative spin–orbit energy.)

The fourth electron goes in the $m=+1$ orbital, so $M_L=+1$ (equivalent to: the subshell is more-than-half full, so M_L aligns with M_S), which means $J=L+S=2$.

Note that many references show the *p*-orbitals filling up with four electrons like this:

- One electron $\uparrow$
- Two electrons $\uparrow\ \uparrow$
- Three electrons $\uparrow\ \uparrow\ \uparrow$
- Four electrons $\uparrow\ \downarrow\ \uparrow\ \uparrow$

This is wrong. As shown previously, if the spin is up, the left-most orbital must be $m=-1$. Then the fourth electron goes into the right-most orbital, with $m=+1$. Furthermore, the two paired electrons do not have definite directions; they are in a superposition of both up and down orientations: $|\uparrow\downarrow\rangle-|\downarrow\uparrow\rangle$.

6.3.4   *Why Hund's Rules?*

Why are Hund's rules true? The rules describe the ground-state of the atom; each rule describes how electrons align for lowest energy. As noted earlier, the three forces at work (in decreasing order of magnitude) are the nuclear potential, the interelectron Coulomb energy, and the spin–orbit energy. (Note that interelectron orbit–orbit and spin–spin energies are below the level of the fine structure, and therefore negligible in this analysis ([5], p. 353m).) We now consider each of Hund's rules in turn:

1. Electrons align for maximum spin. You might think that this is to align the magnetic moments of the electrons, because aligned moments are low energy: you have to do work (add energy) to make magnets oppose each other. But there is a much more important reason: when all the spins align, the spin states are fully symmetric. This means the wave-functions are fully antisymmetric. As described earlier, antisymmetric wave-functions keep the electrons apart (probability of being in the same place is zero), which minimizes the Coulomb energy. Coulomb forces are an order of magnitude or more stronger than magnetic forces, so it is the Coulomb energy which dominates.

Some references also note that antisymmetric wave-functions keep each electron "out of the way" of other electrons, thus reducing nuclear screening. In other words, the reduced interelectron repulsion allows the electrons to be more tightly bound to the nucleus.

2. Orbital angular momenta align for maximum $|M_L|$. Within a subshell of given *l*:

a. Electrons occupy different *m* values to separate themselves from each other and minimize Coulomb energy.
b. An electron's orbital m_l values opposes its spin (m_s) as much as possible, to minimize spin–orbit energy.

To see point (a) more clearly, let us look at the probability densities of the first two electrons for two cases: $l=1$ (*p*-subshell) and $l=2$ (*d*-subshell). As noted earlier, in the Hartree–Fock approximation, the electrons are in hydrogen-like space states that separate into radial and angular parts:

$$\psi_{nlm}(r,\theta,\phi) \propto P_{lm}(\cos\theta)e^{im\phi}, \quad where \quad P_{lm} \equiv \text{the associated Legendre function.}$$

The particle density for each electron is then:

$$\left|\psi_{nlm}(r,\theta,\phi)\right|^2 \propto \left|P_{lm}(\cos\theta)\right|^2 \left|e^{im\phi}\right|^2.$$

However,

$$\left|e^{im\phi}\right|^2 = 1 \qquad \Rightarrow \qquad \left|\psi_{nlm}\right|^2 \propto \left|P_{lm}(\cos\theta)\right|^2.$$

Thus, it is only $P_{lm}(\cos\theta)$ that determines where on a sphere the electrons are likely found, and different *m* have different P_{lm}. We now look at the θ dependence on the surface of a sphere (at fixed *r*).

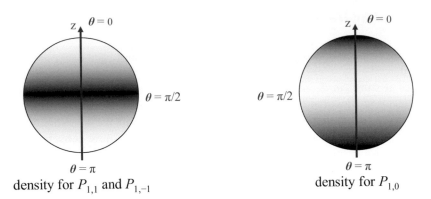

density for $P_{1,1}$ and $P_{1,-1}$ density for $P_{1,0}$

Fig. 6.3 Areal particle density for $l=1$ P_{lm} states (*darker* is more dense)

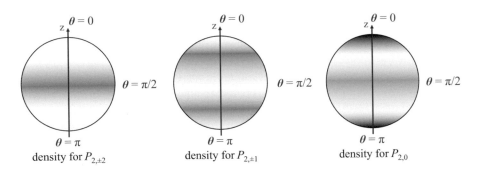

density for $P_{2,\pm 2}$ density for $P_{2,\pm 1}$ density for $P_{2,0}$

Fig. 6.4 Areal particle density for $l=2$ P_{lm} states (*darker* is more dense)

For *l=1* (*p-subshell*): When $m=\pm 1$, $|P_{1,\pm 1}|^2 \sim \sin^2\theta$, density is concentrated at the "equator" (Fig. 6.3, *left*).

When $m=0$, $|P_{10}|^2 \sim \cos^2\theta$, density is concentrated at the z-axis poles (Fig. 6.3, *right*). So two electrons are farthest apart if they are in $m_1=1$ or -1 and $m_2=0$ states. If they were in $m_1=1$ and $m_2=-1$ states, they would be nearly "on top of each other" at the equator. Even though antisymmetrization would keep them from being exactly on top of each other, they would be much closer than in the $m_1=\pm 1$ and $m_2=0$ states. Closer electrons means more Coulomb energy, farther means less energy, and hence the ground state has distinct orbital m values of $m_1=-1$ and $m_2=0$. Furthermore, taking the spin as "up," $m_1=-1$ is lower energy than $m_1=1$ because spin–orbit coupling has lower energy when spin opposes orbital angular momentum.

For *l=2* (*d-subshell*): When $m=\pm 2$, $|P_{2,\pm 2}|^2 \sim \sin^4\theta$, the density is tightly concentrated at the "equator" (Fig. 6.4, *left*):

When $m=\pm 1$, $|P_{21}|^2 \sim \sin^2 2\theta$, density is concentrated in two bands (Fig. 6.4, *middle*). When $m=0$, $|P_{20}|^2 \sim (3\cos^2\theta - 1)^2$, density is concentrated at the poles, and less so around the equator (Fig. 6.4, *right*). So two electrons are farthest apart if they are in $m_1=-2$ or 0 and $m_2=-1$ states. If they were in $m_1=2$ and $m_2=0$ or -2 states, they would have significant overlap. Antisymmetrization would keep them from being exactly on top of each other, but they would be much closer than in the $m_1=-2$ or 0, and $m_2=-1$ states. Closer electrons mean more Coulomb energy, farther means less energy, and hence the ground state. Again, because the *spin* is up, the m values are negative to minimize spin–orbit energy, so $m_1=-2$ and $m_2=-1$.

For all l, the antisymmetry of the multielectron wave-function (space state) forces $L=|M_L|$ (see later for proof).

We now consider the previous point (b): orbital m_l values oppose their own spin (m_s) as much as possible, to minimize spin–orbit energy. Within a subshell, the spins are entirely or predominantly up (Hund's Rule 1). Furthermore, the individual m_l values each oppose their spin as much as possible, meaning they point down. The farther down, the better, to minimize spin–orbit coupling energy. So the first electron goes spin up, and $m_1=-l$ (maximally down). The second electron goes spin up, and $m_2=-(l-1)$, down as low as it can given that $m=-l$ is already occupied.

Examples: Titanium ($Z=22$, $4s^2\,3d^2$) has $l=2$, $m_1=-2$, $m_2=-1$, and $L=3$ (not 4). Similarly, vanadium ($Z=23$, $4s^2\,3d^3$) has $l=2$, $m_1=-2$, $m_2=-1$, $m_3=0$, and $L=3$ (not 4, 5, or 6); praseodymium ($Z=59$, $6s^2\,4f^3$) has $m_1=-3$, $m_2=-2$, $m_3=-1$, and $L=6$ (not 7, 8, or 9); and even heavy tantalum ($Z=73$, $6s^2\,5d^3$) has $L=3$ (not 4, 5, or 6).

[Baym ([1], p. 452) has a misprint, saying that praseodymium ($Z=59$) has $S=3/2$, $L=6$, $J=3/2$ ($^4I_{3/2}$), which is impossible because minimum $J=L-S=9/2$. ([8], p. 427 and [6], p. 308) have it right with $S=3/2$, $L=6$, $J=9/2$ ($^4I_{9/2}$).]

The alignment of orbital angular momentum is sometimes incorrectly explained by claiming that electrons orbiting "in the same direction" do not cross each other's paths as often as electrons orbiting opposite each other; therefore, the electrons are alleged to be farther apart on average, and the Coulomb energy is reduced.

This is not even true classically. If a fast particle orbits in the same direction as a slow one (instead of the opposite direction), it overtakes the slow particle less often, but it stays in the neighborhood of the slow particle for *longer* on each pass. The average distance, classically, is actually *independent* of the speed or direction of the orbits (unless the angular velocities are exactly equal). More to the point: quantum mechanically, all the electrons are in *exactly* stationary states (with no approximations), which means the probabilities of their locations do not change with time. Therefore, there is no "crossing each others' paths" in the classical sense. Coulomb energy is minimized, as described previously, because the electron densities are more widely separated with electrons having m values with different magnitudes.

Another common misconception is the classical view that maximum L spreads the electrons farther apart from each other, because larger L means the electrons are "moving faster" and hence spread around a larger orbit. This is not true.

First, as we have seen, the atoms are *not* in states of maximum L; they are in states of maximum $|M_L|$, and then *minimum* L allowed for that M_L. Second, recall that the electrons are in hydrogen-like space states (orbitals) that separate into radial and angular parts:

$$\psi_{nlm}(r,\theta,\phi) = R_{nl}(r)NP_{lm}(\cos\theta)e^{im\phi},$$

where $P_{lm} \equiv$ the associated Legendre function.

All the electrons in a subshell have the same n and l, and therefore the same radial wave-function $R_{nl}(r)$. The R_{nl} do not change with total L, i.e., how two electrons' angular momenta align does not change $R_{nl}(r)$. (The distance from the nucleus can only increase if the individual electrons' n or l values increase.) The Coulomb energy is minimized by electrons occupying the orbitals (m_l values) that minimize their antisymmetrized spatial overlap due to the spherical harmonics. Furthermore, spreading the electrons farther from the nucleus would cost far more in nuclear Coulomb energy than that reduced by interelectron energy.

   3. *For less than half-full subshells,* $J=|L-S|$, *otherwise* $J=L+S$. This is due to the spin–orbit energy. Recall that L and S are the quantum numbers for the *total* orbital angular momentum and *total* spin for the entire subshell.

a.   An electron's spin interacts with its own orbit around the nucleus ([5] p. 353m). We have seen that each electron's orbital angular momentum opposes its own spin as much as it can, to minimize its spin–orbit coupling energy. We call the spin of the first electron "up," which implies that subshells put the first (spin-up) electrons in $m_l<0$ orbits:

$$\frac{\uparrow}{m=-2}\ \ \frac{}{m=-1}\ \ \frac{}{m=0}\ \ \frac{}{m=+1}\ \ \frac{}{m=+2} \qquad M_S>0,\ M_L<0.$$

b.   By Rule 1, the spins align, so the magnitude of total spin grows while filling the first half of the states, and it is always up. Each orbit opposes its own "up" spin as much as it can, or aligns with it as little as it can. This implies the orbits fill in order of increasing m_l from $-l$ to $+l$:

$$\frac{\uparrow}{m=-2}\ \ \frac{\uparrow}{m=-1}\ \ \frac{\uparrow}{m=0}\ \ \frac{\uparrow}{m=+1}\ \ \frac{}{m=+2} \qquad M_S>0,\ M_L<0.$$

c.   Less-than-half full subshells have total $M_L=m_1+m_2+\ldots+m_k<0$ (down), and total spin M_S up, thus making M_S opposed to M_L, and therefore the total angular momentum quantum number is:

$$J=|L-S| \qquad \text{(less than half full)}.$$

d.   Continuing the example, with five electrons, a d-subshell is half full and all spins are up. $L=M_L=0$, and therefore $J=S$. The crucial step is adding the sixth electron. When added, it must fill one of the orbitals (similar to the fourth electron in oxygen's $2p$ subshell). Since the spatial state for both electrons in that full orbital is the same, the two-particle spatial state is symmetric:

$$\psi(\mathbf{r}_1,\mathbf{r}_2) = \psi_{nlm}(\mathbf{r}_1)\psi_{nlm}(\mathbf{r}_2), \qquad \textit{where} \quad l=2 \text{ in our example.}$$

This means the spin state must be antisymmetric: full orbital spin $|{\uparrow}{\downarrow}\rangle - |{\downarrow}{\uparrow}\rangle$. This spin superposition cancels the spin–orbit energy, which becomes zero. Therefore, the sixth electron goes into the orbital with the highest (unfavorable) spin–orbit energy and cancels it:

$$\underset{m=-2}{\underline{\uparrow}} \quad \underset{m=-1}{\underline{\uparrow}} \quad \underset{m=0}{\underline{\uparrow}} \quad \underset{m=+1}{\underline{\uparrow}} \quad \underset{m=+2}{\underline{\uparrow\downarrow}}$$

$$M_S > 0, M_L > 0. \qquad \uparrow\downarrow \equiv |{\uparrow}{\downarrow}\rangle - |{\downarrow}{\uparrow}\rangle.$$

e. The total spin is still up, but now is shrinking with each new electron. M_L has reversed, and is now up (positive). Thus more-than-half full subshells have both total spin M_S and total orbital M_L pointing up (aligned), and therefore the total angular momentum quantum number is:

$$J = L + S \quad \text{(more than half full)}$$

f. Subsequent orbitals fill in order of m_l shrinking from $+l$ to $-l$, so that each new electron cancels the (then) least favorable spin–orbit energy.

Summarizing:

For less-than-half-full subshells: spins are all up, and individual electron spin–orbit coupling causes $M_L < 0$, i.e., total S opposes total L. Therefore, $J = |L - S|$.

For exactly half-full subshells: Spins are all up, space state is fully antisymmetric, and $L = |M_L| = 0$. Therefore, $J = S$.

For more-than-half full subshells: "Newest" electrons cancel the most unfavorable spin–orbit energies, causing $M_L > 0$, now *aligned* with the total spin. Therefore, $J = L + S$.

Full orbitals: All the spins are canceled, so $S = 0$, and there is zero spin-orbit energy. All the spin-up electrons have antisymmetric spatial states, so $L_{up} = |M_L| = 0$, and similarly for the spin-down electrons. Therefore, total $L = 0$, and $J = L + S = 0$.

[It can be shown that the total spin–orbit energy is proportional to the dot product of total orbital angular momentum and total spin, $E_{so,total} = \zeta_k \mathbf{L} \cdot \mathbf{S}$, where ζ_k depends on k, the number of electrons in the subshell ([1], pp. 462–465). Our analysis here shows that for less-than-half full subshells, $\zeta_k > 0$, but for more-than-half full subshells, $\zeta_k < 0$. Note, though, that each *individual* electron's spin–orbit energy is always negative when its spin opposes its own orbit, and positive when its spin aligns with orbit. As such, there is no obvious physical meaning to the quantity $\mathbf{L}\cdot\mathbf{S}$, other than as a weighted sum of individual electrons' $\mathbf{l}\cdot\mathbf{s}$.]

Hund's third rule is sometimes explained by an appeal to "holes," which behaves like positive particles. This is a subtle argument, but not necessary, as there are no holes in our previous reasoning. [Holes that truly behave like positive particles, even in the Hall effect, exist in crystals (like semiconductors), where some electrons have negative effective mass.]

6.3.5   Symmetry and Slater Determinants for Subshells

Before we prove that $L=|M_L|$, we describe some prerequisites about symmetry and multielectron wave-functions. For a subshell with k electrons, if all the individual $\mathbf{l}_u$ align (uncertain though they are), then total quantum number $L=kl$, and the wave-function must be symmetric. This follows from addition of angular momentum. Consider an example of three electrons with $l=1$ (p subshell). Consider first the $|L, M_L\rangle = |3, 3\rangle$ state. The only way to get it is:

$$\left| l_1, m_1; l_2, m_2; l_3, m_3 \right\rangle = \left| 1,1; 1,1; 1,1 \right\rangle.$$

By inspection, this state is symmetric with respect to interchange of any two particles. We can find all the other $L=3$ states by repeatedly acting on $|3, 3\rangle$ with the total lowering operator:

$$\hat{L}_- = \hat{l}_{1-} + \hat{l}_{2-} + \hat{l}_{3-}$$

Like the $|3, 3\rangle$ state, this operator is symmetric with respect to interchange of any two particles. Acting on a symmetric state with a symmetric operator necessarily produces a new, symmetric ket. Then by induction, all the $|L = kl, M_L\rangle$ states are fully symmetric.

For example, let us look at a few of the orbital states possible from three p-subshell electrons (like nitrogen). p subshell means $l=1$. We start with:

$$|3,3\rangle = |1;1;1\rangle, \quad where \quad \begin{cases} \left| L, M_L \right\rangle \equiv \text{total orbital angular momentum states} \\ \left| m_1; m_2; m_3 \right\rangle \equiv \text{individual electron orbital states.} \end{cases}$$

Now lower both sides:

$$\hat{L}_- |3,3\rangle = L_- |1;1;1\rangle \qquad\qquad \text{Use} \quad \hat{L}_- |L, M_L\rangle = \text{const} |L, M_L - 1\rangle, \text{ etc.}$$
$$|3,2\rangle = \text{const} \left(|0;1;1\rangle + |1;0;1\rangle + |1;1;0\rangle \right) \qquad \text{a symmetric state.}$$

To further illustrate, we can lower again:

$$L_- |3,2\rangle = (const)\left(\hat{l}_{1-} + \hat{l}_{2-} + \hat{l}_{3-} \right)\left(|0;1;1\rangle + |1;0;1\rangle + |1;1;0\rangle \right)$$

$$|3,1\rangle = const \left(\begin{array}{c} \underbrace{|-1;1;1\rangle + |0;0;1\rangle + |0;1;0\rangle}_{\text{from } |0;1;1\rangle} + \underbrace{|0;0;1\rangle + |1;-1;1\rangle + |1;0;0\rangle}_{\text{from } |1;0;1\rangle} \\ + \underbrace{|0;1;0\rangle + |1;0;0\rangle + |1;1;-1\rangle}_{\text{from } |1;1;0\rangle} \end{array} \right)$$

$$= const \left(\begin{array}{c} |-1;1;1\rangle + |1;-1;1\rangle + |1;1;-1\rangle + 2|0;0;1\rangle \\ + 2|0;1;0\rangle + 2|1;0;0\rangle \end{array} \right) \qquad \text{a symmetric state.}$$

Recall that given a set of single-particle wave-functions (or states), we can (usually) construct a fully antisymmetric multiparticle wave-function as a superposition of products of these states, by using a **Slater determinant**. For example, for three particles:

Given $\quad \phi_1(\mathbf{r}), \phi_2(\mathbf{r}), \phi_3(\mathbf{r}), \qquad$ 3 orthonormal single-particle wave-functions,

$$\psi(\mathbf{r}_1,\mathbf{r}_2,\mathbf{r}_3) = \frac{1}{\sqrt{3!}}\begin{vmatrix} \phi_1(\mathbf{r}_1) & \phi_2(\mathbf{r}_1) & \phi_3(\mathbf{r}_1) \\ \phi_1(\mathbf{r}_2) & \phi_2(\mathbf{r}_2) & \phi_3(\mathbf{r}_2) \\ \phi_1(\mathbf{r}_3) & \phi_2(\mathbf{r}_3) & \phi_3(\mathbf{r}_3) \end{vmatrix}$$

$$= \frac{1}{\sqrt{6}}\big[\phi_1(\mathbf{r}_1)\phi_2(\mathbf{r}_2)\phi_3(\mathbf{r}_3) + \phi_2(\mathbf{r}_1)\phi_3(\mathbf{r}_2)\phi_1(\mathbf{r}_3) + \phi_3(\mathbf{r}_1)\phi_1(\mathbf{r}_2)\phi_2(\mathbf{r}_3)$$
$$- \phi_1(\mathbf{r}_1)\phi_3(\mathbf{r}_2)\phi_2(\mathbf{r}_3) - \phi_2(\mathbf{r}_1)\phi_3(\mathbf{r}_2)\phi_1(\mathbf{r}_3) - \phi_3(\mathbf{r}_1)\phi_1(\mathbf{r}_2)\phi_2(\mathbf{r}_3)\big]$$

(antisymmetric).

Note that the determinant of a matrix of functions is itself a function. ψ is antisymmetric under any particle pair interchange, because such an interchange swaps two rows of the determinant, and it is a property of determinants that they are antisymmetric under interchanging any two rows.

You can also easily construct a symmetric state by simply changing the minus signs in the Slater determinant to plus signs:

$$\psi(\mathbf{r}_1,\mathbf{r}_2,\mathbf{r}_3) = \frac{1}{\sqrt{6}}\big[\phi_1(\mathbf{r}_1)\phi_2(\mathbf{r}_2)\phi_3(\mathbf{r}_3) + \phi_2(\mathbf{r}_1)\phi_3(\mathbf{r}_2)\phi_1(\mathbf{r}_3)$$
$$+ \phi_3(\mathbf{r}_1)\phi_1(\mathbf{r}_2)\phi_2(\mathbf{r}_3) + \phi_1(\mathbf{r}_1)\phi_3(\mathbf{r}_2)\phi_2(\mathbf{r}_3) + \phi_2(\mathbf{r}_1)\phi_3(\mathbf{r}_2)\phi_1(\mathbf{r}_3)$$
$$+ \phi_3(\mathbf{r}_1)\phi_1(\mathbf{r}_2)\phi_2(\mathbf{r}_3)\big] \quad \text{(symmetric)}.$$

6.3.6   Why Does $L=|M_L|$?

Consider a d-subshell, $l=2$, with room for ten electrons, but with only four electrons in it:

$$\begin{array}{ccccc} \uparrow & \uparrow & \uparrow & \uparrow & \\ \overline{\quad} & \overline{\quad} & \overline{\quad} & \overline{\quad} & \overline{\quad} \\ m=-2 & m=-1 & m=0 & m=+1 & m=+2 \end{array} \qquad M_S=2, M_L=-2.$$

The spin state is symmetric, so the four-particle wave-function $\psi(\mathbf{r}_1, \mathbf{r}_2, \mathbf{r}_3, \mathbf{r}_4)$ must be antisymmetric with respect to interchange of any two positions $\mathbf{r}_i$ and $\mathbf{r}_j$. This means we can write $\psi(\mathbf{r}_1, \mathbf{r}_2, \mathbf{r}_3, \mathbf{r}_4)$ as a Slater determinant of the four states $|m\rangle: |-2\rangle, |-1\rangle, |0\rangle, |+1\rangle$. Every term in the determinant, i.e., every term in ψ, has each m value exactly once:

$$\psi(\mathbf{r}_1,\mathbf{r}_2,\mathbf{r}_3,\mathbf{r}_4) = |-2;-1;0;+1\rangle - |-1;-2;0;+1\rangle + |-1;0;-2;+1\rangle - \ldots$$

using $\quad |m_1;m_2;m_3;m_4\rangle$ notation.

Now consider lowering this state with the total lowering operator:

$$\hat{L}_- \equiv \hat{l}_{1-} + \hat{l}_{2-} + \hat{l}_{3-} + \hat{l}_{4-}.$$

This is a symmetric operator. A symmetric operator acting on an antisymmetric state produces another antisymmetric state. Now $L = -M_L = 2$ if and only if the total lowering operator annihilates ψ, i.e.,

$$\hat{L}_- \big| L = 2, M_L = -2 \big\rangle = \mathbf{0}_v, \quad \text{or} \qquad \hat{L}_- \psi(\mathbf{r}_1, \mathbf{r}_2, \mathbf{r}_3, \mathbf{r}_4) = \mathbf{0}_v \qquad \text{(the zero vector).}$$

Consider the first operator in $\hat{L}_-$, which is $\hat{l}_{1-}$. This term lowers the first electron state on each term in the Slater determinant. If $m_1 = -2$, then l_{1-} annihilates the term. If $m_1 > -2$, then lowering m_1 makes it duplicate with some other m in the term. E.g.,

$$\hat{l}_{1-} \big| -1; -2; 0; +1 \big\rangle = const \big| -2; -2; 0; +1 \big\rangle. \tag{6.2}$$

This makes the term symmetric with respect to interchange of the electrons with equal m values. But we know that the final lowered state will be fully antisymmetric, so we know that this symmetric term must be cancelled by some other term from the other electron's lowering operator. You can see the canceling explicitly by considering a pair of terms in the original antisymmetric state. For example, the left-side term $\big| -1; -2; 0; +1 \big\rangle$ must appear in $\big| \psi \big\rangle$ along with the negative of the term with m_1 and m_2 swapped:

$$\big| \psi \big\rangle = \ldots + a \big| -1; -2; 0; +1 \big\rangle - a \big| -2; -1; 0; +1 \big\rangle + \ldots \tag{6.3}$$

The total lowering operator $\hat{L}_-$ includes $\hat{l}_{1-} + \hat{l}_{2-}$. $\hat{l}_{1-}$ lowers the first term in Eq. 6.3, and $\hat{l}_{2-}$ lowers the second term, to give:

$$\hat{L}_- \big| \psi \big\rangle = \ldots + a \big| -2; -2; 0; +1 \big\rangle - a \big| -2; -2; 0; +1 \big\rangle + \ldots$$

These two terms cancel. But all other terms in $\big| \psi \big\rangle$ with $m_1 > -2$ also appear in such antisymmetric pairs, and so they all cancel. Thus, every term in $\big| \psi \big\rangle$ is either annihilated outright, when $m_1 = -2$, or is symmetric in two electrons and therefore cancelled by another lowering operator's term. No term survives the total lowering operator: the result of $\hat{L}_- \big| \psi \big\rangle$ is the zero vector, $\mathbf{0}_v$. Hence, $L = -M_L$. QED.

This argument implies that full and half-full subshells have $L = 0$, because

$$M_L = \sum_{u=-l}^{l} m_u = 0.$$

6.4   Multiparticle Entanglement

We have seen multiparticle states that are simple tensor products of single-particle states, e.g., two particles where particle 1 is spin up, and particle 2 is spin down:

$$\big| \chi_{12} \big\rangle = \big| \uparrow_1 \big\rangle \otimes \big| \downarrow_2 \big\rangle \equiv \big| \uparrow_1 \big\rangle \big| \downarrow_2 \big\rangle \equiv \big| \uparrow_1 \downarrow_2 \big\rangle \equiv \big| \uparrow \downarrow \big\rangle.$$

Each particle is in a definite state, and measuring either (or both) particles produces the only possible result: particle 1 measures up and particle 2 measures down.

However, we have also seen that a general two-particle state needs not be a simple (tensor) product of one-particle states. It could be a superposition of such product states, e.g.,

$$|\chi_{12}\rangle = \frac{1}{\sqrt{2}}|\uparrow\downarrow\rangle + \frac{1}{\sqrt{2}}|\downarrow\uparrow\rangle, \quad \text{or} \quad \chi_{12} = \frac{1}{\sqrt{2}}|\uparrow\downarrow\rangle - \frac{1}{\sqrt{2}}|\downarrow\uparrow\rangle. \quad (6.4)$$

Neither of the previous states can be written as a tensor product of single particle states. Neither particle has a definite direction of spin. What happens when we measure individual particles in such states? Such states, where neither particle has a definite value of some property, but the values of the two particles are interdependent, are called **entangled states**, and such particles are said to be **entangled**. This means a measurement of one system provides information about the other. [Entangled particles are sometimes called "correlated," but we avoid this term because it conflicts with the statistical term "correlated."]

6.4.1 Partial Measurements

As always, we use here the term "collapse" of the quantum state as a shortcut to the more complicated process of making a measurement, described in Sect. 1.12. Consider the entanglement of two spin-1/2 particles (in the usual z-basis):

$$|\chi_{12}\rangle = \frac{1}{\sqrt{2}}|\uparrow_1\rangle|\downarrow_2\rangle + \frac{1}{\sqrt{2}}|\downarrow_1\rangle|\uparrow_2\rangle.$$

What if we measure a property of only one of those particles? How does the state then collapse? Answer:

Partial measurement leads to partial collapse.

In the previous state, we easily suppose that there is a 50% chance of measuring particle 1 to be up, but that either way, if we then measure particle 2, it will be opposite to that of particle 1. This is correct, but other superpositions are not so obvious (as shown shortly), so we must develop a mathematical formalism to unambiguously compute the results of measurement on any two-particle state.

To find the probability of measuring a property of *only one* particle of a two-particle system, we use a "partial inner product". Recall that in a one-particle system, an inner product is a scalar. We now extend this idea to a **partial inner product**: Given a ket in the tensor-product space (i.e., a two-particle ket such as $|\uparrow_1\rangle|\uparrow_2\rangle$), we form a partial inner product with a particle 1 bra; the result is a particle 2 *ket*, i.e., a ket in the particle two-ket space. We have,

$$\langle\uparrow_1|\chi_{12}\rangle = \frac{1}{\sqrt{2}}\underbrace{\langle\uparrow_1|\uparrow_1\rangle}_{1}|\uparrow_2\rangle + \frac{1}{\sqrt{2}}\underbrace{\langle\uparrow_1|\downarrow_1\rangle}_{0}|\downarrow_2\rangle = \frac{1}{\sqrt{2}}|\uparrow_2\rangle.$$

In other words, the bra from particle one forms an inner product with the particle 1 piece of the two-particle ket, leaving the particle 2 ket alone. As with our "standard" inner product, the partial inner product is linear in the ket (and antilinear in the bra), so the partial inner product distributes across the superposition of the ket. (If our quantum states were continuous, our inner product would be an integral, and we would say we "integrate out" particle 1.) With this definition of a partial inner product, we find that the probability of obtaining a given measurement follows essentially the same well-known rule as for a single particle:

$$\Pr\left(\uparrow_1\right) = \left|\langle\uparrow_1|\chi_{12}\rangle\right|^2 = \left|\frac{1}{\sqrt{2}}|\uparrow_2\rangle\right|^2 = \frac{1}{2}.$$

The only difference from the single particle case is that for a single particle, the inner product is a scalar, and the probability is its squared magnitude, but in the two-particle case, the inner product is a *vector* (a particle 2 ket), and the probability is the squared magnitude of this vector.

We now consider a more complicated example. Recall that for a single particle, a state of definite spin pointing in the $x+$ direction is:

$$|x+\rangle = \frac{1}{\sqrt{2}}\left(|\uparrow\rangle + |\downarrow\rangle\right).$$

Suppose our two-particle state is:

$$|\chi_{12}\rangle = \frac{1}{\sqrt{2}}|\uparrow_1\uparrow_2\rangle + \frac{1}{\sqrt{2}}|\downarrow_1\downarrow_2\rangle.$$

By inspection, you might think that measuring either particle 1 or particle 2 along the x-axis will give a definite value of $|x+\rangle$ (probability=1). Let us test this by computing the probability for measuring particle 1 in $|x+\rangle$:

$$\Pr\left(|x+_1\rangle\right) = \left|\langle x+_1|\chi_{12}\rangle\right|^2$$

$$\langle x+_1|\chi_{12}\rangle = \frac{1}{\sqrt{2}}\left(\langle\uparrow_1| + \langle\downarrow_1|\right)\frac{1}{\sqrt{2}}\left(|\uparrow_1\uparrow_2\rangle + |\downarrow_1\downarrow_2\rangle\right)$$

$$= \frac{1}{2}\left[\underbrace{\langle\uparrow_1|\uparrow_1\rangle}_{1}|\uparrow_2\rangle + \underbrace{\langle\uparrow_1|\downarrow_1\rangle}_{0}|\downarrow_2\rangle + \underbrace{\langle\downarrow_1|\uparrow_1\rangle}_{0}|\uparrow_2\rangle + \underbrace{\langle\downarrow_1|\downarrow_1\rangle}_{1}|\downarrow_2\rangle\right]$$

$$= \frac{1}{2}\left[|\uparrow_2\rangle + |\downarrow_2\rangle\right] = \frac{1}{\sqrt{2}}|x+_2\rangle. \tag{6.5}$$

As expected, the inner product is a particle 2 ket. The probability of $|x+_1\rangle$ is the squared magnitude of this ket, which is only 1/2! Therefore, despite its look, the state $\frac{1}{\sqrt{2}}(|\uparrow_1\uparrow_2\rangle+|\downarrow_1\downarrow_2\rangle)$ is *not* a state of definite particle 1 spin in the $x+$ direction. Looks can be deceiving. By symmetry, it is also *not* a state of definite particle 2 spin.

We are now ready to address partial collapse. We know that measurements collapse quantum states, so if we measure particle 1 to be $|x+_1\rangle$, what is our resulting system state? The particle 1 state is that consistent with our measurement, i.e., $|\chi_1\rangle=|x+_1\rangle$ (just as with single-particle collapse). The particle 2 state is simply that of the inner product earlier, except that being a quantum state, we must normalize it to unit magnitude:

$$|\chi_2\rangle = normalize\left(\frac{1}{\sqrt{2}}|x+_2\rangle\right)=|x+_2\rangle.$$

Therefore, our two-particle state after measuring particle 1 is:

$$\chi_{12,after} =|x+_1\rangle|x+_2\rangle \quad or \quad |x+_1,x+_2\rangle \quad \text{(after measurement)}.$$

Note that the initial state is entangled, and in this case, measuring particle 1 determines also the state of particle 2.

The probability of measuring $|x+_1,x+_2\rangle$ was only 1/2. What else might we have measured? And with what probabilities? We can compute these by subtracting this known component state from our initial state:

$$|\chi_{12}\rangle = \frac{1}{\sqrt{2}}|\uparrow_1\uparrow_2\rangle+\frac{1}{\sqrt{2}}|\downarrow_1\downarrow_2\rangle = \frac{1}{\sqrt{2}}|x+_1,x+_2\rangle+|others\rangle \quad \Rightarrow$$

$$|others\rangle = \frac{1}{\sqrt{2}}|\uparrow_1\uparrow_2\rangle+\frac{1}{\sqrt{2}}|\downarrow_1\downarrow_2\rangle-\frac{1}{\sqrt{2}}|x+_1,x+_2\rangle. \quad (6.6)$$

Now we expand the entangled state $|x+_1,x+_2\rangle$ in our z basis:

$$|x+_1,x+_2\rangle=\underbrace{\frac{1}{\sqrt{2}}(|\uparrow_1\rangle+|\downarrow_1\rangle)}_{|x+_1\rangle}\otimes\underbrace{\frac{1}{\sqrt{2}}(|\uparrow_2\rangle+|\downarrow_2\rangle)}_{|x+_2\rangle}$$

$$=\frac{1}{2}(|\uparrow_1\uparrow_2\rangle+|\uparrow_1\downarrow_2\rangle+|\downarrow_1\uparrow_2\rangle+|\downarrow_1\downarrow_2\rangle). \quad (6.7)$$

Then,

$$|others\rangle = \frac{1}{\sqrt{2}}|\uparrow_1\uparrow_2\rangle+\frac{1}{\sqrt{2}}|\downarrow_1\downarrow_2\rangle-\frac{1}{\sqrt{2}}\frac{1}{2}(\uparrow_1\uparrow_2 + \uparrow_1\downarrow_2 + \downarrow_1\uparrow_2 + \downarrow_1\downarrow_2)$$

$$=\frac{1}{2\sqrt{2}}[|\uparrow_1\uparrow_2\rangle-|\uparrow_1\downarrow_2\rangle-|\downarrow_1\uparrow_2\rangle+|\downarrow_1\downarrow_2\rangle].$$

The ket in square brackets factors into a tensor product of single-particle states:

$$|others\rangle = \frac{1}{2\sqrt{2}}\left[\underbrace{\left(|\uparrow_1\rangle - |\downarrow_1\rangle\right)}_{\sqrt{2}|x_{-1}\rangle} \otimes \underbrace{\left(|\uparrow_2\rangle - |\downarrow_2\rangle\right)}_{\sqrt{2}|x_{-2}\rangle}\right] = \frac{1}{\sqrt{2}}|x-_1, x-_2\rangle.$$

Plugging in to Eq. 6.6 we see that our initial state is a superposition of two simple product states:

$$|\chi_{12}\rangle = \frac{1}{\sqrt{2}}|\uparrow_1 \uparrow_2\rangle + \frac{1}{\sqrt{2}}|\downarrow_1 \downarrow_2\rangle = \frac{1}{\sqrt{2}}|x+_1, x+_2\rangle + \frac{1}{\sqrt{2}}|x-_1, x-_2\rangle.$$

It is interesting that $|\chi_{12}\rangle$ has the same form in both the z basis and the x basis. We see this by rewriting our shorthand for the z basis in the notation we used for the x basis:

$$|\chi_{12}\rangle = \frac{1}{\sqrt{2}}|\uparrow_1 \uparrow_2\rangle + \frac{1}{\sqrt{2}}|\downarrow_1 \downarrow_2\rangle \equiv \frac{1}{\sqrt{2}}|z+_1, z+_2\rangle + \frac{1}{\sqrt{2}}|z-_1, z-_2\rangle.$$

In contrast to this superposition of product states, let us consider a simple tensor product state, $|x+_1\rangle \otimes |x+_2\rangle$, which we worked out in Eq. 6.7. By construction, it *is* a state of definite particle 1 spin in the $x+$ direction, *and separately* definite particle 2 spin. In expanded form, as on the RHS of Eq. 6.7, it is hard to see by inspection that this state has definite spin for both particles.

These principles of partial measurements, partial inner products, probabilities of partial measurements, and partial collapse apply to any tensor product state, including space $\otimes$ spin. Recall the space $\otimes$ spin example of the Stern–Gerlach device we considered in Sect. 4.4.3: position is entangled with spin, so position is a *proxy* for spin. Measuring the position tells us also the spin and collapses the state to one of definite position *and* definite spin.

6.4.2   The EPR Paradox (Not)

Multiparticle QM provides a way to decide whether nature is truly probabilistic, or QM is incomplete. The effect we examine is called the "EPR paradox," after Einstein, Podolski, and Rosen, who first brought attention to it with a thought experiment. We show that, though the result may be unexpected and counterintuitive, there is no contradiction in the physics, and nature is indeed probabilistic. This result is quite important: you can now buy commercial quantum encryption systems that use this principle. Such systems are essentially unbreakable (so long as QM is correct). Our development uses only the simple QM we have already developed, and also shows that the axioms of quantization, entanglement, and the correspondence principle together imply that QM must be nonlocal (i.e., must have effects

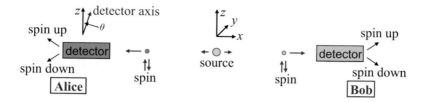

Fig. 6.5 Schematic of the EPR experiment. Alice's detector axis is off by θ

that span distances in a time faster than light can traverse them). Our analysis is inspired by a section of Heinz R. Pagels' book *In Search of Schrödinger's Cat*.

Is QM is incomplete? Are there "hidden variables", which quantum theory is neglecting, that would eliminate the uncertainty of QM? In the EPR experiment, a spin-0 source emits a pair of particles, A and B, in opposite directions. The total spin of the pair is zero, but they are emitted in the entangled state $|\uparrow\downarrow\rangle + |\downarrow\uparrow\rangle$ (the first arrow refers to particle A, the second to particle B). The experiment measures the spins of both particles, and is repeated over many pairs, with varying directions of measurement. (In principle, the experiment can use any of a variety of two-state properties, such as polarization of photons, etc.)

We start by being skeptical of QM, and suppose that each pair is actually emitted in a state with *definite* orientation, but QM is too ignorant to know about it. In other words, we suppose that the true orientation of each pair has a definite value from its creation, but which QM is "too dumb" to quantify. QM is only smart enough to predict probabilities. We suppose there is no essential uncertainty in nature, just ignorance of some of the state variables.

Bob starts with a perfect z-spin measuring device, which measures the spin direction (up or down along z) of particle B. Alice measures the spin of particle A. She tries to measure along the z-axis, but her experimental setup is imperfect, and is actually slightly rotated around the x-axis by an angle θ with respect to the z-axis. If she measures spin down, then knowing that the total spin is zero, she predicts Bob's perfect device measures B spin up. Thus, she decides the two-particle state is $|\downarrow\uparrow\rangle$. However, because of her angular error, θ, she sometimes predicts incorrectly.

We now compute the fraction of the time that Alice and Bob make various errors, and how often they agree on results. We compare the predictions of QM, where the particle state is *not determined even in principle* until Alice makes her measurement, against the predictions of our skeptical supposition that the state is determined at pair creation, and we just do not know which state it is until Alice measures it. We show that the two theories predict different results, and are therefore distinguishable by experiment. Many such experiments have been done, and QM wins every time.

a.   With Alice's imperfect measurement, and Bob's perfect measurement, what fraction of her decisions are in error (call it f)? We find f from Eq. (5.5), to wit:

$$\Pr(z+) = \cos^2 \frac{\theta}{2} \quad \Rightarrow \quad f = 1 - \cos^2 \frac{\theta}{2} = \sin^2 \frac{\theta}{2}.$$

This fraction is a QM result which has been tested, and is correct. We note that her error rate f is a maximum when $\theta = \pi/2$, and $f = 0.5$.

b.   We now replace Bob's perfect measuring device with one that is angled in error by $-\theta$, and we replace Alice by a perfectly aligned measuring device. By symmetry, what fraction of Bob's decisions is in error? Since physics is rotationally invariant, we must again have $f = \sin^2(\theta/2)$.

c.   Now consider that Alice and Bob both have alignment errors: Alice with $+\theta$ and Bob with $-\theta$ misalignment. Each decides the state $|AB\rangle$ from his/her own measurements. Each makes independent errors with respect to a hypothetical perfect z-axis device. Given f, then from simple probabilities, without using QM, what fraction of the time will Alice and Bob have different decisions for the state $|AB\rangle$, in terms of f? Note that in this case, two wrongs make an agreement. So the probability of agreement is the chance that Alice and Bob are either both right or both wrong:

$$\Pr(agree) = \Pr(\text{Alice right})\Pr(\text{Bob right}) + \Pr(\text{Alice wrong})\Pr(\text{Bob wrong})$$
$$= (1-f)(1-f) + f^2 = 1 - 2f + 2f^2.$$

d.   Substitute from part (a), to eliminate f in favor of θ. In terms of θ, under the assumption of independent errors, for what fraction will Alice and Bob agree? This is what would happen if the particles had some definite, but hidden, spin orientation, and if one person's measurement did not affect the other's.

$$\Pr(agree) = \left(\cos^2\frac{\theta}{2}\right)\left(\cos^2\frac{\theta}{2}\right) + \left(\sin^2\frac{\theta}{2}\right)\left(\sin^2\frac{\theta}{2}\right) = \cos^4\frac{\theta}{2} + \sin^4\frac{\theta}{2}. \qquad (6.8)$$

Classical probabilities predict that when $\theta = \pi/2$, they agree half the time.

e.   Now, using QM, and noting that Alice and Bob's devices are separated by an angle of 2θ, what does QM predict as the fraction that Alice and Bob agree?

$$\Pr(agree) = \cos^2\frac{2\theta}{2} = \cos^2\theta.$$

This fully quantum prediction is different than classical probabilities (part (d), Eq. 6.8). In reality, when $\theta = \pi/2$, they *never* agree: when one measures "up," the other also measures "up." In contrast, the classical Eq. 6.8 predicts 50% agreement, which does not, in fact, occur.

This difference in the predictions of (d): independent errors with hidden variables and (e): those of QM for entangled states, where Alice's state is *not determined until she makes a measurement*, has been experimentally confirmed to over 30σ. By this measure, the probability that QM is wrong is less than 1 chance in 10^{197}.

(Realistically, though, the chance that the experimenters made a mistake would be vastly higher than this.) No experiment has ever contradicted QM. QM works.

Conclusions: This experiment shows that quantum superpositions and entanglement are fundamental aspects of nature and not a result of our ignorance. Hidden variables do not explain quantum uncertainty.

In some sense, it seems the partial collapse of the wave-function brought on by Alice's measurement occurs instantly, across arbitrarily large distances. At first, this might seem to violate Special Relativity (SR): some kind of "signal" travels from Alice to Bob, collapsing the wave-function near him. The facts are subtle, and defy such a simple classical explanation. We here note a few important points, but will not digress into speculations about their interpretation; the facts of QM are clear, indisputable, and accurately predict the outcomes of experiment. That is good enough for us.

By itself, the EPR effect does *not* allow Alice to communicate with Bob. Only a comparison of their results, including their measurement angles, reveals the interdependence of their measurements. Such a comparison requires an independent communication channel, which is itself subject to the laws of physics, precluding communication faster than light. Here is a valid quantum model to describe the process from Bob's point of view (other models are possible): for Bob, quantum theory says that when Alice takes a measurement, she and her measuring equipment evolve into a superposition state of her measuring "up" and her measuring "down." (Recall that in our model of QM, Alice cannot collapse Bob's wave-function; see Sect. 1.4.) Only when Bob finally sees her later communication does Alice and all her measuring equipment collapse into a single state of either "up" or "down." Nothing needs travel faster than light. [Note that since Alice is macroscopic, she and her measurement state decohere essentially instantly, preventing Bob from making an interference pattern from Alice and her measurement. In other words, in this case, the situation is indistinguishable from Alice collapsing the wave-function by her measurement.]

Interestingly, SR implies a restriction on QM: the no cloning theorem. If Bob could make many exact copies of his particle's state, he could make hundreds of copies, and measure the polarization of all the copies many times at each of slightly different angles. This would determine very well what the original particle's incoming polarization eigendirection had been. This would allow Alice to *choose* a polarization direction for her measurement as a signal to Bob, who could then (quickly) determine which direction Alice chose from arbitrarily far away, and thus receive her "message" faster than light. By this circuitous route, SR demands that arbitrary quantum states be impossible to duplicate exactly. One can also prove the no-cloning theorem directly from QM theory.

Chapter 7
Quantum Electromagnetic Radiation

Electromagnetic (EM) interactions are probably the most fundamental and important in our lives. Most of our everyday experience is dominated by EM interactions (and gravity). Quantum electromagnetism is the first quantum field theory (QFT), on which all following QFTs are built. Thus, quantum electromagnetism not only explains much of the world around us, but is a prerequisite for the QFTs that explain the rest.

We present here an overview of quantized radiation and matter–radiation interaction. This is a huge field, and an active area of research. [10] provides an accessible, but fairly thorough, tutorial on the subject.

In many cases, one can reasonably approximate some features of quantized matter interactions with radiation by using the "semiclassical approximation:" one treats the charged particles as quantized, but the radiation field as a classical potential in which the particle acts. Such an approximation describes EM absorption by matter and stimulated emission. This can also be thought of as a semiclassical EM field: the photon has a definite energy and a simple wave-function, just like ordinary quantum mechanics (QM). For a semiclassical EM field, the vector potential $A(t, r)$ is essentially the wave-function of the photon. $|A(t, r)|^2$ is proportional to the particle density of the photon at (t, r). We do not address this semiclassical approximation here. We also do not address spin interactions.

However, many phenomena cannot be described by a semiclassical approximation, the most important being spontaneous emission. It requires a fully quantized EM field, which allows for the creation of a photon even in the absence of any prior EM excitation, i.e., even into the vacuum. The quantized EM field is the main topic of this chapter. It also quantitatively describes the Lamb shift of atomic spectra, the gyromagnetic ratio of the electron, the Casimir force, and multiple photon detections.

In general, a quantized field allows for particle creation and annihilation, and is the essence on which all QFTs are built.

We discuss the following aspects of quantum EM radiation:

1. Quantized EM field: A failed attempt at photons.
2. Quantized EM field: the simple harmonic oscillator (SHO) analogy.
3. Quantized EM fields conceptually introduce a new kind of wave-function (the quantized field), but we find that we never need to explicitly use it.
4. Differences between a one-dimensional (1D) harmonic oscillator and the EM field.
5. Combining matter states and photon (radiation) states.
6. Example of a simplified matter-photon operator in action, evaluating an inner product.

E. L. Michelsen, *Quirky Quantum Concepts*, Undergraduate Lecture Notes in Physics, DOI 10.1007/978-1-4614-9305-1_7, © Springer Science+Business Media New York 2014

7. Example of the complete $\hat{\mathbf{A}}$ operator (vector potential)in action, evaluating an inner product.
8. Spontaneous radiation.
9. Photons have all phases.
10. Photons in other bases (other modes).
11. The wave-function of a photon.
12. Quasiclassical states, average E-field is not quantized, and interference (number operators).

On a first reading, one can skip the more advanced photon and quasiclassical theory.

This section requires a thorough understanding of classical EM propagation, including phasors (the complex representation of sinusoids), wave-vectors, elementary polarization, and the three-vector potential, $\mathbf{A}(t, \mathbf{r})$. You must also understand Dirac notation, the quantum SHO, ladder operators, multiparticle QM, and tensor product states and their inner products.

Before going further, we again caution against stating what is "really" happening at the microscopic, quantum scale. None of QM is directly observable. Therefore, we present here the generally accepted model, as simply as we can. The model is based on now-familiar QM principles, and it quantitatively predicts the outcomes of experiments.

7.1 Quantized EM Field: A Failed Attempt at Photons

Imagine an essentially classical EM wave of frequency ω in a volume of space, but quantized in amplitude such that its energy is $E = \hbar\omega$. Then:

$$\mathbf{A}(t,\mathbf{r}) = \mathrm{Re}\left\{\mathbf{A}_0 e^{i(\mathbf{k}\cdot\mathbf{r}-\omega t)}\right\} \quad \text{where} \quad \mathbf{A}_0 \equiv \text{complex vector};$$

$$E = \hbar\omega \infty \int_\infty d^3\mathbf{r} \, \mathbf{A}^2(t,\mathbf{r}) = \frac{1}{2}(Vol)|\mathbf{A}_0|^2.$$

We might propose this as a crude model for a "photon," a quantized EM field (though we will see it is not a very good model). We usually separate out the overall magnitude and phase of the wave from its polarization, using a phasor, A_0 (a complex number representing a sinusoid) [10, p. 354], such that,

$$\mathbf{A}_0 = A_0\boldsymbol{\varepsilon}.$$

Then,

$$\mathbf{A}(t,\mathbf{r}) = \mathrm{Re}\left\{A_0\boldsymbol{\varepsilon}e^{i(\mathbf{k}\cdot\mathbf{r}-\omega t)}\right\}$$

$$\textit{where} \quad A_0 \equiv \text{phasor}; \qquad \boldsymbol{\varepsilon} \equiv \text{complex polarization vector.}$$

$$|\boldsymbol{\varepsilon}|^2 = 1 \Rightarrow E = \hbar\omega \propto \frac{1}{2}(Vol)|A_0|^2.$$

This simple model of a photon explains the well-known photoelectric effect: it takes radiation of a certain frequency or higher to ionize a substance. No amount of light of lower frequency, no matter how intense, will ionize the substance. We explain this frequency cutoff by noting that for ionization, our incident radiation must provide enough energy to unbind an electron. Since we have supposed that radiation is quantized into bundles of energy called photons, ionization requires a photon of sufficient energy. Finally, since $E=\hbar\omega$, sufficient energy for ionization requires sufficient frequency ω.

However, this simple model does not explain several phenomena, such as spontaneous emission, the Lamb shift of atomic spectra, the gyromagnetic ratio of the electron, and the Casimir force.

7.2   Photon Number States

Despite the failure of our simple model, it does suggest a useful concept for the quantum state of an EM field. We know that some kind of quantized EM field exists, since photons of energy $E=\hbar\omega$ are seen experimentally. Furthermore, the existence of coherent classical EM fields suggests that multiple photons of a single mode (single frequency, phase, and polarization) can exist. Also, multiple modes of excitation can simultaneously exist. We, therefore, suppose that the quantum state of an EM field can be given by a list of each mode, and the number of photons in that mode. For example, given modes characterized by wave-vectors $\mathbf{k}_j$, and polarization vectors $\boldsymbol{\varepsilon}_1(\mathbf{k}_j)$ and $\boldsymbol{\varepsilon}_2(\mathbf{k}_j)$, we might have EM states such as:

$$\left|N_1(\mathbf{k}_1)\right\rangle; \qquad \left|N_2(\mathbf{k}_1)\right\rangle; \qquad \left|N_1(\mathbf{k}_2)\right\rangle; \quad \ldots \quad \left|N_\lambda(\mathbf{k}_n)\right\rangle,$$

$$\lambda = 1, 2; \quad N_\lambda = 0, 1, \ldots \infty; \quad n = 1, 2, \ldots \infty.$$

Note that the subscript on N and $\boldsymbol{\varepsilon}$ refers to one of the two polarization states, whereas the subscript on $\mathbf{k}$ refers to one of an infinite number of wave-vectors.

In the momentum basis, a photon is described by its wave-vector, $\mathbf{k}$ (aka propagation vector) and its polarization vector, $\boldsymbol{\varepsilon}$. Therefore, a photon state is a (polarization-vector, wave-vector)pair, written as $\boldsymbol{\varepsilon}(\mathbf{k})$. The wave-vector tells you the state's EM propagation direction and spatial frequency $|\mathbf{k}|=2\pi/\lambda$ rad/m, and therefore also temporal frequency $\omega=c|\mathbf{k}|$. The polarization vector tells you how the photon state is polarized. For each $\mathbf{k}$, two independent polarizations exist, $\boldsymbol{\varepsilon}_1(\mathbf{k})$ and $\boldsymbol{\varepsilon}_2(\mathbf{k})$. A general photon state is then written $\boldsymbol{\varepsilon}_\lambda(\mathbf{k})$, where $\lambda=1$ or 2. (Many references use the alternate notation $\boldsymbol{\varepsilon}_{\mathbf{k}\lambda}$.)

> For every wave-vector $\mathbf{k}$ (aka propagation vector), there are two independent polarizations that can propagate with that wave-vector.

Here, we do not need to know the details of a polarization vector; we need only that for any given propagation vector **k**, there are two independent polarization modes, say horizontal and vertical, or right-hand circular (RHC) and left-hand circular (LHC). Each polarization mode is described by the presence of a distinct photon. Either photon or both may exist in space.

Photons are bosons, so a single state can be occupied by any number of photons, from 0 on up. A complete photon state therefore includes a photon count (aka **occupation number**) for each $(\mathbf{k}, \boldsymbol{\varepsilon})$ pair, written in general as $N_\lambda(\mathbf{k}_j)$.

We write an EM state of multiple mode excitations (multiple photons) as a single ket, e.g.,

$$\left| N_2(\mathbf{k}_1) = 3, N_1(\mathbf{k}_2) = 2, N_1(\mathbf{k}_3) = 1 \right\rangle \quad \text{(multimode quantum EM state).} \quad (7.1)$$

Of course there are an infinite number of possible modes, and most of them are unoccupied [$N_\lambda(\mathbf{k}) = 0$], so we omit those from the notation: any modes not listed in the ket are defined to be unoccupied. Such a state is said to be written in the **number basis**.

  A two-photon state is *not* the vector sum of two one-photon states.

For example:

$$\left| N_1(\mathbf{k}_1) = 1, N_1(\mathbf{k}_2) = 1 \right\rangle \neq \left| N_1(\mathbf{k}_1) = 1 \right\rangle + \left| N_1(\mathbf{k}_2) = 1 \right\rangle.$$

For one thing, the RHS is not normalized. Even if we normalized it, though, it would be a superposition of two one-photon states. We might detect either one of the component photons (thus collapsing the wave-function), but not both. In contrast, a two-photon state of the EM field has two whole photons in it, and both can be detected.

The state with no photons at all is called the *vacuum state*, and written:

$$\left| 0 \right\rangle \equiv \text{vacuum state.} \qquad \text{Note: } \left| 0 \right\rangle \neq \mathbf{0_v}.$$

NB: the vacuum state is *not* the zero-vector, $\mathbf{0}_v$ (which is also known as the "null ket").

For brevity, we may write modes of a single excitation ($N=1$) as:

$$\left| \boldsymbol{\varepsilon}_1(\mathbf{k}) \right\rangle \equiv \left| N_1(\mathbf{k}) = 1 \right\rangle, \qquad \text{or more generally,} \qquad \left| \boldsymbol{\varepsilon}_\lambda(\mathbf{k}) \right\rangle \equiv \left| N_\lambda(\mathbf{k}) = 1 \right\rangle.$$

Since the occupation number is 1, we do not write it explicitly.

In a state of multiple modes, the order does not matter. For example, a state with two photons, one of wave-vector $\mathbf{k}_1$ and polarization $\boldsymbol{\varepsilon}_1(\mathbf{k}_1)$, and another of $\boldsymbol{\varepsilon}_1(\mathbf{k}_2)$, can be written: $\left| \boldsymbol{\varepsilon}_1(\mathbf{k}_1), \boldsymbol{\varepsilon}_1(\mathbf{k}_2) \right\rangle$ or $\left| \boldsymbol{\varepsilon}_1(\mathbf{k}_2), \boldsymbol{\varepsilon}_1(\mathbf{k}_1) \right\rangle$. Again, since the occupation numbersare 1, we do not write them explicitly.

Note that higher excitations, $N \geq 2$, describe *independent* photons, each with completely indeterminate phase [11, p. 253b].

7.2.1 *Aside on Polarization Vectors*

The polarization vector completely describes the polarization of the state, given as the polarization of its vector-potential, **A**. Classically, the polarization vector is a vector of three phasors: one for each component of the A-field: A_x, A_y, and A_z. In QM, for a single photon, the polarization vector is a complex-valued vector in ordinary three-space:

$$\boldsymbol{\varepsilon} = (\varepsilon_x, \varepsilon_y, \varepsilon_z) \quad \text{where} \quad \varepsilon_x, \varepsilon_y, \varepsilon_z \text{ are complex components,}$$

giving the QM amplitudefor the x, y, and z components of **A**. Polarization vectors are dimensionless, and normalized to unit magnitude:

$$|\boldsymbol{\varepsilon}|^2 = \boldsymbol{\varepsilon} \cdot \boldsymbol{\varepsilon} = \boldsymbol{\varepsilon}^* \boldsymbol{\varepsilon} = |\varepsilon_x|^2 + |\varepsilon_y|^2 + |\varepsilon_z|^2 = \varepsilon_x^* \varepsilon_x + \varepsilon_y^* \varepsilon_y + \varepsilon_z^* \varepsilon_z = 1.$$

(It is not necessary here to understand the details of what a polarization vector means, however *Funky Electromagnetic Concepts* explains them completely for **E** fields; the transition to **A** fields is straightforward.) All that matters right now is that for every wave-vector **k**, there are two independent polarizations that can propagate with that wave-vector.

As with all quantum states, the polarization state can be a superposition of $\boldsymbol{\varepsilon}_1$ and $\boldsymbol{\varepsilon}_2$. Therefore, for a given propagation direction, $\mathbf{e_k} \equiv \mathbf{k}/|\mathbf{k}|$, any polarization vector (aka polarization "state") can be written as a linear combination of two basis polarization vectors, $\boldsymbol{\varepsilon}_1$ and $\boldsymbol{\varepsilon}_2$. The most common basis polarization vectors are RHC and LHC, which we could write as $\boldsymbol{\varepsilon}_R(\mathbf{k})$ and $\boldsymbol{\varepsilon}_L(\mathbf{k})$. (These are handy because they are angular momentum eigenstates.)However, for most of our calculations, the basis is irrelevant. Note that the polarization vector $\boldsymbol{\varepsilon}_R(\mathbf{k})$ *is a function of the direction of* **k**, so $\boldsymbol{\varepsilon}_R(z$-direction) is a *different* vector than $\boldsymbol{\varepsilon}_R(x$-direction). [In fact, $\boldsymbol{\varepsilon}_R(\mathbf{e}_z) = (1, i, 0)/\sqrt{2}$ and $\boldsymbol{\varepsilon}_R(\mathbf{e}_x) = (0, 1, i)/\sqrt{2}$, but we do not need to know that here.]

7.3 Quantized EM Field: The SHO Analogy

When we describe the interaction of radiation and matter, we must talk about photons being created (radiated) and destroyed (absorbed). Thus, we introduce a new concept to QM: particle creation and destruction (aka **annihilation**). It is this concept that demands QFT (aka "second quantization"). In this section, we take an approach known as "canonical quantization," where the classical vector-potential (**A**-field) becomes an operator $\hat{\mathbf{A}}$, acting on photon states.

(A different approach uses Feynman Path Integrals (FPIs), where the **A**-field remains a complex number function of space (as in classical E&M when working in Fourier space). FPI have some advantages over canonical quantization, but we do not address that further here.)

In classical EM, we define a plane wave with a definite vector potential $\mathbf{A}(t, \mathbf{r})$:

$$\mathbf{A}(t,\mathbf{r}) = \text{Re}\left\{A_0\ \boldsymbol{\varepsilon}\ e^{i(\mathbf{k}\cdot\mathbf{r}-\omega t)}\right\} \quad \text{where} \quad A_0 \text{ is a phasor; } \boldsymbol{\varepsilon} \text{ is the polarization vector,}$$

with no uncertainty. Note that the classical $\mathbf{A}$-field oscillate in both space and time.

However, when we quantize the EM field, i.e., when we consider photons as quantum particles, the vector-potential, $\mathbf{A}$, is no longer definite. A state of definite photon count, say $|N_\lambda(\mathbf{k})=1\rangle$, has uncertain $\mathbf{A}$. $\mathbf{A}$ is a probabilistic function of the photon state $|N_\lambda(\mathbf{k})=1\rangle$. This is analogous to a 1D harmonic oscillator: a particle in a state of definite energy, say $|n\rangle \equiv \psi_n(x)$, has uncertain position. The position x is a probabilistic function of the state.

Why Should the EM Field Behave Like an SHO? The SHO for a particle has a potential energy proportional to x^2. For EM waves, both the E- and B-fields contain energy proportional to E^2 (and B^2). In the SHO, to move a particle to position x takes energy proportional to x^2; in EM waves, to create an E-field with magnitude E takes energy proportional to E^2. Thus, it is reasonable that EM fields quantize similarly to the SHO. Furthermore, the analogy between harmonic oscillator and vector-potential continues. For a given $\mathbf{k}$ and λ [which selects $\boldsymbol{\varepsilon}_\lambda(\mathbf{k})$]:

$$\hat{x} \quad \leftrightarrow \quad \hat{\mathbf{A}}_\lambda(\mathbf{k})$$

$$\hat{x} \sim \left(\hat{a}+\hat{a}^\dagger\right) \quad \leftrightarrow \quad \hat{\mathbf{A}}_\lambda(\mathbf{k}) \sim \left(\hat{a}_\lambda(\mathbf{k})+\hat{a}_\lambda^\dagger(\mathbf{k})\right)$$

$$\text{energy level } n \quad \leftrightarrow \quad \text{photon count } N_\lambda(\mathbf{k})$$

Note that like the SHO quantum number n, the photon count, $N_\lambda(\mathbf{k})$, for a plane-wave mode is also a measure of the energy level of the EM field for that mode (i.e., for that value of $\mathbf{k}$ and $\boldsymbol{\varepsilon}_\lambda$).

An SHO particle's position is a wave-function (superposition) of many possible positions, whose probabilities are determined by the excitation level, n, of the oscillator. For each position x, $\psi_n(x)$ gives the probability amplitude for the particle to have that position x (Fig. 7.1, *left*). As usual, $\hat{x}$ is an operator which acts on the oscillator state $\psi_n(x)$ and multiplies each point of the wave-function by its position, x:

$$\hat{x}\psi_n(x) = x\psi_n(x).$$

The x-hat operator weights each possible position x by the complex amplitude, $\psi(x)$, for the particle to have that position.

The analog for an EM field is this: consider a single point in time and space $(t, \mathbf{r})$: at this point, it is not a particle *position* which is uncertain, it is the vector potential instantaneous amplitude $\mathbf{A}(t, \mathbf{r})$ which is uncertain (Fig. 7.1, right). Therefore, it is equivalent to there being a new kind of wave-function for the vector-potential *at that one point* [25, p. 18b]. We here call it $\phi_N(\mathbf{A})$ ("phi" and "photon" both start with "ph") and N is the photon count (for some given $\mathbf{k}$ and $\boldsymbol{\varepsilon}$). This new wave-function is a function of the three-dimensional (3D) *vector* $\mathbf{A}$:

$$\phi_N(\mathbf{A}) \equiv \phi_N(A_x, A_y, A_z).$$

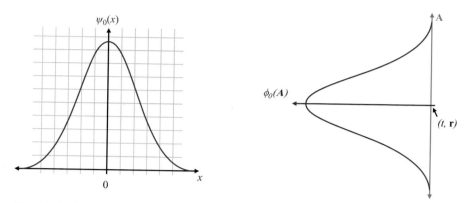

Fig. 7.1 Analogy between (*left*) particle wave-function $\psi_0(x)$ and (*right*) photon vector-potential wave-function $\phi_0(\mathbf{A})$ at a spacetime point $(t, \mathbf{r})$

For each possible value of vector potential $\mathbf{A}$, $\phi_N(\mathbf{A})$ gives the probability amplitude for the field to have that vector potential $\mathbf{A}$. The probabilities of the possible $\mathbf{A}$-values vary with photon count, N. Then, $\hat{\mathbf{A}}$ is an operator which acts [at the point $(t, \mathbf{r})$] on the EM field state, and (much like SHO operators) multiplies each point of $\phi_N(\mathbf{A})$, in the 3D space of $\mathbf{A}$, by its vector potential $\mathbf{A}$:

$$\hat{\mathbf{A}}\phi_N(\mathbf{A}) = \mathbf{A}\phi_N(\mathbf{A}).$$

This is a vector-valued function of the 3D $\mathbf{A}$-space. Note that a different instance of $\hat{\mathbf{A}}\phi_N(\mathbf{A})$ exists at every point in real space.

We will see, though, that in calculations we never need to explicitly use this new kind of wave-function, $\phi_N(\mathbf{A})$, because we use raising and lowering operator sinstead. However, it is important to understand its meaning conceptually. The vacuum state, or ground state, is $\left|N_\lambda(\mathbf{k}) = 0\right\rangle$ for all λ, $\mathbf{k}$. But just like the SHO, the vacuum has some probability of producing an $\mathbf{A}$ field, *even though there are no photons present!* We sketch one dimension of $\phi_0(\mathbf{A})$ in Fig. 7.2, *left* (for some arbitrary $\mathbf{k}$, $\boldsymbol{\varepsilon}$), along with the probability density for the possible values of $\mathbf{A}$ over some distance in the (say) x-direction of real space.

> [We note in passing that, since the $\mathbf{E}$ and $\mathbf{B}$ fields derive from $\mathbf{A}$, the vacuum has some probability of producing $\mathbf{E}$ and $\mathbf{B}$ fields, too! The *average* values of $\mathbf{A}$, $\mathbf{E}$, and $\mathbf{B}$ are all zero, since they are equally likely to be positive or negative. But, perhaps more significantly, the *average* value of $\mathbf{E}^2$ and $\mathbf{B}^2$ are positive! Which, if we take this at face value, means the vacuum has energy: an infinite amount of energy. Some physicists believe this has something to do with dark energy and the cosmological constant of General Relativity. Though these vacuum states can be used to explain the Casimir force, their full nature is, in my opinion, not clear.]

In Fig. 7.2, *right*, we sketch one dimension of $\phi_1(\mathbf{A})$, where $N = 1$ (for some arbitrary λ, $\mathbf{k}$) and the probability density for the possible values of $\mathbf{A}$. Here we find the

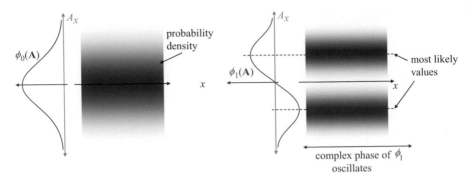

Fig. 7.2 Probability densities of values of **A** at a point. (*Left*) For vacuum, $N=0$. (*Right*) For one photon, $N=1$. $\phi(\mathbf{A})$ oscillates in space

most likely values of **A** are nonzero, consistent with the presence of an EM wave. However, the average is still zero, because the EM phase is completely uncertain.

7.3.1   SHO vs. EM Wave: Similar, but Different

Though there is an analogy between a 1D SHO and the oscillations of an EM field, the field is more complicated than the SHO. We here briefly compare the two, then discuss each point in more detail:

1. The SHO $\hat{x}$ acts on oscillator states such as $|n\rangle$. The vector potential operator $\hat{\mathbf{A}}$ acts on EM field states such as $|N_\lambda(\mathbf{k})\rangle$.
2. For a 1D harmonic oscillator, position is a scalar, but the vector-potential is a vector. Therefore, the position operator, $\hat{x}$, produces a scalar function of space, but the vector-potential operator, $\hat{\mathbf{A}}$, produces a vector function of A-space. As with any vector operator, you can think of $\hat{\mathbf{A}}$ as a set of three operators:

$$\hat{\mathbf{A}} \equiv (\hat{A}_x, \hat{A}_y, \hat{A}_z) \equiv \hat{A}_x \mathbf{e}_x + \hat{A}_y \mathbf{e}_y + \hat{A}_z \mathbf{e}_z.$$

3. $\hat{x}$ describes a *single* SHO position value, but $\hat{\mathbf{A}}$ is an infinite set of coupled oscillators, spread through space. Therefore, there is a different $\hat{\mathbf{A}}$ operator at each point of time and space. In other words, the $\hat{\mathbf{A}}$ operator is itself a function of time and space, thus $\hat{\mathbf{A}}$ (for a given $\mathbf{k}$ and $\mathbf{\varepsilon}$) is sometimes written $\hat{\mathbf{A}}(t,\mathbf{r})$.
4. Furthermore, every possible mode (combination of $\mathbf{k}$ and $\mathbf{\varepsilon}$) has its own $\hat{\mathbf{A}}$. The total vector potential is the sum of all possible modes; each mode is its own harmonic oscillator. Therefore, at each point in time and space, there is a contribution from the infinite set of harmonic oscillators covering all modes.

More detail:
1. The SHO $\hat{x}$ acts on oscillator states such as $|n\rangle$. The vector potential operator $\hat{\mathbf{A}}$ acts on EM field states such as $|N_\lambda(\mathbf{k})\rangle$.

For a harmonic oscillator, we can write the $\hat{x}$ operator in the energy basis:

$$\hat{x} = \sqrt{\frac{\hbar}{2m\omega}}\left(\hat{a} + \hat{a}^\dagger\right) \text{ where } \hat{a} \text{ and } \hat{a}^\dagger \text{ are energy annihilation and creation operators.}$$

(Note that $\hat{a}$ and $\hat{a}^\dagger$ are operators, but are often written in the literature without a "hat".) Similarly, we can write the vector-potential operator in the photon-number basis for a given $(\mathbf{k}, \boldsymbol{\varepsilon})$ mode as:

$$\hat{\mathbf{A}}_\lambda(\mathbf{k}) = something\left(\hat{a}_\lambda(\mathbf{k}) + \hat{a}_\lambda^\dagger(\mathbf{k})\right). \tag{7.2}$$

$\hat{\mathbf{A}}$ is Hermitian, because the vector potential is real (recall that Hermitian operators produce real observables). Therefore, we have:

$$\hat{\mathbf{A}}_\lambda(\mathbf{k}) = \mathbf{f}(t,\mathbf{r})\hat{a}_\lambda(\mathbf{k}) + \mathbf{f}^*(t,\mathbf{r})\hat{a}_\lambda^\dagger(\mathbf{k}) \quad \text{where} \quad \mathbf{f}(t,\mathbf{r}) \text{ is a complex vector function.}$$

2. For a 1D harmonic oscillator, position is a scalar, but the vector-potential is a vector. The vector-potential operator, $\hat{\mathbf{A}}$ is a vector operator: it produces a vector function. We can think of a vector operator as three scalar operators: $\hat{\mathbf{A}} \equiv \left(\hat{A}_x, \hat{A}_y, \hat{A}_z\right)$. However, $\hat{A}_x$, $\hat{A}_y$, and $\hat{A}_z$ are tied together at each wave-vector $\mathbf{k}$ by the polarization vector $\boldsymbol{\varepsilon}(\mathbf{k})$, which does two things: (1) since $\boldsymbol{\varepsilon} \cdot \mathbf{k} = 0$, $\mathbf{A} \cdot \mathbf{k} = 0$ as well; and (2) the vector potential components A_x, A_y, and A_z have a coherent phase relationship between them that maintains the polarization at all times (e.g., RHC). Because of these tight constraints on $\hat{A}_x$, $\hat{A}_y$, and $\hat{A}_z$, it is clearer to write them as a single vector-valued operator, $\hat{\mathbf{A}}$. The vector for $\hat{\mathbf{A}}$ for a *single* mode has two terms: one proportional to the polarization vector $\boldsymbol{\varepsilon}$ and the other proportional to its complex conjugate $\boldsymbol{\varepsilon}^*$. But for a given $\mathbf{k}$, there are also two independent polarization vectors, $\boldsymbol{\varepsilon}_1$ and $\boldsymbol{\varepsilon}_2$, so $\hat{\mathbf{A}}$ looks like this:

$$\hat{\mathbf{A}}(t,\mathbf{r}) \sim \sum_{\lambda=1}^{2}\left[(something)\boldsymbol{\varepsilon}_\lambda + (something\text{-}else)\boldsymbol{\varepsilon}_\lambda^*\right].$$

Note that the sum of a complex number and its conjugate is real, so pairing $\boldsymbol{\varepsilon}$ with $\boldsymbol{\varepsilon}^*$ is necessary to make $\hat{\mathbf{A}}$ Hermitian. Combining with the prior requirement for $\hat{\mathbf{A}}$ to be a sum of creation and annihilation operators, Eq. (7.2), we get (for a given $(\mathbf{k}, \boldsymbol{\varepsilon})$ mode):

$$\hat{\mathbf{A}}(t,\mathbf{r}) \sim \sum_{\lambda=1}^{2}\left(g(t,\mathbf{r})\boldsymbol{\varepsilon}_\lambda(\mathbf{k})\hat{a}_\lambda(\mathbf{k}) + g^*(t,\mathbf{r})\boldsymbol{\varepsilon}_\lambda^*(\mathbf{k})\hat{a}_\lambda^\dagger(\mathbf{k})\right) \tag{7.3}$$

where $g(t,\mathbf{r})$ is a complex scalar function.

3. $\hat{x}$ describes a *single* position value, but the vector-potential field has an infinite set of coupled values, spread through time and space.

Therefore, there is a different $\hat{\mathbf{A}}$ operator at each point of time and space. In other words:

The $\hat{\mathbf{A}}$ operator is itself a function of time and 3D space. Thus, $\hat{\mathbf{A}}$ is sometimes written $\hat{\mathbf{A}}$ $(t, \mathbf{r})$.

This means our photon oscillator wave-function $\phi_N(\mathbf{A})$ is really $\phi_N(\mathbf{A}; t, \mathbf{r})$ because each point in space has its own ϕ_N for $\mathbf{A}$. However, these spatially distinct ϕ_N are coupled to each other by Maxwell's equations. In other words, $\hat{\mathbf{A}}(t, \mathbf{r})$ is an operator which acts on a photon state $|N_\lambda(\mathbf{k})\rangle$, to produce a *superposition of EM waves*, each of which is spread over space and time. Still, at a given point $(t, \mathbf{r})$, $\hat{\mathbf{A}}(t, \mathbf{r})$ weights each possible vector potential value by the complex amplitude for the EM field to have that value.

For the plane-wave basis, the spatial part of $\hat{\mathbf{A}}$ includes just the plane-wave propagation factor. For a single wave vector $\mathbf{k}$:

$$g(t,\mathbf{r}) = e^{i(\mathbf{k}\cdot\mathbf{r}-\omega t)} \;\Rightarrow\; \mathbf{A}(t,\mathbf{r}) \sim \sum_{\lambda=1}^{2}\left(\boldsymbol{\varepsilon}_\lambda(\mathbf{k})e^{+i(\mathbf{k}\cdot\mathbf{r}-\omega t)}\hat{a}_\lambda(\mathbf{k}) + \boldsymbol{\varepsilon}_\lambda^*(\mathbf{k})e^{-i(\mathbf{k}\cdot\mathbf{r}-\omega t)}\hat{a}_\lambda^\dagger(\mathbf{k})\right)$$

where $\omega \equiv c|\mathbf{k}|$.

Note that we have used the factor $\exp[i(\mathbf{k}\cdot\mathbf{r} - \omega t)]$ to essentially convert us from momentum space, $\mathbf{k}$, to position space, $\mathbf{r}$. The spatial squared-magnitude is always 1. Therefore, the photon particle density is uniform throughout space. This is not true for other bases, such as spherical waves. We consider later the issue of a wave-function for a photon.

4. The vector potential operator is the sum of all possible photon frequencies.

There are potentially photons of all frequencies, from 0 to infinity. Each frequency is a separate harmonic oscillator. Therefore, *at each point in time and space*, there are an infinite number of creation and annihilation operators, one pair of operators for each propagation vector $\mathbf{k}$, and each polarization $\boldsymbol{\varepsilon}_\lambda$. Therefore, the full $\hat{\mathbf{A}}$ operator includes all these modes. In other words, for all $(\mathbf{k}, \boldsymbol{\varepsilon})$ pairs, $\hat{\mathbf{A}}$ produces an EM field. Thus, the complete $\hat{\mathbf{A}}$ is an integral over all wave-vectors $\mathbf{k}$, with a creation and annihilation operator for each $(\mathbf{k}, \boldsymbol{\varepsilon})$ pair:

$$\hat{\mathbf{A}}(t,\mathbf{r}) \sim \sum_{\lambda=1}^{2}\int_{-\infty}^{\infty} d^3\mathbf{k}\,(\text{something})\hat{a}_\lambda(\mathbf{k}) + (\text{something-else})\hat{a}_\lambda^\dagger(\mathbf{k}).$$

(Some references use notation such as "$\hat{a}_{\mathbf{k},\lambda}$" for the creation/annihilation operators.)

Any real EM wave is a superposition of many frequencies, with a **spectral density**: the number of photons per unit frequency. This spectral density is handled in the system state, by making the photon *state* a superposition of a continuum of frequencies. Therefore, the existence of spectral density does *not* affect the *operator* $\hat{\mathbf{A}}$.

Putting all these properties together we get the full vector potential operator $\hat{\mathbf{A}}$, in the plane-wave basis [1, 13.63 p. 274]:

$$\hat{\mathbf{A}}(t,\mathbf{r}) = \sum_{\lambda=1}^{2} \int_{-\infty}^{\infty} d^3\mathbf{k} \; A_0(\mathbf{k}) \left(\underbrace{\boldsymbol{\varepsilon}_\lambda(\mathbf{k})e^{+i(\mathbf{k}\cdot\mathbf{r}-\omega t)}}_{\text{spacetime}} \underbrace{\hat{a}_\lambda(\mathbf{k})}_{\text{photon}} + \underbrace{\boldsymbol{\varepsilon}_\lambda^*(\mathbf{k})e^{-i(\mathbf{k}\cdot\mathbf{r}-\omega t)}}_{\text{spacetime}} \underbrace{\hat{a}_\lambda^\dagger(\mathbf{k})}_{\text{photon}} \right)$$

where $A_0(\mathbf{k}) \equiv$ real amplitude for a single photon of mode $\mathbf{k}$.   (7.4)

We do not need the details here, but we can find the amplitude constant $A_0(\mathbf{k})$ by setting the energy of a single photon (above the nonzero vacuum energy) to $\hbar\omega$. This must be done carefully, since quantum effects give a nonclassical result [13, pp. 13–16]. The answer is [1, 13.60 and 63 pp. 273–274]:

$$A_0(\mathbf{k}) = \sqrt{\frac{2\pi\hbar c}{(Vol)|\mathbf{k}|}}.$$

We take A_0 as real, since number basis states are superpositions of all phases, and different modes have all possible phase relationships.

7.3.2   The A-field from State and Operator

We frequently encounter inner products of the forms:

$$\langle 0 | \hat{\mathbf{A}} | \boldsymbol{\varepsilon}(\mathbf{k}) \rangle \quad \text{or} \quad \langle \boldsymbol{\varepsilon}(\mathbf{k}) | \hat{\mathbf{A}} | 0 \rangle. \tag{7.5}$$

How can we think of these? The first represents absorption: a transition of the EM field from a single-photon state to the vacuum. The second represents radiation: a transition from the vacuum EM field to a single-photon state. We consider each in turn.

Absorption: The absorption inner product expands by substituting the full $\hat{\mathbf{A}}$ operator, Eq. (7.4):

$$\langle 0 | \hat{\mathbf{A}} | \boldsymbol{\varepsilon}(\mathbf{k}) \rangle = \langle 0 | \sum_{\lambda=1}^{2} \int_{-\infty}^{\infty} d^3\mathbf{k} \; A_0(\mathbf{k})$$

$$\left(\underbrace{\boldsymbol{\varepsilon}_\lambda(\mathbf{k})e^{+i(\mathbf{k}\cdot\mathbf{r}-\omega t)}}_{\text{spacetime}} \underbrace{\hat{a}_\lambda(\mathbf{k})}_{\text{photon}} + \underbrace{\boldsymbol{\varepsilon}_\lambda^*(\mathbf{k})e^{-i(\mathbf{k}\cdot\mathbf{r}-\omega t)}}_{\text{spacetime}} \underbrace{\hat{a}_\lambda^\dagger(\mathbf{k})}_{\text{photon}} \right) | \boldsymbol{\varepsilon}(\mathbf{k}) \rangle.$$

Of the integral over $\mathbf{k}$, and the sum over λ, only the $\mathbf{k},\lambda$ pair matching $\boldsymbol{\varepsilon}(\mathbf{k})$ survives, because the raising and lowering operators for all other modes are between the vacuum ket $|0\rangle$ and the vacuum bra $\langle 0|$, which contribute zero. Therefore,

$$\langle 0|\hat{\mathbf{A}}|\boldsymbol{\varepsilon}(\mathbf{k})\rangle = \langle 0|A_0(\mathbf{k})\boldsymbol{\varepsilon}(\mathbf{k})e^{+i(\mathbf{k}\cdot\mathbf{r}-\omega t)}\hat{a}_\varepsilon(\mathbf{k})|\boldsymbol{\varepsilon}(\mathbf{k})\rangle.$$

We can think of $\hat{a}$ acting to the *left* as *raising* the bra, so:

$$\langle 0|\hat{\mathbf{A}}|\boldsymbol{\varepsilon}(\mathbf{k})\rangle = \langle \boldsymbol{\varepsilon}(\mathbf{k})|A_0(\mathbf{k})\boldsymbol{\varepsilon}(\mathbf{k})e^{+i(\mathbf{k}\cdot\mathbf{r}-\omega t)}|\boldsymbol{\varepsilon}(\mathbf{k})\rangle.$$

This is similar to the standard form for the average value of an operator in a given quantum state, Eq. (1.5), but the "operator" is just a complex function of $(t, \mathbf{r})$. Thus, the inner product $\langle 0|\hat{\mathbf{A}}|\boldsymbol{\varepsilon}(\mathbf{k})\rangle$ is essentially the *effective* A-field for the given single-photon quantum state. This is a vector-valued function of spacetime given by:

$$\begin{aligned}\langle 0|\hat{\mathbf{A}}|\boldsymbol{\varepsilon}(\mathbf{k})\rangle &= \langle \boldsymbol{\varepsilon}(\mathbf{k})|A_0(\mathbf{k})\boldsymbol{\varepsilon}(\mathbf{k})e^{+i(\mathbf{k}\cdot\mathbf{r}-\omega t)}|\boldsymbol{\varepsilon}(\mathbf{k})\rangle = A_0(\mathbf{k})\boldsymbol{\varepsilon}(\mathbf{k})e^{+i(\mathbf{k}\cdot\mathbf{r}-\omega t)}\langle \boldsymbol{\varepsilon}(\mathbf{k})|\boldsymbol{\varepsilon}(\mathbf{k})\rangle \\ &= A_0(\mathbf{k})\boldsymbol{\varepsilon}(\mathbf{k})e^{+i(\mathbf{k}\cdot\mathbf{r}-\omega t)} \quad \text{where} \quad A_0(\mathbf{k}) \equiv \sqrt{\frac{2\pi\hbar c}{(Vol)|\mathbf{k}|}}. \end{aligned} \quad (7.6)$$

Thus, we see that $\langle 0|\hat{\mathbf{A}}|\boldsymbol{\varepsilon}(\mathbf{k})\rangle$ is identical to the classical A-field of energy $\hbar\omega$.

[Instead of having $\hat{a}$ acting to the left, we can equivalently think of it as *lowering* the state on the right, which reduces the number-state inner product to $\langle 0|0\rangle$, instead of $\langle \boldsymbol{\varepsilon}(\mathbf{k})|\boldsymbol{\varepsilon}(\mathbf{k})\rangle$, but either way, the inner product is 1.]

Radiation: What, then, is $\langle \boldsymbol{\varepsilon}(\mathbf{k})|\hat{\mathbf{A}}|0\rangle$ (the matrix element for emission of a photon)? Is it the average A-field of the vacuum? Not quite, because that is zero: $\langle 0|\hat{\mathbf{A}}|0\rangle = 0$. However, expanding $\hat{\mathbf{A}}$ with Eq. (7.4), we again find only one term survives:

$$\langle \boldsymbol{\varepsilon}(\mathbf{k})|\hat{\mathbf{A}}|0\rangle = \langle \boldsymbol{\varepsilon}(\mathbf{k})|A_0(\mathbf{k})\boldsymbol{\varepsilon}^*(\mathbf{k})e^{-i(\mathbf{k}\cdot\mathbf{r}-\omega t)}\hat{a}_\varepsilon^\dagger(\mathbf{k})|0\rangle = A_0(\mathbf{k})\boldsymbol{\varepsilon}^*(\mathbf{k})e^{-i(\mathbf{k}\cdot\mathbf{r}-\omega t)}. \quad (7.7)$$

Again, the inner product reduces to an *equivalent* classical A-field, but *not* zero. Some call this the "vacuum fluctuations" of the EM field, and that this provides an ever-present EM field of the vacuum for "spontaneous" radiation. In this view, the emitted photon is actually "stimulated" by the vacuum EM field. This view correctly predicts experiments, and is therefore a valid interpretation, but is not required.

Higher Excitations: For higher excitations, the matrix elements, Eq. (7.5), become:

$$\langle N_\lambda(\mathbf{k})-1|\hat{\mathbf{A}}|N_\lambda(\mathbf{k})\rangle \quad \text{or} \quad \langle N_\lambda(\mathbf{k})+1|\hat{\mathbf{A}}|N_\lambda(\mathbf{k})\rangle, \quad (7.8)$$

which again correspond to absorption and radiation, respectively. Using the same expansion as for the single-photon cases previously, we find:

$$\begin{aligned}\langle N_\lambda(\mathbf{k})-1|\hat{\mathbf{A}}|N_\lambda(\mathbf{k})\rangle &= \sqrt{N_\lambda(\mathbf{k})}A_0(\mathbf{k})\boldsymbol{\varepsilon}_\lambda(\mathbf{k})e^{+i(\mathbf{k}\cdot\mathbf{r}-\omega t)} \quad \text{or} \\ \langle N_\lambda(\mathbf{k})+1|\hat{\mathbf{A}}|N_\lambda(\mathbf{k})\rangle &= \sqrt{N_\lambda(\mathbf{k})+1}A_0(\mathbf{k})\boldsymbol{\varepsilon}_\lambda^*(\mathbf{k})e^{-i(\mathbf{k}\cdot\mathbf{r}-\omega t)}. \end{aligned}$$

These correspond to the classical A-fields for the given EM excitation modes. Notice, though, that for the *same* EM field excitation, the effective A-field for radiation is slightly higher than that for absorption. In other words, other things being equal, radiation is more likely than absorption. This is consistent with the view that radiationincludes a stimulated part and a spontaneous part, whereas absorption only has a stimulated part.

7.4   Quantum Interaction of Radiation and Matter

We now develop the QM of spontaneous radiation, e.g., an excited atom radiates (creates) a photon, and thus decays to a lower energy state. To compute radiation probabilities and metastable state lifetimes (or equivalently, decay rates), we must evaluate a matrix element of the form:

$$\langle \psi_f, \boldsymbol{\varepsilon}(\mathbf{k}) | \hat{\mathbf{A}} \cdot \hat{\mathbf{p}} | \psi_i, 0 \rangle \quad \text{where} \quad | \psi_i, 0 \rangle \text{ is the initial matter state with no photons;}$$

$$\langle \psi_f, \boldsymbol{\varepsilon}(\mathbf{k}) | \text{ is the final matter state plus one photon.}$$

In other words, we evaluate the complex amplitude to go from an initial matter state with no photons, to a final matter state and one photon. The probability of such a transition is the squared-magnitude of the amplitude, as usual. We now describe the meaning of all the pieces of this inner product, and how to evaluate it, in general.

7.4.1   Multiparticle Matter and Photon States

Quantum states for a system of interacting matter and radiation are combination states (tensor product states): they include a matter piece and an EM field piece. The matter states are the usual wave-functions (plus spin-states), that we already know and love. For example, the combined matter-EM state:

$$| \psi, N_\lambda(\mathbf{k}) \rangle \equiv | \psi \rangle | N_\lambda(\mathbf{k}) \rangle \quad \text{has}$$

$$| \psi \rangle \equiv \text{matter state of a charged particle;}$$

$$| N_\lambda(\mathbf{k}) \rangle \equiv \text{photon state with wave-vector } \mathbf{k},$$

$$\text{polarization } \boldsymbol{\varepsilon}\lambda(\mathbf{k}), \text{ and } N_\lambda(\mathbf{k}) \text{ photons present.}$$

A hydrogen atom in the $|100\rangle$ state with a photon in the $\boldsymbol{\varepsilon}_1(\mathbf{k})$ state can be written:

$$| 100, \boldsymbol{\varepsilon}_1(\mathbf{k}) \rangle \quad \text{or} \quad | 100 \rangle | \boldsymbol{\varepsilon}_1(\mathbf{k}) \rangle \quad \text{(combined matter/photon state).}$$

(Both of these are shorthand for $|100\rangle \otimes |\boldsymbol{\varepsilon}_1(\mathbf{k})\rangle$, which is the tensor-product (aka direct-product) of the two states.) Note that the matter-states exist in one Hilbert space, and the photon states exist in a *different* Hilbert space. Therefore, some operators act on the matter-state alone, some on the photon-state alone, and some act on both the matter state and the photon state.

Because we have chosen that $\mathbf{k}$ has a definite value for the photons, they are plane-wave eigenstates of momentum. Later, we write them in the position basis, so we can take inner products with $\psi(\mathbf{r})$, the matter state in the position basis.

In this section, our photon counts $N_\lambda(\mathbf{k})$ are always 1 or 0.

7.4.2 $\hat{A}$ in Action

Now we look more closely at $\hat{A}$ in operation. We can think of the set of operators $\hat{A}(t, \mathbf{r})$ as acting on the photon number counts to produce a (quantum) superposition of EM waves spread over time and space. A typical matrix element is an inner product of matter and EM tensor-product states, e.g., for radiation:

$$\left\langle \psi_f, \boldsymbol{\varepsilon}(\mathbf{k}') \middle| \hat{A}(t,\mathbf{r}) \cdot \hat{\mathbf{p}} \middle| \psi_i, 0 \right\rangle \qquad \text{(typical matter-EM inner product).}$$

The result is a complex number. This is the matrix element connecting an initial state of matter in $|\psi_i\rangle$ with no photons, to a final state of matter in $|\psi_f\rangle$ with one photon of state $|\boldsymbol{\varepsilon}(\mathbf{k})\rangle$. Thus, it is the matrix element relevant to spontaneous emission of a photon with wave-vector $\mathbf{k}$ and polarization $\boldsymbol{\varepsilon}(\mathbf{k})$.

To evaluate such a matrix element, we must explicitly take account of the fact that $\hat{A}(t, \mathbf{r})$ produces a vector function of space time. If we expand the inner product between $\langle \psi_f |$ and $|\psi_i\rangle$ we get a 3D space integral:

$$\left\langle \psi_f, \boldsymbol{\varepsilon}(\mathbf{k}') \middle| \hat{A}(t,\mathbf{r}) \cdot \hat{\mathbf{p}} \middle| \psi_i, 0 \right\rangle = \int_\infty \psi_f^*(\mathbf{r}) \underbrace{\left\langle \boldsymbol{\varepsilon}(\mathbf{k}') \middle| \hat{A}(t,\mathbf{r}) \middle| 0 \right\rangle}_{\text{3D vector}} \cdot \underbrace{\hat{\mathbf{p}} \psi_i(\mathbf{r})}_{\text{3D vector}} \, d^3 r.$$

We recognize the first inner product in the integrand as the effective A-field for the given EM transition, Eq. (7.7).

A Simplified Example: Before tackling this full spontaneous emission matrix element, we illustrate some concepts with a simpler example. Suppose we have a hypothetical operator that acts on both matter and EM states, for just one mode ($\mathbf{k}'$, $\boldsymbol{\varepsilon}$), given as:

$$\hat{B}(t,\mathbf{r}) \equiv \underbrace{\boldsymbol{\varepsilon}^*(\mathbf{k}') e^{-i(\mathbf{k}' \cdot \mathbf{r} - \omega t)}}_{\text{spacetime}} \underbrace{\hat{a}_\varepsilon^\dagger(\mathbf{k}')}_{\text{photon \#}}$$

$$\text{where} \quad \omega \equiv c|\mathbf{k}'| \qquad \text{(simplified hypothetical operator).}$$

Note that is a non-Hermitian vector operator, i.e., when it acts on a ket, it multiplies the ket by a complex vector function of space. $\hat{B}$ includes a spacetime dependence and a photon number operator.

Suppose we wish to find $\left\langle \psi_f, \boldsymbol{\varepsilon}(\mathbf{k}') \middle| \hat{\mathbf{B}} \middle| \psi_i, 0 \right\rangle$, a vector-valued matrix element. To evaluate it, we evaluate the photon number part first, which gives a function of spacetime. This function then becomes part of the inner product between matter states. In other words, separate $\hat{\mathbf{B}}$ and the states into their spacetime parts (matter and EM) and photon number parts:

$$\mathbf{M} \equiv \left\langle \psi_f, \boldsymbol{\varepsilon}(\mathbf{k}') \middle| \hat{\mathbf{B}} \middle| \psi_i, 0 \right\rangle$$

$$= \Big\langle \psi_f \Big| \underbrace{\langle \boldsymbol{\varepsilon}(\mathbf{k}') \big|}_{\text{photon}} \underbrace{\boldsymbol{\varepsilon}^*(\mathbf{k}') e^{-i(\mathbf{k}' \cdot \mathbf{r} - \omega t)}}_{\text{spacetime}} \underbrace{\hat{a}^\dagger_{\boldsymbol{\varepsilon}}(\mathbf{k}')}_{\text{photon}} \underbrace{\big| \psi_i \big\rangle}_{\text{matter}} \underbrace{\big| 0 \big\rangle}_{\text{photon}} \,.$$

The braand ket are both tensor productsof matter states and photon number states. We group the spacetime factors together and the photon number parts together. The photon number parts now evaluate to a simple constant (in this case, 1):

$$\mathbf{M} = \underbrace{\left\langle \psi_f \middle| \boldsymbol{\varepsilon}^*(\mathbf{k}') e^{-i(\mathbf{k}' \cdot \mathbf{r} - \omega t)} \middle| \psi_i \right\rangle}_{\text{spacetime}} \underbrace{\left\langle \boldsymbol{\varepsilon}(\mathbf{k}') \middle| \hat{a}^\dagger(\mathbf{k}') \middle| 0 \right\rangle}_{\text{photon part} = 1}$$

$$= \left(\int_\infty \psi_f^*(\mathbf{r}) \boldsymbol{\varepsilon}^*(\mathbf{k}') e^{-i(\mathbf{k}' \cdot \mathbf{r} - \omega t)} \psi_i(\mathbf{r}) \, d^3 r \right)(1),$$

where we used $\left\langle \boldsymbol{\varepsilon}(\mathbf{k}') \middle| \hat{a}^\dagger_{\boldsymbol{\varepsilon}}(\mathbf{k}') \middle| 0 \right\rangle = \left\langle \boldsymbol{\varepsilon}(\mathbf{k}') \middle| \boldsymbol{\varepsilon}(\mathbf{k}') \right\rangle = 1$, which is the matrix element between zero and one photons. We can now pull out some nonspatial factors, and (in principle) evaluate the space integral:

$$\mathbf{M} = \int_\infty d^3 r \, \psi_f^*(\mathbf{r}) \boldsymbol{\varepsilon}^*(\mathbf{k}') e^{-i(\mathbf{k}' \cdot \mathbf{r} - \omega t)} \psi_i(\mathbf{r}) = \boldsymbol{\varepsilon}^*(\mathbf{k}') e^{+i\omega t} \int_\infty d^3 r \, \psi_f^*(\mathbf{r}) e^{-i\mathbf{k}' \cdot \mathbf{r}} \psi_i(\mathbf{r}).$$

Again, note that $\mathbf{M}$ is a vector, and a function of time only through its complex phase. This phase is usually irrelevant, as we will see when we consider Fermi's Golden Rule (FGR). It typically contributes only to an energy-conserving δ-function [1, p. 251, 16, p. 329], and therefore any realistic calculation will depend only on the magnitude of $\mathbf{M}$.

The Real Deal: Returning now to the more realistic (and more complicated) case:

$$\left\langle \psi_f, \boldsymbol{\varepsilon}(\mathbf{k}') \middle| \hat{\mathbf{A}}(t, \mathbf{r}) \cdot \hat{\mathbf{p}} \middle| \psi_i, 0 \right\rangle \qquad \text{(typical matter-EM inner product)}.$$

To evaluate this inner product, which is a scalar, we write $\hat{\mathbf{A}} \cdot \hat{\mathbf{p}}$ as a sum of $\boldsymbol{\varepsilon}\hat{a}$ and $\boldsymbol{\varepsilon}^*\hat{a}^\dagger$ terms, then (as before) separate each term into spacetime and photon number parts. The photon number parts reduce to simple constants, and then we can integrate over space. Since $\hat{\mathbf{A}} \cdot \hat{\mathbf{p}}$ is a scalar operator, the final inner product is a single complex number. We now proceed.

First, we separate the $\boldsymbol{\varepsilon}\hat{a}$ and $\boldsymbol{\varepsilon}^*\hat{a}^\dagger$ terms, and use the full $\hat{\mathbf{A}}$ from (7.4):

$$M \equiv \left\langle \psi_f, \boldsymbol{\varepsilon}_1(\mathbf{k}') \middle| \hat{\mathbf{A}} \cdot \hat{\mathbf{p}} \middle| \psi_i, 0 \right\rangle$$

$$\text{where} \quad \left\langle \boldsymbol{\varepsilon}_1(\mathbf{k}') \middle| \equiv \left\langle N_1(\mathbf{k}') = 1 \middle| \,, \quad \omega \equiv c \middle| \mathbf{k}' \middle| \,.$$

$$M = \left\langle \psi_f, \boldsymbol{\varepsilon}_1(\mathbf{k}') \middle| \sum_{\lambda=1}^{2} \int_{-\infty}^{\infty} d^3\mathbf{k}\, A_0(\mathbf{k}) \right.$$

$$\left. \left(\underbrace{\boldsymbol{\varepsilon}_\lambda(\mathbf{k}) e^{+i(\mathbf{k}\cdot\mathbf{r}-\omega t)}}_{\text{spacetime}} \underbrace{\hat{a}_\lambda(\mathbf{k})}_{\text{photon}} + \underbrace{\boldsymbol{\varepsilon}_\lambda^*(\mathbf{k}) e^{-i(\mathbf{k}\cdot\mathbf{r}-\omega t)}}_{\text{spacetime}} \underbrace{\hat{a}_\lambda^\dagger(\mathbf{k})}_{\text{photon}} \right) \cdot \hat{\mathbf{p}} \middle| \psi_i, 0 \right\rangle .$$

Separate the positive and negative frequency component operators, and interchange the order of integration over $\mathbf{k}$ with the spatial inner product (integration over $\mathbf{r}$):

$$M = \sum_{\lambda=1}^{2} \int_{-\infty}^{\infty} d^3\mathbf{k}\, A_0(\mathbf{k}) \left[\left\langle \psi_f, \boldsymbol{\varepsilon}_1(\mathbf{k}') \middle| \boldsymbol{\varepsilon}_\lambda(\mathbf{k}) e^{+i(\mathbf{k}\cdot\mathbf{r}-\omega t)} \underbrace{\hat{a}_\lambda(\mathbf{k})}_{\text{photon}} \cdot \hat{\mathbf{p}} \middle| \psi_i, 0 \right\rangle \right.$$

$$\left. + \left\langle \psi_f, \boldsymbol{\varepsilon}_1(\mathbf{k}') \middle| \boldsymbol{\varepsilon}_\lambda^*(\mathbf{k}) e^{-i(\mathbf{k}\cdot\mathbf{r}-\omega t)} \underbrace{\hat{a}_\lambda^\dagger(\mathbf{k})}_{\text{photon}} \cdot \hat{\mathbf{p}} \middle| \psi_i, 0 \right\rangle \right] .$$

We separate each term into spacetime and photon number parts, and see that the first term is 0:

$$M = \sum_{\lambda=1}^{2} \int_{-\infty}^{\infty} d^3\mathbf{k}\, A_0(\mathbf{k}) \left[\underbrace{\left\langle \psi_f \middle| e^{+i(\mathbf{k}\cdot\mathbf{r}-\omega t)} \left(\boldsymbol{\varepsilon}_\lambda(\mathbf{k}) \cdot \hat{\mathbf{p}} \right) \middle| \psi_i \right\rangle}_{\text{spacetime}} \underbrace{\cancel{\left\langle \boldsymbol{\varepsilon}_1(\mathbf{k}') \middle| \hat{a}_\lambda(\mathbf{k}) \middle| 0 \right\rangle}}_{\text{photon part} = 0} \right.$$

$$\left. + \underbrace{\left\langle \psi_f \middle| e^{-i(\mathbf{k}\cdot\mathbf{r}-\omega t)} \left(\boldsymbol{\varepsilon}_\lambda^*(\mathbf{k}) \cdot \hat{\mathbf{p}} \right) \middle| \psi_i \right\rangle}_{\text{spacetime}} \underbrace{\left\langle \boldsymbol{\varepsilon}_1(\mathbf{k}') \middle| \hat{a}_\lambda^\dagger(\mathbf{k}) \middle| 0 \right\rangle}_{\text{photon part} = 1} \right] .$$

In the second term, we have a sum of two photon parts, one for each polarization ($\lambda=1$ and $\lambda=2$). Considering just that factor, we have:

$$\sum_{\lambda=1}^{2} \left\langle \boldsymbol{\varepsilon}_1(\mathbf{k}') \middle| \hat{a}_\lambda^\dagger(\mathbf{k}) \middle| 0 \right\rangle = \left\langle \boldsymbol{\varepsilon}_1(\mathbf{k}') \middle| \boldsymbol{\varepsilon}_1(\mathbf{k}) \right\rangle + \underbrace{\cancel{\left\langle \boldsymbol{\varepsilon}_1(\mathbf{k}') \middle| \boldsymbol{\varepsilon}_2(\mathbf{k}) \right\rangle}}_{0} = 1,$$

where we used the fact that $\boldsymbol{\varepsilon}_1$ and $\boldsymbol{\varepsilon}_2$ are orthogonal. This eliminates the sum on λ. Notice that the photon inner products reduced to two trivial values: 0 and 1. This always happens if all states involved have occupation numbers $N_\lambda(\mathbf{k}') = 0$ or 1. If we recognize this ahead of time, we can eliminate those parts right from the start, and simplify the algebra.

Also, for our given emission wave-vector $\mathbf{k}'$, the photon part is nonzero only for $\mathbf{k} = \mathbf{k}'$, which eliminates the integral over $\mathbf{k}$. This leaves just the spacetime part of the inner product:

$$M = A_0(\mathbf{k}')\langle \psi_f | e^{-i(\mathbf{k}'\cdot\mathbf{r}-\omega t)} \left(\boldsymbol{\varepsilon}_1^*(\mathbf{k}')\cdot\hat{\mathbf{p}} \right) | \psi_i \rangle$$

$$= A_0(\mathbf{k}')e^{+i\omega t}\boldsymbol{\varepsilon}_1^*(\mathbf{k}') \cdot \left(\int_\infty d^3\mathbf{r}\ \psi_f^*(\mathbf{r})e^{-i\mathbf{k}'\cdot\mathbf{r}}\frac{\hbar}{i}\nabla\psi_i(\mathbf{r}) \right).$$

The $\boldsymbol{\varepsilon}^*\cdot\hat{\mathbf{p}} \sim \boldsymbol{\varepsilon}^*\cdot\nabla$ is a scalar operator, so the final inner product is a scalar complex number.

Photon Number Inner Products: In general, for a given EM mode, when we say that $|N=0\rangle$ "couples to" $|N=1\rangle$, it means that $|N=0\rangle$ can time evolve into $|N=1\rangle$. In this example, the zero photon state can time evolve into the one photon state through the process of matter radiating the photon (which implies a change in the matter state, as well). Since the Hamiltonian is the generator of time evolution, saying that $|N=0\rangle$ "couples to" $|N=1\rangle$ implies:

$$\langle N=1 | \hat{H} | N=0 \rangle \neq 0.$$

7.4.3 Bigger and Better QFTs

A matter particle wave-function oscillates in time according to its energy, by the factor $\exp(-i\omega t)$. Therefore, matter particle wave-functions can be thought of as spatially distributed harmonic oscillators and second quantized similarly to photons. This process leads inevitably to antimatter, and matter/antimatter particle creation and annihilation. Many further complications also arise. Such QFTs are the only known way to calculate the results of modern particle experiments.

7.5 Photons Have All Phases, Semiclassical Approximation

An individual photon is a superposition of all possible EM field phases [11, pp. 186–187]. This is completely analogous to a stationary state of a quantum particle harmonic oscillator, which is in a superposition of all possible positions (or "phases" of its oscillation). Therefore, the photon number states we have been discussing, e.g., $|N_1(\mathbf{k}_1),N_1(\mathbf{k}_2),...\rangle$, are called **incoherent states**, since the different modes have no particular phase relationship to each other. (We discuss coherent states later).

The matrix elements we have been computing, cannot distinguish between a *superposition* and a *mix* of all phases, so based on experimental results, we could equally well say a single photon state is a *mixed state* of all possible phases [GAF p. 359t]. However, our model of an A-space "wave-function" (Sect. 7.3) for the A-field is more consistent with a superposition than a mixed state.

A single-photon state is a superposition of all possible phases, and therefore has no definite phase.

The fact that single-photons have all phases leads to an important distinction between the effective A-field of a photon, a complex function of space given by Eq. (7.6):

$$\langle 0|\hat{A}|\varepsilon(\mathbf{k})\rangle = A_0(\mathbf{k})\varepsilon(\mathbf{k})e^{+i(\mathbf{k}\cdot\mathbf{r}-\omega t)} \equiv \mathbf{A}_{eff}(t,\mathbf{r}) \quad \text{effective A-field of photon,}$$

and the classical EM "analytic" (or phasor) A-field, also a complex function of space, that looks very similar:

$$\mathbf{A}_{analytic}(\mathbf{r}) = A_0\varepsilon e^{+i\mathbf{k}\cdot\mathbf{r}} \quad \text{classical analytic (phasor) A-field.}$$

The classical *true* A-field is a real-valued function of space, defined by the classical phasor A-field as:

$$\mathbf{A}_{true}(t,\mathbf{r}) = \sqrt{2}\,\text{Re}\left\{\mathbf{A}_{analytic}(\mathbf{r})e^{-i\omega t}\right\} = \sqrt{2}\,\text{Re}\left\{A_0\varepsilon e^{+i(\mathbf{k}\cdot\mathbf{r}-i\omega t)}\right\} = \sqrt{2}A_0\varepsilon\cos(\mathbf{k}\cdot\mathbf{r}-\omega t).$$

(Other normalizations are often used, as well.) The distinction is that the classical true A-field amplitude varies in space: it has positive peaks, negative troughs, and is zero in places. Where it is zero, there is no chance it interacts with a particle there. In contrast, the effective A-field of the photon is complex, and equal magnitude everywhere. It is equally likely to interact at all points. It has no "zeros." This reflects the fact that the photon is a superposition of all phases, and is equally likely to have any given amplitude at all points.

Note that the factor of $2^{1/2}$ in the formula for $\mathbf{A}_{true}$ gives the semiclassical "photon" the same energy as a true photon, by making their squared magnitudes, summed over space, equal at every point in time:

$$\int_\infty \mathbf{A}_{true}^2 \, d^3r = \int_\infty 2A_0^2 \cos^2(\mathbf{r}) \, d^3r = A_0^2(Vol) \quad \text{and}$$

$$\int_\infty \mathbf{A}_{eff}^2 \, d^3r = \int_\infty A_0^2(\mathbf{r})|\varepsilon|^2 \left|e^{i(\mathbf{k}\cdot\mathbf{r})}\right|^2 d^3r = A_0^2(Vol).$$

7.6   Realistic Photons: Multimode Single Photon States

We have taken our basis states for the EM field to be integer excitations of all possible plane-wave EM modes, which for simplicity we take here to be discrete (indexed by l for $\varepsilon_l\mathbf{k}_l$):

$$|0\rangle, \quad |n_1=1\rangle, \ |n_1=2\rangle, \ |n_1=3\rangle, \ \ldots$$
$$|n_2=1\rangle, \ |n_2=2\rangle, \ |n_2=3\rangle, \ \ldots$$
$$\vdots \qquad\qquad\qquad where \quad l \equiv \varepsilon_l(\mathbf{k}_l).$$

However, we know from experiment that atoms radiate single photons, and also that they have some uncertainty in their energy (and therefore in $\mathbf{k}$). We must conclude that the EM state for single-photon radiation is a (normalized) superposition of single-photon states of different modes:

$$\left|\psi_{EM}\right\rangle = \sum_{l} c_l \left|\boldsymbol{\varepsilon}_l(\mathbf{k}_l)\right\rangle = \sum_{l} c_l \left|N_l = 1\right\rangle \tag{7.9}$$

where $l \equiv$ mode of the plane-wave state.

When all the $\mathbf{k}_l$ are "nearby," this is a wave-packet. Compared to an infinite plane-wave mode, it is a more realistic single-photon state. This is the EM state after an atom has radiated a photon. It has an envelope of essentially finite size in space.

Such a state is *not* an eigenstate of any single-mode number operator, $\hat{N}_{l'} = \hat{a}_{l'}^{\dagger}\hat{a}_{l'}$, but *is* an eigenstate of the total number operator, with eigenvalue 1:

$$\hat{N}_{tot} \equiv \sum_{l'=1}^{\infty} \hat{N}_{l'} \quad \Rightarrow \quad \hat{N}_{tot}\left|\psi_{EM}\right\rangle = \sum_{l'=1}^{\infty}\sum_{l=1}^{n} \hat{N}_{l'}\, c_l \left|N_l = 1\right\rangle = \sum_{l=1}^{n} 1 \cdot c_l \left|N_l = 1\right\rangle = 1\left|\psi_{EM}\right\rangle.$$

In the double sum, we used the fact that only terms with $l'=l$ contribute. Since the total photon number observable is 1, it is a single-photon state, even though it is a superposition of many modes. It can be called a "multimode single-photon state." Such states could be propagating wave-packets, localized in space, and moving in time.

The inner product for detecting a photon of mode l' will be of the form:

$$\left\langle 0\right|\hat{a}_{l'}\left|\psi_{EM}\right\rangle = \left\langle 0\right|\sum_{l=0}^{\infty} c_l\hat{a}_{l'}\left|N_l = 1\right\rangle = c_{l'}\left\langle 0\,|\,0\right\rangle = c_{l'},$$

where we have used:

$$\hat{a}_{l'}\left|N_{l'} = 1\right\rangle = \left|0\right\rangle, \quad \text{and} \quad \hat{a}_{l'}\left|N_l = 1\right\rangle = \mathbf{0}_v, \quad (l' \neq l) \quad \text{and therefore:}$$

$$\hat{a}_{l'}\left|\psi_{EM}\right\rangle = \hat{a}_{l'}c_0\left|N_0 = 1\right\rangle + \hat{a}_{l'}c_1\left|N_1 = 1\right\rangle + \hat{a}_{l'}c_2\left|N_2 = 1\right\rangle + \dots \hat{a}_{l'}c_{l'}\left|N_{l'} = 1\right\rangle + \dots$$

$$= \mathbf{0}_v + \mathbf{0}_v + \mathbf{0}_v + \dots c_{l'}\left|0\right\rangle + \dots = c_{l'}\left|0\right\rangle.$$

> The multimode single-photon state is the EM state most like a single matter particle.

Note that the probability of detecting two photons from such a state is zero, because such a detection requires *two* lowering operators, one for each photon detected [10, p. 376]. As just shown, any one lowering operator produces a ket proportional to the vacuum state $\left|0\right\rangle$. Therefore, any second lowering operator returns the zero vector $\mathbf{0}_v$. Thus,

$$\left\langle 0\right|\hat{a}_{l'}\hat{a}_{l''}\left|\psi_{EM}\right\rangle = 0 \qquad \forall\, l,l',$$

and the probability of any such double detection is 0.

7.7 Photons in Other Bases

So far, we have chosen plane waves of a given $(\varepsilon, \mathbf{k})$ mode as our basis states. Such states are eigenstate of energy and momentum, and are single excitations of the number operator in the plane-wave basis. For simplicity, we now describe our plane-wave basis as a discrete basis labeled by modes l, where each l defines an $(\varepsilon_l, \mathbf{k}_l)$ pair. Then a single photon in mode l is written:

$$\left| N_l = 1 \right\rangle \equiv \left| \varepsilon_l(\mathbf{k}_l) \right\rangle \quad \text{(single-photon plane-wave state)}.$$

This is appropriate for a resonant cavity, where there exist only discrete modes. However, the same principles apply to both continuous and discrete bases [11, Chap. 6]. We follow the notation of [10] and a similar method.

All of the properties of the quantized EM field derive from three facts: (1) the commutator of the annihilation and creation operators for a single mode is the identity operator:

$$\left[\hat{a}_l, \hat{a}_l^\dagger \right] = \mathbf{1}_{op};$$

(2) different modes do not interact, so:

$$\left[\hat{a}_l, \hat{a}_{l'} \right] = \left[\hat{a}_l^\dagger, \hat{a}_{l'}^\dagger \right] = \mathbf{0}_{op} \tag{7.10}$$

and (3) photons are bosons, which means the quantized EM field can have from 0 to ∞ photons in a single mode (i.e., in any given single-particle state).

These facts have been used for plane-wave modes, however, we saw previously that a photon can exist as a single-photon wave-*packet*: a superposition of many modes. How would we "create" such a state with creation operators? We simply superpose the plane-wave component creation operators to create the wave-packet state Eq. (7.9). Define the single-photon wave-packet state as $|b\rangle$. Then:

$$|b\rangle = \sum_{l=0}^{\infty} c_l |N_l = 1\rangle = \left(\sum_l c_l \hat{a}_l^\dagger \right) |0\rangle \quad \Rightarrow \quad \sum_l c_l \hat{a}_l^\dagger \equiv \hat{b}^\dagger,$$

where $\hat{b}^\dagger$ is a creation operator for the wave-packet. More generally, we can construct an entire basis of orthogonal wave-packet states, $|b_m\rangle$, where m denotes the wave-packet "mode." Each mode has its own set of coefficients c_l, so we replace the c_l with a notation for a unique set of coefficients for each m:

$$|b_m\rangle = \sum_l U_{ml} |N_l = 1\rangle = \left(\sum_l U_{ml} \hat{a}_l^\dagger \right) |0\rangle \quad \Rightarrow \quad \sum_l U_{ml} \hat{a}_l^\dagger \equiv \hat{b}_m^\dagger.$$

U_{ml} is a unitary matrix, because it transforms from the a-to-b basis: given a column vector of a-mode coefficients, multiplying by U_{ml} gives the coefficients in the b

basis. Similarly, we create b-mode annihilation operators:

$$\hat{b}_m \equiv \left(\hat{b}_m^\dagger\right)^\dagger = \sum_l U_{ml}^* \hat{a}_l.$$

Now what are the commutation relations of the b-mode creation and annihilation operators? Using the fact that different a-modes commute, Eq. (7.10), that the commutator is linear in both arguments, Eq. (1.8), and that every row of U_{ml} is a unit-magnitude vector:

$$\left[\hat{b}_m, \hat{b}_m^\dagger\right] = \left[\left(\sum_l U_{ml}^* \hat{a}_l\right), \left(\sum_{l'} U_{ml'} \hat{a}_{l'}^\dagger\right)\right] = \sum_l \left[U_{ml}^* \hat{a}_l, U_{ml} \hat{a}_l^\dagger\right] = \sum_l U_{ml}^* U_{ml} \left[\hat{a}_l, \hat{a}_l^\dagger\right] = 1_{op}.$$

Similarly, because the rows of U_{ml} are orthogonal, if $m \neq m'$:

$$\left[\hat{b}_m, \hat{b}_{m'}^\dagger\right] = \sum_l U_{ml}^* U_{m'l} \left[\hat{a}_l, \hat{a}_l^\dagger\right] = 0_{op}.$$

Thus, the b-modes, the nonplane-wave modes, satisfy the same commutation relations as the plane-wave a-modes. Since all the number-basis properties derive from this commutation relation, there are then excitations in the b-number-basis with all the same number-basis properties. These excitations are eigenstates of the b-basis number operator $\hat{b}^\dagger \hat{b}$, and are therefore deserving of the name "photons." In fact, the b-mode number states are the multimode photon states.

When an atom radiates, we usually take the a-mode basis states as infinite spherical waves of definite k. However, the radiation is in fact a spherical wave-packet (a shell of radiation of finite thickness) radiating outward. Therefore, it may be considered a wave-packet superposition of infinite spherical waves. Equivalently, it may be considered a single excitation in a spherical wave-packet basis. In either basis, it is an eigenstate of the total photon operator, with eigenvalue 1; in other words, in either basis, it is a single photon.

7.8   The Wave-Function of a Photon?

If a photon is a particle, does it have a traditional wave-function? (Not the A-space "wave-function" of Sect. 7.3) This is a very important question, because it provides the link between nonrelativistic QM, and the more complete, relativistic QFT. Because photons are massless, they are easily created and absorbed, unlike (say) electrons, which are more durable throughout interactions. It is the creation and absorption of photons, and their bosonic nature, that demands we use a QFT, rather than ordinary QM. Another complication is that photons, being massless, cannot be at rest; they are always moving. And finally, photons are vector particles, somewhat more complicated than our nonrelativistic view of quantum particles with a scalar wave-function. Nonetheless, it is possible to define a useful "wave-function" for a single photon state, though "with some reservations" [10, p. 384t]. Such a photon wave-function is useful in our later discussion of the quantum eraser.

We define our photon wave-function, $\psi_{EM}(\mathbf{r})$, to have the usual properties for a quantum particle, namely that the probability of detecting a photon is proportional to ψ_{EM^2}, therefore it is normalized, and it is complex-valued so that it produces the required interference:

$$\mathrm{pdf}_\psi(\mathbf{r}) = \left|\psi_{EM}(\mathbf{r})\right|^2, \quad \int_\infty \left|\psi_{EM}(\mathbf{r})\right|^2 d^3r = 1, \quad \psi_{EM}(\mathbf{r}) \text{ complex.}$$

In fact, if normalized, the classical analytic (phasor)A-field has these properties. So our wave-function can be the *effective* A-field of the photon, but normalized.

A single photon state in a single mode is written $\left|\boldsymbol{\varepsilon}(\mathbf{k})\right\rangle$. One might think that we can simply find the average value of the A-field due to this photon in the usual QM way:

$$\left\langle \mathbf{A} \right\rangle = \left\langle \boldsymbol{\varepsilon}(\mathbf{k}) \left| \hat{\mathbf{A}} \right| \boldsymbol{\varepsilon}(\mathbf{k}) \right\rangle,$$

but we have already seen that this average is zero. The photon is a superposition of all phases, and each component of $\mathbf{A}$ is equally positive and negative. Instead, we already know the effective A-field, from Eq. (7.6):

$$\left\langle 0 \left| \hat{\mathbf{A}} \right| \boldsymbol{\varepsilon}(\mathbf{k}) \right\rangle = A_0(\mathbf{k})\boldsymbol{\varepsilon}(\mathbf{k})e^{+i(\mathbf{k}\cdot\mathbf{r}-\omega t)}.$$

It is the coupling of the EM state by $\hat{\mathbf{A}}$ *to the vacuum* that gives the effective A-field. We define our wave-function by simply replacing the amplitude $A_0(\mathbf{k})$ and polarization $\boldsymbol{\varepsilon}(\mathbf{k})$ with a normalization constant:

$$\psi_{EM}(\mathbf{r}) = Ne^{+i(\mathbf{k}\cdot\mathbf{r}-\omega t)}$$

Since this is an infinite plane-wave state, we have the same options for normalizing it as we did for incident plane waves in scattering (Sect. 2.5).

A more realistic single photon state is the multimode state, Eq. (7.9). We construct our wave-function the same way as for a single-mode photon, by coupling the EM state to the vacuum (and assuming all the $\boldsymbol{\varepsilon}_l$ are the same, and dropping them):

$$\psi_{EM}(\mathbf{r}) = N\left\langle 0 \left| e^{+i(\mathbf{k}_l\cdot\mathbf{r}-\omega_l t)}\hat{a}_l \right| \psi_{EM} \right\rangle = N\sum_l c_l \left\langle 0 \left| e^{+i(\mathbf{k}_l\cdot\mathbf{r}-\omega_l t)}\hat{a}_l \right| N_l = 1 \right\rangle$$

where N is chosen such that $\int_\infty \left|\psi_{EM}(\mathbf{r})\right|^2 d^3r = 1.$

Because the wave-packet is finite in extent, we have used ordinary wave-function normalization.

> The effective A-field of a photon, as defined by its "wave-function," follows Maxwell's equations.

So all the classical laws of radiation (e.g., reflection, diffraction) apply, even to a single photon.

Like any quantum particle, a single photon can be spread out over an arbitrarily large area. In my graduate research, a single photon (loosely speaking) returning from the moon is spread over many square kilometers. Our telescope's ~ 10 m^2 cross section then has only a small probability of detecting such a photon, because ultimately the photon detection is a quantized event: either we detect it or we do not. A detailed consideration of "detection" leads us to the interaction of photons (and radiation in general) with matter.

This method of defining a wave-function from a field excitation by coupling it to the vacuum is fairly general, and can be used (with similar reservations) for many other kinds of particles [1, p. 422, 17, 2 3.376 p. 147b, 23 2013 p. xvi top, 14 p. 24b].

7.9 Quasiclassical States

The closest quantum analog of a classical field is called a *quasiclassical state*, or a "coherent state,", or a "Glauber state," after Nobel laureate Roy J. Glauber who first described them in detail (Phys. Rev. 130, 2529–2539 (1963)). The *average* field of a quasiclassical state is a classical EM field. For example, the long-time limit of a classical oscillating current produces an EM field that is a quasi-classical state, as does a strong-beam laser [11, p. 190m].

Let us construct a quasiclassical state. Consider a single-mode classical wave that is essentially infinite in space. In the usual complex notation, the wave is given by:

$$\mathbf{A}(t,\mathbf{r}) = 2\,\mathrm{Re}\left[\alpha A_0 \boldsymbol{\varepsilon} e^{+i(\mathbf{k}\cdot\mathbf{r}-\omega t)}\right] \quad where \quad A_0 = \sqrt{\frac{2\pi\hbar c}{(Vol)|\mathbf{k}|}},$$

where we have written the phasor for the wave as αA_0 for later convenience, with $\langle\alpha|$ a complex number and A_0 taken as real. We seek a normalized quantum state $\langle\alpha|$ which reproduces the previous classical wave as closely as possible. Therefore, its average A-field equals the classical wave:

$$\langle \mathbf{A}(t,\mathbf{r})\rangle = \langle\alpha|\,\hat{\mathbf{A}}\,|\alpha\rangle = 2\,\mathrm{Re}\left[\alpha A_0 \boldsymbol{\varepsilon} e^{+i(\mathbf{k}\cdot\mathbf{r}-\omega t)}\right]. \tag{7.11}$$

$\hat{\mathbf{A}}$ is Hermitian, consisting of two terms proportional to $\hat{a}$ and $\hat{a}^\dagger$. If we make the average of the $\hat{a}$ term equal to the complex-valued signal (the previous bracketed quantity), then the hermiticity of $\hat{\mathbf{A}}$ guarantees the classical correspondence, Eq. (7.11). Thus,

$$\langle\alpha|\,A_0 \boldsymbol{\varepsilon} e^{+i(\mathbf{k}\cdot\mathbf{r}-\omega t)}\hat{a}\,|\alpha\rangle = \alpha A_0 \boldsymbol{\varepsilon} e^{+i(\mathbf{k}\cdot\mathbf{r}-\omega t)} \quad where \quad \hat{a} \equiv \hat{a}_{\boldsymbol{\varepsilon}}(\mathbf{k}). \tag{7.12}$$

(We used the fact that only one mode is present to eliminate from $\hat{\mathbf{A}}$ the sum over λ, and the integral over $\mathbf{k}$.) Now recall that $\hat{a}$ acts only on the number part of $|\alpha\rangle$, and not on the spacetime part. Therefore Eq. (7.12) becomes:

$$\langle\alpha| A_0 \boldsymbol{\varepsilon} e^{+i(\mathbf{k}\cdot\mathbf{r}-\omega t)}\hat{a}|\alpha\rangle = A_0\boldsymbol{\varepsilon} e^{+i(\mathbf{k}\cdot\mathbf{r}-\omega t)}\langle\alpha|\hat{a}|\alpha\rangle$$

$$= \alpha A_0\boldsymbol{\varepsilon} e^{+i(\mathbf{k}\cdot\mathbf{r}-\omega t)} \quad\Rightarrow\quad \langle\alpha|\hat{a}|\alpha\rangle = \alpha. \qquad (7.13)$$

The simplest way to satisfy the last equality is:

$$\hat{a}|\alpha\rangle = \alpha|\alpha\rangle \qquad \text{(quasi-classical state)}.$$

In fact, if different modes $\boldsymbol{\varepsilon}(\mathbf{k})$ are to remain independent (not intermix in $\langle\alpha|\hat{\mathbf{A}}|\alpha\rangle$), then this eigenvalue equation is the only one allowed. Thus, we have the general form for a quasiclassical state: $|\alpha\rangle$ is an eigenstate of the lowering operator $\hat{a}$.

We write $|\alpha\rangle$ in the number basis, with as-yet undetermined coefficients, b_N:

$$|\alpha\rangle \equiv \sum_{N=0}^{\infty} b_N|N\rangle \quad \text{for a given mode, } \boldsymbol{\varepsilon} \text{ and } \mathbf{k}.$$

The eigenvalue α is a given, but arbitrary, complex number, and is used to label the state $|\alpha\rangle$. We now show that α alone determines the state. Recall that α is proportional to the phasor for the average value of the EM field in the quasiclassical state $|\alpha\rangle$.

First, we find a recurrence relation for the coefficients in the superposition, b_N, using the fact that $|\alpha\rangle$ is an eigenstate of $\hat{a}$:

$$\hat{a}b_N|N\rangle = \sqrt{N}b_N|N-1\rangle = \alpha b_{N-1}|N-1\rangle \quad\Rightarrow\quad b_N = \frac{\alpha b_{N-1}}{\sqrt{N}}.$$

Thus, $|\alpha\rangle$ is a superposition of all photon occupation numbers of a quantum state, from 0 to ∞. By induction, the entire state is determined by no more than b_0, the first coefficient, and the eigenvalue α, since (by induction):

$$b_N = \frac{\alpha^N}{\sqrt{N!}}b_0.$$

However, this general form determines the normalization, and $|\alpha\rangle$ is normalized. Then:

$$\sum_{N=0}^{\infty}|b_N|^2 = \sum_{N=0}^{\infty}\frac{|\alpha|^{2N}}{N!}|b_0|^2 = \exp(|\alpha|^2)|b_0|^2 = 1 \quad\Rightarrow\quad |b_0| = \exp(-|\alpha|^2/2).$$

Therefore, choosing our phase such that b_0 is real, the entire state is defined by the complex eigenvalue α:

$$|\alpha\rangle \equiv \exp(-|\alpha|^2/2)\sum_{N=0}^{\infty}\frac{\alpha^N}{\sqrt{N!}}|N\rangle \qquad \text{(quasi-classical state)}.$$

As a check on our derivation, we compute the average A-field for $|\alpha\rangle$ in the usual way, with an inner product of $\hat{\mathbf{A}}$, using $\hat{a}|\alpha\rangle = \alpha|\alpha\rangle$ and its adjoint, $\langle\alpha|\hat{a}^\dagger = \alpha^*\langle\alpha|$:

$$\langle \mathbf{A}(t,\mathbf{r})\rangle = \langle\alpha|\hat{\mathbf{A}}|\alpha\rangle = \langle\alpha|\left[A_0\boldsymbol{\varepsilon} e^{+i(\mathbf{k}\cdot\mathbf{r}-\omega t)}\hat{a} + A_0\boldsymbol{\varepsilon}^* e^{-i(\mathbf{k}\cdot\mathbf{r}-\omega t)}\hat{a}^\dagger \right]|\alpha\rangle$$

$$= A_0\boldsymbol{\varepsilon} e^{+i(\mathbf{k}\cdot\mathbf{r}-\omega t)}\langle\alpha|\hat{a}|\alpha\rangle + A_0\boldsymbol{\varepsilon}^* e^{-i(\mathbf{k}\cdot\mathbf{r}-\omega t)}\langle\alpha|\hat{a}^\dagger|\alpha\rangle$$

$$= \alpha A_0\boldsymbol{\varepsilon} e^{+i(\mathbf{k}\cdot\mathbf{r}-\omega t)} + \alpha^* A_0\boldsymbol{\varepsilon}^* e^{-i(\mathbf{k}\cdot\mathbf{r}-\omega t)} = 2\,\mathrm{Re}\left[\alpha A_0\boldsymbol{\varepsilon} e^{+i(\mathbf{k}\cdot\mathbf{r}-\omega t)} \right].$$

As expected, this reproduces the earlier classical correspondence, Eq. (7.11).

Further Properties of the Quasiclassical State: The eigenvalue α is arbitrary, so the *average* A-field amplitude is any one of a *continuum* of amplitudes, i.e., it is *not* quantized. Also, $|\alpha\rangle$ is *not* an eigenstate of the photon number operator $\hat{N}$. Therefore, the number of photons in a quasiclassical state is uncertain. This may be surprising, since it is often incorrectly assumed that a general EM state must comprise an integer number of photons.

A general quantum EM state may have an uncertain number of photons in it, and the average number of photons in a state can be fractional.

The average A-field amplitude is αA_0, where $A_0 \equiv A_0(\mathbf{k})$ is the average A-field amplitude of a single-photon state $|\boldsymbol{\varepsilon}(\mathbf{k})\rangle$. Therefore, α is dimensionless. A quasiclassical state is *not* an eigenstate of the Hamiltonian, and therefore has uncertain energy. The energy of a single photon is $\hbar\omega$, and the average energy of the EM state $|\alpha\rangle$ is (relative to the vacuum energy):

$$\hat{H} = \hbar\omega\hat{a}^\dagger\hat{a} \quad\Rightarrow\quad \langle E_\alpha\rangle = \hbar\omega\langle\alpha|\hat{a}^\dagger\hat{a}|\alpha\rangle = \hbar\omega|\alpha|^2.$$

So $|\alpha|^2$ can be thought of as the average number of photons in the state $|\alpha\rangle$, which need not be an integer.

Quasiclassical states are also states of minimum uncertainty: the EM field amplitude is as precisely defined as QM allows.

Possibly the most widespread misunderstanding about quantized EM fields is the belief that they must comprise an integer number of photons.

7.10   Classical EM Superposition vs. Quantum Superposition: Interference of Distinct Photons?

Classical superposition of EM fields is a somewhat different concept than quantum superposition of states and kets. Classical superposition (aka interference) simply says that if two (or more) sources create EM fields at the same point in spacetime, the total field is the vector sum of the component fields:

$$\mathbf{A}_{tot}(t,\mathbf{r}) = \mathbf{A}_1(t,\mathbf{r}) + \mathbf{A}_2(t,\mathbf{r}), \quad \text{and similar for } \mathbf{E}(t,\mathbf{r}) \text{ and } \mathbf{B}(t,\mathbf{r}).$$

Since *independent* classical EM fields interfere, and because all classical results are explained by QM, it must be true that independent *quantized* EM fields also interfere.

Dirac made a famous, though widely misunderstood, statement that "... each photon only interferes with itself. Interference between different photons never occurs" [4, p. 9]. We show here the mechanism by which independent quantum EM fields *do* interfere. (If we say, loosely, that a quantum EM field comprises "photons," then classical EM interference forces us to accept that distinct photons do, in fact, interfere. However, see the earlier discussion on the uncertainty of photon count in a quantum EM state, Sect. 7.8.)

To study independent photon interference, Pfleegor and Mandel performed a famous experiment in 1967 [28] where two independent lasers were attenuated to such a low intensity that the probability of detecting two photons at the same time was negligible. Nonetheless, the two lasers produce a distinct interference pattern, thus confirming that two "photons" that can reasonably be considered independent do, in fact, interfere.

The quantum explanation for both classical superposition, and the Pfleegor–Mandel results, is found in the full $\hat{\mathbf{A}}$ operator already developed in Eq. (7.4):

$$\hat{\mathbf{A}}(t,\mathbf{r}) = \sum_{\lambda=1}^{2} \int_{-\infty}^{\infty} d^3\mathbf{k}\; A_0(\mathbf{k}) \left(\underbrace{\boldsymbol{\varepsilon}_\lambda(\mathbf{k})e^{+i(\mathbf{k}\cdot\mathbf{r}-\omega t)}}_{\text{spacetime}} \underbrace{\hat{a}_\lambda(\mathbf{k})}_{\text{photon}} + \underbrace{\boldsymbol{\varepsilon}_\lambda^*(\mathbf{k})e^{-i(\mathbf{k}\cdot\mathbf{r}-\omega t)}}_{\text{spacetime}} \underbrace{\hat{a}_\lambda^\dagger(\mathbf{k})}_{\text{photon}} \right).$$

The effective A-field is the coherent sum of all the modes. The resulting A-field becomes part of an inner product, whose magnitude is squared to produce a measurable result. Thus the A-field follows the usual quantum rule: first sum coherently, then square the magnitude.

We see, then, that the Fleegor–Mandel interference is essentially classical interference. As noted in the section on quasiclassical states, the average amplitude of the field is *not* quantized, and is therefore essentially classical. It is the *interaction* of the EM field with matter that is quantized (through the raising and lowering operators in $\hat{\mathbf{A}}$). In the experiment, even though the probability of detecting a given photon is low, the EM field is still excited to some level, and interference occurs.

We can now consider Dirac's statement in a new light. Recall that a single photon is *incoherent* (a superposition of all phases), i.e., it has no definite phase. Therefore, two such photons have no definite phase relationship and cannot interfere. However, a single photon (even though incoherent) *can* interfere with itself, because each *component* of the superposition *has* a definite phase. In the two-slit experiment, each component reinforces and cancels in the same way, so the complete wavefunction (which is the sum of its components) exhibits interference. In contrast:

Quasiclassical states describe fields with a well-defined sinusoidal phase of small uncertainty, and therefore, independent sources can interfere in essentially the classical way.

This is in contrast to single-photon states, which are a superposition of all phases, so independent single-photon sources cannot interfere.

In Summary: A significant point of confusion about Dirac's statement concerns the meaning of the word "photon." The naive view that EM fields comprise a definite number of "photons" is incorrect. EM fields are not quantized like electrons or marbles. Any single photon state is incoherent, meaning a superposition of all phases. The more complete theory of the quantized EM field defines quasiclassical states, which interfere in an essentially classical way, through the summation over modes in the $\hat{A}$ operator. However, even in such quasiclassical states, the EM field interactions with matter are quantized in units we call "photons." (We note in passing that the general results of our development apply to all boson fields, not just photons [10, p. 384b].)

7.11   Don't Be Dense

Density of states is a very important concept for radiation and other particle creation/annihilation operations. Density of states is often used, and much abused. The term "density of states" comes in many forms, but in QM, "density of states" always refers to number density of stationary quantum states per unit of some parameter space. Recall that stationary quantum states are energy eigenstates. We consider these common densities of states:

- 1D harmonic oscillator
- 1D particle in a box
- 1D free particle
- 3D free particle

We consider only the density of spatial states (wave-functions). Fermions, e.g., electrons, would have twice as many states, because each spatial state has two spin-states. Similarly, photon states are twice as dense, because each spatial state has two polarization states.

Our analysis refers to three different "spaces": real physical space (1D, two-dimensional, or 3D), k-space, and phase space. Since $p = \hbar k$, we take momentum space to be essentially equivalent to k-space. Phase space is a "tensor product" of real-space and momentum space:

$$V_{\text{phase space}} = (Vol)(V_{\text{momentum space}}) \quad \Rightarrow \quad dV_{\text{phase space}} = d^3r \, d^3p.$$

1D Harmonic Oscillator: Perhaps the simplest density of states is that for a 1D harmonic oscillator. We write the number of states per unit energy as $\rho(E)$. Since the energy spacing is constant, $\hbar\omega$, the number of states per unit energy is (Fig. 7.3, *left*):

$$\rho(E) = \frac{1}{\hbar\omega} \quad \text{(1D harmonic oscillator).}$$

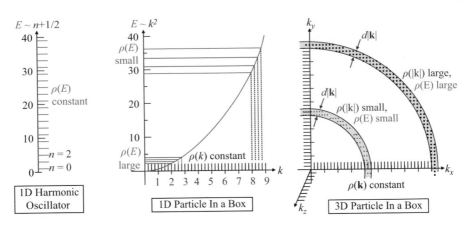

Fig. 7.3 Densities of states

The units are $[E]^{-1}$, i.e., inverse energy. The momentum (and therefore, wave vector, k) of the particle in any stationary state is uncertain, and does not figure into the density of states.

1D Particle in a Box: Another simple density of states is 1D particle in a box. We will see that the result per unit energy is the same as for a free particle, and use this as a warm up for 3D. Each state has a purely real wave-function, which must be 0 at the boundaries. This leads to quantized wave-functions (ignoring normalization):

$$\psi(x) = \sin(kx) = \frac{1}{2i}\left(e^{ikx} - e^{-ikx}\right)$$

where $k = n\pi / L$, $L =$ length of box, $n =$ quantum number.

Each energy eigenstate is a superposition of $+$ and $-$ momentum eigenstates. Because k is easily quantized (an integer number of 1/2 wavelengths: $kL = n\pi$), we can write the density of states parameterized by k. At first, we let k take only positive values, knowing that each positive value of k describes a state which is a standing-wave superposition of both $+k$ and $-k$ momenta. We wrote k as a function of n previously, but for a density of states, we must invert this relation to get number density per unit of positive k:

$$n(+k) = \frac{kL}{\pi} \quad \Rightarrow \quad \rho_+(k) = \frac{dn}{dk} = \frac{L}{\pi}, \qquad \text{in units of } [\text{rad/m}]^{-1} = [\text{m}]. \qquad (7.14)$$

But taking into account that each state consumes a $+k$ and $-k$ value, the final (two-sided) density per unit k (of either sign) is only half the previous equation:

$$\rho(k) = \frac{1}{2}\frac{dn}{dk} = \frac{L}{2\pi} \qquad \text{(1D particle in box).} \qquad (7.15)$$

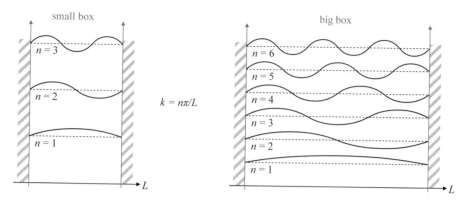

Fig. 7.4 Density of states, $\rho(k)$, for particle-in-a-box is proportional to the box size

Thus, for a given size box, the density of energy eigenstates per unit k is constant. Note that there is no upper bound to k (or E).

Rewriting our two-sided formula for $n(k)$ in terms of momentum introduces a general principle:

$$n(k) = \frac{kL}{2\pi} = \frac{pL}{2\pi\hbar} = \frac{\text{phase space volume}}{2\pi\hbar}.$$

> For large systems, the number of quantum states enclosed by boundaries of p and the real-space volume is closely given by the "volume" of the phase-space, divided by $2\pi\hbar \, (=h)$.

It is as if each quantum state takes up h units of phase space volume [8, p. 177]. In this case, phase space extends over positions 0 to L, and momenta $-\hbar k$ to $\hbar k$. The number of states in this phase-space volume is:

$$\# \, states \equiv n(k) = \frac{V_{\text{phase-space}}}{2\pi\hbar} = (L-0)\big(\hbar k - (-\hbar k)\big) = \frac{2L\hbar k}{2\pi\hbar} = \frac{Lk}{\pi},$$

in agreement with $n(k)=kL/\pi$, Eq. (7.14) previously, since the quantum number n counts the total number of states up to k.

We also notice that a bigger box allows for more states per unit k, since more k values are allowed (Fig. 7.4). In fact, the density of states per unit k is proportional to the size of the box. So sometimes we say, generally, that the density of states per unit k, per unit size L, is a constant $1/2\pi$ (still in 1D):

$$\frac{\rho(k)}{L} = \frac{1}{2\pi} \qquad \text{in units of } [k]^{-1}[L]^{-1} \; = \; [\text{rad/m}]^{-1}[\text{m}]^{-1} = \text{dimensionless}.$$

It is usually convenient to have the density of states as a function of energy, $\rho(E)$, rather than $\rho(k)$. This allows for easily integrating things like $\int dE (\ldots)\rho(E)\delta(E_f - E_i)$, which appears in FGR. We can find $\rho(E)$ from $\rho(k)$ by a change of variables:

$$\rho(E)\, dE = \rho(k)\, dk \quad \Rightarrow \quad \rho(E) = \rho(k)\frac{dk}{dE}.$$

Then:

$$E = \frac{(\hbar k)^2}{2m} \qquad \Rightarrow \qquad k(E) = \frac{(2mE)^{1/2}}{\hbar}, \qquad \frac{dk}{dE} = \left(\frac{1}{2}\right)\frac{\sqrt{2m}}{\hbar} E^{-1/2}.$$

$$\rho(E) = \underbrace{\frac{L}{2\pi}}_{\rho(k)}\underbrace{\frac{1}{2}\frac{\sqrt{2m}}{\hbar} E^{-1/2}}_{dk/dE} = \frac{L}{2\pi\hbar}\sqrt{\frac{m}{2E}} \qquad \text{(1D).} \qquad (7.16)$$

We see (Fig. 7.3, *center*) that in 1D, while $\rho(k)$ is constant, $\rho(E)$ is large at low E and decreases with E (opposite of the 3D case, later). The units of $\rho(E)$ are $[E]^{-1}$, i.e., inverse energy.

For a massless particle, e.g., a photon, $\rho(k)$ is the same, but the E-k relation (dispersion relation) is now:

$$E(k) = \hbar\omega = \hbar ck \quad \Rightarrow \quad k(E) = \frac{1}{\hbar c}E,$$

$$\frac{dk}{dE} = \frac{1}{\hbar c} \quad \Rightarrow \quad \rho(E) = \rho(k)\frac{dk}{dE} = \frac{L}{\pi\hbar c}.$$

Notice that since E is always positive (kinetic energy), we used the one-sided $\rho(k)$ to find $\rho(E)$. Equivalently, we could have used the two-sided $\rho(k)$, and then doubled the result since $E(k)=E(-k)$ are two states of the same energy.

1D Free Particle: Here, the particle is essentially free, but we consider it to be in a large box. Since it has well-defined momentum (k is known), its position is therefore spread widely (uncertainty principle). Since there is no potential, the energy eigenstatesare also momentum eigenstates. We simply use the density of states for a particle in a box of length L, but at the end of any calculation, we take the limit as $L \to \infty$. For any realistic calculation, this limit must be well-defined. Thus, from Eqs. (7.15) and (7.16):

$$\rho(k) = \lim_{L\to\infty}\frac{L}{2\pi} \qquad \rho(E) = \lim_{L\to\infty}\frac{L}{2\pi\hbar}\sqrt{\frac{m}{2E}} \qquad \text{(1D particle in large box).}$$

We take the two-sided view of k ($+k$ and $-k$ are separate states), since left-going and right-going free particles are separate states.

Many references [1, p. 264t, 16, p. 339m, 14, p. 4m, 406m, 21, p. 406m] find the density of states by considering a box of size L, and imposing *periodic* boundary conditions: the complex wave-function at the right and left edges must be the same. We are unaware of any physical justification for this requirement, but it leads to the same result. Periodic boundary conditions means an integer number of cycles fits in the box. This yields a different quantization condition for k:

$$kL = n2\pi \quad \Rightarrow \quad n(k) = \frac{kL}{2\pi}, \quad \rho(k) = \frac{dn}{dk} = \frac{L}{2\pi}, \quad \text{units} = [\text{m}] \quad (\text{periodic BC}).$$

Notice that now both $+k$ and $-k$ are separate states, so the number of states up to a given *magnitude* of k is the same as a particle-in-a-box. This implies that the (phase-space volume)/$(2\pi\hbar)$ is again the number of quantum states enclosed by a given magnitude of k. Then $\rho(E)$ is also the same as for particle in a box. For massless particles, $\rho(E)$ is also the same: $\rho(E) = L/\pi\hbar c$.

3D Free Particle: Here, each of k_x, k_y, and k_z are quantized separately according to the 1D free particle. The 3D wave-vector $\mathbf{k}$ then must sit on a lattice point in $\mathbf{k}$ space (Fig. 7.3, *right*). The k-space volumetric density of these points, i.e., points per unit k-volume, is:

$$\rho(\mathbf{k}) = \rho_x(\mathbf{k})\rho_y(\mathbf{k})\rho_z(\mathbf{k}) = \frac{L_x}{2\pi}\frac{L_y}{2\pi}\frac{L_z}{2\pi} = \frac{Vol}{(2\pi)^3} \quad \text{in units of } [\text{rad/m}]^{-3} = [\text{m}]^3.$$

where *Vol* is the volume of the box in *real* 3D space (not k-space). Much like the 1D case, we usually quote the density of states per unit k-volume per unit (real-space) volume:

$$\frac{\rho(\mathbf{k})}{Vol} = \frac{1}{(2\pi)^3} \quad \text{in units of } [\text{rad/m}]^{-3}[\text{m}]^{-3} = \text{dimensionless}.$$

Though derived for a rectangular volume, this result is quite general for any shape volume [15, p. 4m].

The "volume of phase space" relation extends to higher dimensions as:

$$\text{Number of states} = \frac{V_{\text{phase-space}}}{h^N}$$

where $N \equiv$ number of generalized coordinates of the system.

In such a case, the accessible phase space goes over *Vol*, and over the k-space hypersphere up to radius $k \equiv |\mathbf{k}|$.

In 3D, for large k (enclosing many lattice points), we closely approximate the number of states by the k-space volume times the k-space density of states:

$$\# states \equiv n(k) = \int_0^k d^3k'\, \rho(k') = \int_0^k dk'\, 4\pi k'^2 \frac{Vol}{(2\pi)^3} = \frac{4}{3}\pi k^3 \frac{Vol}{(2\pi)^3}.$$

For small k, this approximation for number of states is less accurate. In terms of momentum:

$$\text{Number of states} = \frac{4}{3}\pi \frac{p^3}{(2\pi\hbar)^3} Vol = \frac{\left(V_{momentum\ space}\right)(Vol)}{h^3}$$

$$= \frac{V_{phase\ space}}{h^3} \quad \text{where} \quad p \equiv \hbar|k|.$$

To find the energy density of states, $\rho(E)$, for free (spinless) particles (or particles in a 3D box), we change variables as before, but now there are many $\mathbf{k}$ vectors with the same energy (all the $\mathbf{k}$ on a sphere of radius $k \equiv |\mathbf{k}|$):

$$\rho(E) = \rho(|\mathbf{k}|)\, 4\pi|\mathbf{k}|^2 \frac{d|\mathbf{k}|}{dE}, \quad \text{where} \quad |\mathbf{k}| = \frac{(2mE)^{1/2}}{\hbar}, \quad \frac{d|\mathbf{k}|}{dE} = \frac{(2m)^{1/2}}{2\hbar} E^{-1/2}.$$

$$\rho(E) = \frac{Vol}{(2\pi)^3}\, 4\pi \left(\frac{2mE}{\hbar^2}\right)\frac{(2m)^{1/2}}{2\hbar} E^{-1/2} = Vol\, \frac{4\pi\sqrt{2}}{(2\pi\hbar)^3} m^{3/2} E^{1/2} \quad \text{in units of } [E]^{-1}.$$

In 3D we see that $\rho(E)$ increases with E, opposite to the 1D case (Fig. 7.3).

For massless particles, the k-space density is the same, but the E-k relation is $E = c\hbar k$:

$$\rho(E) = \rho(|\mathbf{k}|)\, 4\pi|\mathbf{k}|^2 \frac{d|\mathbf{k}|}{dE}, \quad \text{where} \quad |\mathbf{k}| = \frac{E}{\hbar c}, \quad \frac{d|\mathbf{k}|}{dE} = \frac{1}{\hbar c}.$$

$$\rho(E) = \frac{Vol}{(2\pi)^3}\, 4\pi \left(\frac{E^2}{\hbar^2 c^2}\right)\frac{1}{\hbar c} = Vol\, \frac{4\pi}{(2\pi\hbar)^3 c^3} E^2 \quad \text{in units of } [E]^{-1}.$$

This is very different behavior than the massive density of states, increasing much more rapidly with E.

It can be proven that:

For large volumes or high energies, the density of states per unit k per unit volume is essentially $\rho(k)/Vol = 1/(2\pi)^3$, regardless of the shape of the volume or the boundary conditions applied.

This shape-independence follows from short-wavelength considerations similar to the WKB approximation.

Note The k-volume density of states per unit *Vol* (real-space volume) is a constant $1/(2\pi)^3$, leading to integrals of the form $d^3k/(2\pi)^3$. Another related situation looks similar: the inverse Fourier transform (from k basis to x basis) also leads to integrals of the form $d^3k/(2\pi)^3$ (depending on the normalization choice). However, the integrands and limits are usually different, and you should be careful to understand the distinction.

7.12 Perturb Unto Others: FGR

FGR often first comes up when considering interaction of matter with EM fields. This section assumes you have been through the derivation in a standard text [1, p. 248+, 16, p. 327+], and want clarification.

For an EM transition, FGR allows us to compute the transition rate, in transitions/s. The rule is often confusingly written with a δ-function, such as (for absorption):

$$R = \frac{2\pi}{\hbar}\left|\langle f|\hat{H}_{int}|i\rangle\right|^2 \delta(E_f - E_i - \hbar\omega) \qquad \text{(not rigorous)}$$

where $|i\rangle \equiv$ initial state; $\langle f| \equiv$ final state,

           $\hat{H}_{int}$ is the perturbing interaction hamiltonian,

           $E_f \equiv$ final energy of matter, $E_i \equiv$ initial energy of matter,

           $\hbar\omega \equiv$ energy of incident photon.

This clearly requires further description: a "rate" is a number and cannot have a delta function in it. In fact, the previous formula only becomes a rate when it is integrated over a density of states, which removes the δ-function ("…in all practical applications, the δ-function will get integrated over…." [20, p. 483b]; "…to get actual numbers from this formula we must sum over a continuous group of… states…" [1, p. 251m]). What the earlier δ-function means is that the probability is very high that the final state will lie in a narrow range of energies. This narrow range insures energy conservation, and so the δ-function is sometimes called an *energy-conserving δ-function*. Therefore, we should rewrite FGR as:

$$R = \frac{2\pi}{\hbar}\left|\langle f|\hat{H}_{int}|i\rangle\right|^2 \int \delta(E_f - E_i - \hbar\omega)\rho(something)\,dE$$

$$= \frac{2\pi}{\hbar}\left|\langle f|\hat{H}_{int}|i\rangle\right|^2 \rho(something),$$

but we must be careful to describe what is $\rho(something)$. In fact, this density of states can appear due to any *one* of the three entities in the matrix element $\langle f|\hat{H}_{int}|i\rangle$: the density of initial states, ρ_i; density of final states, ρ_f; and in the interaction

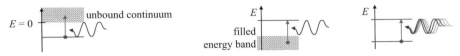

Fig. 7.5 (*Left*) Density of final state. (*Middle*) Density of initial states. (*Right*) Density of $\hat{H}_{int}$

Hamiltonian, H_{int}. Figure 7.5 gives examples of all three cases. The left case is probably the most common: photoionizing an atom moves an electron from a discrete energy state to a continuum of states. The middle case excites one electron from a continuous band of electrons (e.g., in a valence band) into a single state above the band (e.g., due to an impurity atom). The right case excites an electron from a lower to a higher energy state, with radiation having a photon density $\rho(E)$.

(In an idealized case where the initial and final states are both discrete, and the radiation monochromatic, then the transition probability simply oscillates in time with period $2\pi\hbar/|E_i - E_j|$, and there is no density of states [1, p. 250t]. Therefore, there is no "transition rate.")

In some cases, it is more convenient to change variables from E to k, and then we have, in 1D:

$$R = \frac{2\pi}{\hbar}\left|\langle f|\hat{H}_{int}|i\rangle\right|^2 \rho(k)\frac{dk}{dE} \qquad \text{(1D)}.$$

For a free particle density of states in 3D, where $E = (\hbar k)^2/2m$:

$$R = \frac{2\pi}{\hbar}\left|\langle f|\hat{H}_{int}|i\rangle\right|^2 \rho(k)4\pi k^2 \frac{dk}{dE} \qquad \text{where} \quad k \equiv |\mathbf{k}| \quad \text{(3D)}.$$

As an example, consider a system which decays by radiating a photon. Since photons can radiate into any energy, there are many (a continuum of) final photon states. Therefore, it is the photon density of states in the final state that is integrated with the δ-function to produce an actual rate.

For example, to compute an infinitesimal rate of radiation of an atom into a differential solid angle, we have [1, p. 252]:

$$dR(\theta, \phi) = \frac{2\pi}{\hbar}\left|\langle f|\hat{H}_{int}|i\rangle\right|^2 d\Omega \int_0^\infty dE\ \rho(E, \theta, \phi)\delta(E_f - E_i - \hbar\omega)$$

$$= \frac{2\pi}{\hbar}\left|\langle f|\hat{H}_{int}|i\rangle\right|^2 d\Omega\ \rho(\hbar\omega, \theta, \phi).$$

To compute the total decay rate, we must integrate over all radiated photon angles:

$$R = \int_0^\pi \sin\theta\ d\theta \int_0^{2\pi} d\phi\ dR(\theta, \phi).$$

See [1, p. 252] for more details.

How to Remember FGR: One can produce FGR (up to the 2π) with dimensional analysis and basic QM concepts.

1. FGR describes a *probability* (per unit time) of measuring a final state given an initial state, so it must involve an inner product of the form $\left|\langle f|\text{ something }|i\rangle\right|^2$.
2. FGR describes time evolution, which is generated by the Hamiltonian, so $R \propto \left|\langle f|\hat{H}_{\text{int}}|i\rangle\right|^2$.
3. Energy must be conserved, so we need an energy conserving δ-function: $R \propto \left|\langle f|\hat{H}_{\text{int}}|i\rangle\right|^2 \delta\left(E_f - E_i\right)$.
   Remember to include the photon energy, $\hbar\omega$, if this is an absorption or emission transition.
4. We must have an $\hbar$ (because this is QM), and dimensions show it must be in the denominator, thus: $R \propto \left|\langle f|\hat{H}_{\text{int}}|i\rangle\right|^2 \delta\left(E_f - E_i\right)/\hbar$.

The only thing missing is the 2π, which you just have to memorize, or derive in detail.

Validity of Approximation: FGR is only a first-order approximation, and only valid in a range of times [1, p. 251b]. We quote the following results without derivation. A detailed analysis [1, p. 248+] shows that for a range of final-state energies of ΔE, FGR becomes valid when the time is long enough for the main peaks of the probability curve fully contained within the band of energy states [1, 12.24a p. 251b]:

$$t > \frac{2\pi\hbar}{\Delta E}.$$

Further, time must be *short enough* that the energy spacing of the final states, δE, is very fine compared to the time [1, 12.24b p. 251b]:

$$t \ll \frac{2\pi\hbar}{\delta E}.$$

This is usually easily satisfied in realistic situations.

Second Order: As with other perturbation concepts, the second-order transition rate is evaluated by considering the probability that the system evolved temporarily into an unobserved intermediate state, times the probability that it then evolved into the given final state. Thus, the second-order transition complex-amplitude is [1, 12.47 p. 258t]:

$$\lim_{\varepsilon \to 0} \sum_{m \neq i, f} \frac{\langle f|\hat{H}_{\text{int}}|m\rangle\langle m|\hat{H}_{\text{int}}|i\rangle}{E_i - E_m + i\varepsilon\hbar}.$$

Note that energy is *not* conserved on the unobserved intermediate transitions themselves, but the final observed transition $i \to f$ *does* conserve energy [16, p. 333m]. If the first-order amplitude is zero, i.e., $\langle f|\hat{H}_{int}|i\rangle = 0$, then the second-order transition rate is [1, 12.46 p. 257]:

$$R_{i\to f}^{(2)} = \frac{2\pi}{\hbar}\left|\sum_{m\neq i,f}\frac{\langle f|\hat{H}_{int}|m\rangle\langle m|\hat{H}_{int}|i\rangle}{E_i - E_m + i\varepsilon\hbar}\right|^2 \delta(E_f - E_i), \quad \text{when } \langle f|\hat{H}_{int}|i\rangle = 0,$$

where we sum over all possible unobserved intermediate states. However, if $\langle f|\hat{H}_{int}|i\rangle \neq 0$, there may be interference between the first-order and second-order amplitudes. Then as always in QM, we must first add the amplitudes and then square to get the total rate to second order [14, 5.6.37 p. 333]:

$$R_{i\to f}^{(1+2)} = \frac{2\pi}{\hbar}\left|\langle f|\hat{H}_{int}|i\rangle + \sum_{m\neq i,f}\frac{\langle f|\hat{H}_{int}|m\rangle\langle m|\hat{H}_{int}|i\rangle}{E_i - E_m + i\varepsilon\hbar}\right|^2 \delta(E_f - E_i).$$

Chapter 8
Desultory Topics in Quantum Mechanics

8.1 Short Topics

8.1.1 Parity

A parity transformation of a system reflects each point through the origin. It is equivalent to three reflections in three mirrors: $x \to -x$, $y \to -y$, $z \to -z$, or $\mathbf{r} \to -\mathbf{r}$. You might think that you could achieve that effect with one mirror, if you held it at the proper angle. We see that is not true by supposing such a mirror placement exists, and considering what happens to a vector nearly parallel to the plane of the mirror. Its reflection is barely different than the original vector, and clearly *not* reflected through the origin. You must use three mirrors.

> All the laws of physics, except the weak force, are invariant under a parity transformation of the system.

8.1.2 Massive Particle Frequency

Unlike a photon, a massive or charged particle frequency has an *arbitrary* reference point. Therefore, absolute frequency of a massive particle has no physical meaning. Only *differences* in frequency are meaningful. Recall that particle frequency is proportional to *total* energy, kinetic (T) + potential (V):

$$E = T + V = \hbar\omega,$$

but V has an *arbitrary* reference point. In particular, E and therefore ω, can be negative. Negative frequency arises in many areas of physics, engineering, and mathematics, especially when the zero-point of a frequency is arbitrary. Negative fre-

The original version of this chapter was revised. An erratum can be found at
https://doi.org/10.1007/978-1-4614-9305-1_9

E. L. Michelsen, *Quirky Quantum Concepts,* Undergraduate Lecture Notes in Physics,
DOI 10.1007/978-1-4614-9305-1_8, © Springer Science+Business Media New York 2014

quency is essential to quantum mechanics (QM) and quantum field theory (QFT), but what is negative frequency?

An ordinary frequency counts oscillations of something, per unit time, which is necessarily non-negative. The oscillating parameter is some one-dimensional (1D) quantity. As described in Sect. 1.6.2, to allow for negative frequency, we imagine something rotating about an axis (rather than just oscillating). If it rotates clockwise, physicists call it positive frequency. In this way, energy differences are proportional to frequency differences.

Note that phase velocity $v_p = \omega/k$, and thus v_p also has no physically meaningful absolute zero point. In contrast, the group velocity, $v_g = d\omega/dk$, is independent of an arbitrary frequency (or energy) offset. This is consistent with its interpretation as the physically measurable particle velocity.

The previous frequency considerations are in direct contrast to a photon, or other possible massless-chargeless particles, where ω has an absolute value and $E = \hbar\omega$ is an absolute equation for the total energy of the particle. v_p then also has an absolute value.

Note also that massive particles *do* have an absolute wavelength, which can be measured by diffraction. The absolute wavelength is given by the kinetic momentum, and is gauge invariant. However, the spatial frequency k that appears in the wave-function is gauge dependent. In other words, the wave-function itself is gauge dependent, but all observables are, as in all of physics, gauge invariant.

8.1.3   Uncertainty is not From Measurement Clumsiness

The uncertainty principle is often misleadingly described as: "You can't simultaneously know the position and momentum of an object, because when you measure its position, you disturb its momentum, and when you measure its momentum, you disturb its position." This description belies the quantum theory, because it implies that a particle *has* a well-defined position and momentum, but you just cannot know what they both are. In fact, a particle has a wave function, which gives probability densities for all possible values of position and momentum. Due to the nature of the position and momentum operators, which act on the wave function to produce the probability density functions [position probability distribution function (PDF) and momentum PDF], it is impossible to construct a wave function which yields arbitrarily narrow ranges for both position and momentum. Therefore:

> Uncertainty arises not because you cannot *know* both position and momentum, but because a particle cannot *have* precise values of both position and momentum.

One significant consequence of this more accurate statement of uncertainty is this: do not bother looking for "gentler" measurement methods, which disturb the particle *less* than existing measurements, in the hopes of beating Heisenberg. The "disturbance" of a previously known value is not some measurement side effect; it is fundamental to the nature of wave functions and measurements.

Now the measurement postulate *does* say that when you measure position, you "disturb" momentum, because you collapse the wave function to one of a precise position. But if you make a precise position measurement, you do not change the momentum to some precise, but unknown, value. When you make a precise position measurement, you change the wave function, making the position PDF narrow, and the momentum PDF wide. The problem is not that now you do not *know* the new momentum; the problem is that now the particle does not *have* a definite momentum. The particle is in a superposition of many different momenta.

8.1.4 Generalized Uncertainty

In general, the uncertainty product of two observables depends on the state of the system. Many people are confused by this because in the special case of position–momentum uncertainty, the uncertainty product happens to be a constant for any state. However,

> In the general case, the uncertainty product depends on the state of the system.

Consider a set of systems in identical quantum states, each of which is the same superposition of eigenstates of an observable. Recall that if we measure this observable for every system, then we will get a variety of measurements, and they will have some standard deviation, σ. It is easy to show that, if $\hat{A}$ and $\hat{B}$ are two noncommuting observables, then in the state $|\psi\rangle$ ([16], 1.4.63, p. 36):

Using: $\langle \hat{O} \rangle = \langle \psi | \hat{O} | \psi \rangle$

$$\sigma_A^2 \sigma_B^2 \geq \frac{1}{4}\left|\langle \psi | \left[\hat{A},\hat{B}\right] | \psi \rangle\right|^2 + \frac{1}{4}\left|\langle \psi | \left\{\hat{A},\hat{B}\right\} | \psi \rangle\right|^2$$

where $\left[\hat{A},\hat{B}\right] \equiv \left(\hat{A}\hat{B} - \hat{B}\hat{A}\right); \quad \left\{\hat{A},\hat{B}\right\} \equiv \left(\hat{A}\hat{B} + \hat{B}\hat{A}\right).$

Since the second term in the inequality is ≥ 0, the inequality still holds without it, so sometimes people abbreviate ([16], 1.4.53, p. 35):

$$\sigma_A^2 \sigma_B^2 \geq \frac{1}{4}\left|\langle \psi | \left[\hat{A},\hat{B}\right] | \psi \rangle\right|^2$$

or by simply taking square roots:

$$\sigma_A \sigma_B \geq \frac{1}{2}\left|\langle \psi |\left[\hat{A}, \hat{B}\right]| \psi \rangle\right|.$$

For example, the uncertainty product of the x and y components of angular momentum is zero in the state where the total angular momentum is zero, $|J = 0, M = 0\rangle$. However, the same uncertainty product is nonzero for states with $M \neq 0$. Recalling that $\left[\hat{J}_x, \hat{J}_y\right] = i\hbar \hat{J}_z$:

$$\sigma_x^2 \sigma_y^2 \geq \frac{1}{4}\left|\langle J, M |\left[\hat{J}_x, \hat{J}_y\right]| J, M \rangle\right|^2 = \frac{1}{4}\left|\langle J, M | i\hbar \hat{J}_z | J, M \rangle\right|^2$$

$$= \frac{\hbar^2}{4} M^2 \left|\langle J, M | J, M \rangle\right|^2 = \frac{\hbar^4}{4} M^2.$$

The uncertainty product increases as the z component of angular momentum increases. The uncertainty in measurements of the system depends on the state of the system.

Back to position–momentum uncertainty: it happens that the commutator of $\hat{x}$ and $\hat{p}$ is a constant for all states, hence the uncertainty product is a constant for any state $|\psi\rangle$:

$$\sigma_x \sigma_p \geq \frac{1}{2}\left|\langle \psi | i\hbar | \psi \rangle\right| = \frac{\hbar}{2}\left|\langle \psi | \psi \rangle\right| = \frac{\hbar}{2}.$$

Finally, there is often talk of "energy-time" uncertainty. However, though energy *is* an observable, time *is not an observable*. Time is a parameter; time marches on; we know what time it is, but we do not measure the "time" of a system (as we do its energy); time has no corresponding Hermitian operator. Therefore,

---

The energy-time "uncertainty relation" is fundamentally a different phenomenon than the uncertainty product of two observables.

According to ([16], p. 80), "this time-energy uncertainty relation is of a very different nature from the uncertainty relation between two incompatible observables."

8.1.5   Is ∇^2 the Square or Composition of Anything?

"∇^2" denotes the Laplacian operator, sometimes called "grad squared." Usually, the superscript ^{2} notation means either the square of a number, or the composition of an operator on itself. For example, x^2 means simply the square of the number "x." $\hat{p}^2$ means $\hat{p}\hat{p}$. However, contrary to such uses of the superscript ^{2} notation, ∇^2 is neither the square of a number nor the composition of some ∇ operator on itself. ∇^2 is actually the composition of two *different* "∇" operators. We consider those operators now.

First, "∇" is the **gradient** operator. It operates on a scalar field, and produces a vector field. For example, if $\Phi(x, y, z)$ is the electric potential (a scalar field), $\mathbf{E} = -\nabla\Phi(x, y, z)$ is the electric (vector) field. Recall that the gradient $\nabla\Phi$ (pronounced "grad fie" or "del fie") tells how Φ varies when moving in any direction from a point:

$$\Delta\Phi(\text{in direction } \mathbf{s}) \approx \nabla\Phi \cdot \mathbf{s} \quad \text{(to first order)}.$$

The gradient operator is often written in bold, "∇," indicating its result is a vector.

Second, "$\nabla\cdot$" is the **divergence** operator (pronounced "del dot"). It operates on a vector field and produces a scalar field. The divergence operator is never written in bold, because its result is a scalar field.

The divergence of the gradient is a composition of operators that arises very frequently. It can be written "$\nabla\cdot\nabla$", e.g.,

$$\nabla \cdot \nabla\Phi = \text{the divergence of the gradient of the scalar field } \Phi,$$

whose result is another scaler field.

This composition arises so frequently that we use the special notation "∇^2" as shorthand for "$\nabla\cdot\nabla$". But the two "dels" in "$\nabla\cdot\nabla$" are two different things: the right (bold) ∇ is the gradient operator, and the left "$\nabla\cdot$" is the divergence operator. So:

> ∇^2 is not the composition of any operator on itself. It is inconsistent notation, but universally used.

Note that (in simple nontensor mathematics) it is meaningless to square either the gradient or the divergence operators:

$\nabla(\nabla\varphi)$ has the left gradient operator acting on a vector field, which is undefined.

$\nabla\cdot(\nabla\cdot\mathbf{v})$ has the left divergence operator acting on a scalar field, which is undefined.

All that said, there is a sense in which ∇^2 is roughly the square of *something*. It is common to write the square of a vector $\mathbf{p}$ as $p^2 \equiv \mathbf{p}\cdot\mathbf{p}$. Following this idea, we write the definition of the gradient operator ∇ as:

$$\nabla \equiv \left(\frac{\partial}{\partial x} \mathbf{i} + \frac{\partial}{\partial y} \mathbf{j} + \frac{\partial}{\partial z} \mathbf{k} \right) \Rightarrow$$

$$\nabla^2 \equiv \nabla \cdot \nabla = \left(\frac{\partial}{\partial x} \mathbf{i} + \frac{\partial}{\partial y} \mathbf{j} + \frac{\partial}{\partial z} \mathbf{k} \right) \cdot \left(\frac{\partial}{\partial x} \mathbf{i} + \frac{\partial}{\partial y} \mathbf{j} + \frac{\partial}{\partial z} \mathbf{k} \right) = \frac{\partial^2}{\partial x^2} + \frac{\partial^2}{\partial y^2} + \frac{\partial^2}{\partial z^2}.$$

In this loose interpretation, ∇^2 could be called the square of the gradient operator, but it is not the composition of the gradient operator with itself (which can only be defined as a rank-2 tensor).

All of these comments on ∇^2 apply directly to the three-dimensional momentum operator:

$$\hat{\mathbf{p}} \equiv \frac{\hbar}{i} \nabla \Rightarrow \quad \hat{\mathbf{p}}^2 \equiv \hat{\mathbf{p}} \cdot \hat{\mathbf{p}} = -\hbar^2 \nabla^2,$$

which is a *scalar* operator.

8.1.6   The Only Stationary States are Eigenstates of Energy

The only stationary states are eigenstates of energy, because every state evolves in time by the operation of $\exp(i\hat{H}t/\hbar)$, and the only states of fixed energy are eigenstates of $\hat{H}$. That is, for $\hat{H}$ independent of time:

$$\hat{U}(t, t_0) = \exp(i\hat{H}(t - t_0)/\hbar).$$

For a stationary state, time evolution must introduce only a (time-dependent) complex phase factor to the wave-function, which means $\hat{H}$ must produce a constant. This is the definition of an energy eigenstate:

$$\hat{H} |\psi\rangle = E |\psi\rangle.$$

Thus, the only stationary states are energy eigenstates.

8.1.7   g Whiz

The letter "g" has at least three different meanings in QM, the first two of which are especially confusing.

First, when counting quantum states in statistical mechanics, the multiplicity of an energy level, or multiplicity of a spatial state, is called "g." For an electron, or any other fermion, its spin is 1/2, and therefore, in the absence of spin-dependent energy (e.g., no magnetic field), each spatial state has multiplicity $g = 2$ *exactly*: one state spin up, the other spin down.

A second (completely different) use of "g" refers to the gyromagnetic ratio: the ratio of the electron's spin-related magnetic dipole moment to its Bohr magneton. (Other disciplines use the term "gyromagnetic ratio" differently.) Recall the Bohr magneton is the magnetic dipole moment created by an electron in an *orbit* of angular momentum $\hbar$. One can easily show that the Bohr magneton is:

$$\mu_B = \frac{e\hbar}{2mc} \quad \text{(gaussian)} \quad \text{or} \quad \mu_B = \frac{e\hbar}{2m} \text{(SI)}$$

where $e \equiv$ particle charge, $m \equiv$ particle mass.

For an electron, $g \approx -2$, but *not exactly*. In fact, $g_e = -2.002\ 319\ 304\ 362\ 2(15)$ [Nis, "electron g factor"]. And therefore, the actual spin-related dipole moment for an electron is $\mu_e \approx -2.002\mu_B$. For a general particle, $\mu = g\mu_{\text{magneton}}$.

> When referring to the spin-multiplicity of a spatial state, $g = 2$ *exactly*. When referring to gyromagnetic ratio, $g \approx -2$.

The third use of g extends the gyromagnetic ratio to the total angular momentum, $\mathbf{J} = \mathbf{L} + \mathbf{S}$, which is a combination of orbital and spin angular momentum. This is called the **Landé g-factor** (described in detail in Sect. 5.8). Part of $\mathbf{J}$ is due to $\mathbf{L}$, and part due to $\mathbf{S}$. Therefore, part of the magnetic dipole moment is due to $\mathbf{L}$ (with weight 1), and part due to $\mathbf{S}$ (with weight approximately 2). Thus, g is a weighted average such that:

$$\mu = g m_j \mu_B \quad \text{where} \quad m_j \text{ is the } J_z \text{ quantum number.}$$

8.1.8   Why Are Photons Said to be Massless?

Photons have energy, and $m = E/c^2$, so why are photons said to be massless? In the old days, to try to retain $F = ma$, physicists said that a moving particle's mass increased by a factor:

$$\gamma \equiv \frac{1}{\sqrt{1-(v/c)^2}}.$$

This "variable mass" turned out to be a bad idea, because the "mass" parallel to the direction of motion is then different than the "mass" perpendicular to the motion. To avoid this, and have only one kind of mass, modern relativity defines mass as a scalar (all observers measure the same mass): "mass" is always the "rest mass" of the particle. This section assumes you understand four-vectors.

Instead of changing the mass with motion, we use the relativistic laws of motion, which include the γ factor. They are written in three-vector and four-vector form as:

(Three-vector): $\mathbf{p} \equiv \gamma m \mathbf{v}, \quad \mathbf{F} = \dfrac{d\mathbf{p}}{dt}$.

(Four-vector): $p^{\mu} = (E/c, \mathbf{p}) = (\gamma mc, \mathbf{p}), \quad F^{\mu} = \dfrac{dp^{\mu}}{d\tau}, \quad$ where $\tau \equiv$ proper time.

From this, we can write an invariant equation for a particle's mass from its energy-momentum four-vector:

$$p^{\mu} p_{\mu} = \gamma^2 m^2 c^2 - \gamma^2 m^2 v^2 = m^2 c^2 \gamma^2 \left(1 - \dfrac{v^2}{c^2}\right) = m^2 c^2 \quad \Rightarrow \quad m = \sqrt{\dfrac{p^{\mu} p_{\mu}}{c^2}}.$$

For a photon,

$$E = |\mathbf{p}|c \quad \Rightarrow \quad p^{\mu} = |\mathbf{p}|, \mathbf{p} \quad \Rightarrow \quad m = \sqrt{\dfrac{p^{\mu} p_{\mu}}{c^2}} = \sqrt{\dfrac{\mathbf{p}^2 - \mathbf{p}^2}{c^2}} = 0.$$

Particles that move at the speed of light *must* be massless (to have finite energy), and therefore must satisfy $E = |\mathbf{p}|c$.

8.1.9   The Terrible Electromagnetic Hamiltonian

Some references use a confusing notation for the interaction of an electron with a magnetic field. Before we describe this, recall the general description of a charged particle, with charge q, in a magnetic field. Here, q is the charge of the particle (not a universal physical constant), and is negative for electrons, and positive for positrons and protons:

$$\mathbf{p}_{kin} = m\mathbf{v}, \qquad \mathbf{p} = \mathbf{p}_{kin} + \dfrac{q}{c}\mathbf{A} \quad \Rightarrow \quad \hat{H} = \dfrac{\hat{\mathbf{p}}_{kin}^2}{2m} + V(\mathbf{r})$$

$$= \dfrac{\left(\hat{\mathbf{p}} - \dfrac{q}{c}\mathbf{A}(\mathbf{r})\right)^2}{2m} + V(\mathbf{r}) \tag{8.1}$$

where $\mathbf{p} \equiv$ canonical momentum.

Note that $\mathbf{p}$ is $\mathbf{p}_{kin}$ *plus* $(q/c)\mathbf{A}$, and therefore the kinetic energy in the Hamiltonian *subtracts* $(q/c)\mathbf{A}$ to get $\mathbf{p}_{kin}$.

Some references specialize to the particle being an *electron*, and call its charge "$-e$," where e is a physical constant (the charge of a proton). This *reverses* the plus and minus signs in the formulas for the canonical momentum and the Hamiltonian. This notation is both less general (since it only works for negative particles), and it conflicts with standard notation in classical mechanics (which is not obsessed with electrons).

Do not do it. Just let charge be charge: q takes the sign of the particle in question, and Eq. 8.1 holds.

8.2   Current Events: Probability Current and Electric Current

Many QM references discuss a concept called "probability current," more accurately called probability current density, measured in particles/s/m². This is a mathematical construction, which has limited physical meaning and application. However, comparing it to true particle current illuminates the fundamental postulates and workings of QM. Nearly all references give an incomplete equation for probability current. Note that a single-particle *probability* density is actually a *particle* density, for all practical purposes (see "The Meaning of the Wave-Function," Sect. 1.9). However, probability *current* is *different* from the classical currents which generate things like magnetic fields. We proceed as follows:

- the meaning of a general continuity equation
- the quantum continuity equation
- an example of the failure of the "standard" equation
- correcting the failure
- the physical meaning of probability current and other currents, such as electric current and its coupling to magnetics.

This section assumes you are familiar with basic QM, such as operators and the Schrödinger equation, with the vector calculus of divergence, and 1D tunneling.

> Some popular texts incorrectly say, or suggest, that the electromagnetic (EM) field couples to the probability current. This is wrong, as described later.

8.2.1   General Continuity

The most familiar continuity equation is probably that of classical electric current, which expresses that electric charge cannot be created or destroyed. Electric charge can move or flow through space, but total charge is *conserved*:

$$\nabla \cdot \mathbf{J}(t,\mathbf{r}) = -\frac{\partial \rho(t,\mathbf{r})}{\partial t} \quad \text{(electric charge)}$$

where $\mathbf{J} \equiv$ electric current density, C/s/m², $\rho \equiv$ charge density, C/m³.

This means that the net outflow of charge from any volume equals the decrease in charge within that volume. The continuity equation holds for all points in time and space. A similar **continuity equation** can express the conservation of anything which can flow (e.g., mass).

The quantum continuity equation: In QM, consider the (probability) density of a single particle:

$$\rho(t,\mathbf{r}) = \psi^*(t,\mathbf{r})\psi(t,\mathbf{r}).$$

We can visualize this density as a fluid, which deforms and flows through time-evolution of the wave function. This view leads to conservation of particle number (sometimes called "conservation of probability"): we start with one particle, so at all future times, the integral of the particle density must still equal one full particle:

$$\int_\infty \rho(t,\mathbf{r})\, d^3r = \int_\infty \psi^*(t,\mathbf{r})\psi(t,\mathbf{r})\, d^3r = 1, \qquad \forall\, t.$$

This is just normalization. Therefore, when the particle "fluid" (density) flows, we must have a continuity equation that describes the conservation of the total fluid, which always sums to one particle, i.e., at every point in space, and at every time, we must have:

$$\nabla \cdot \mathbf{J}(t,\mathbf{r}) = -\frac{\partial \rho(t,\mathbf{r})}{\partial t} \qquad \text{(quantum particles)}$$

where $\mathbf{J} \equiv$ particle current, particles/s/m^2, $\rho \equiv$ particle density, particles/m^3,
$\mathbf{J}$ and ρ are real.

Since a particle's momentum says something about where the particle is going, we expect the *probability* current to be related to momentum. We show later that indeed, it is. However, first we start with a traditional, though somewhat tedious, derivation directly from the Schrödinger equation. Since the wave function of a particle tells us everything there is to know about the particle (except spin), we can write this continuity equation directly in terms of the wave function. Let us start by writing $\partial\rho/\partial t$ in terms of ψ:

$$\frac{\partial \rho}{\partial t} = \frac{\partial}{\partial t}(\psi^*\psi) = \psi^*\frac{\partial \psi}{\partial t} + \frac{\partial \psi^*}{\partial t}\psi \quad \psi \equiv \psi(t,\mathbf{r}). \tag{8.2}$$

The Schrödinger equation gives us $\partial\psi/\partial t$, and the conjugate Schrödinger equation gives $\partial\psi^*/\partial t$:

$$i\hbar\frac{\partial \psi}{\partial t} = -\frac{\hbar^2}{2m}\nabla^2\psi + V(t,\mathbf{r})\psi \quad \text{and} \quad -i\hbar\frac{\partial \psi^*}{\partial t} = -\frac{\hbar^2}{2m}\nabla^2\psi^* + V(t,\mathbf{r})\psi^*.$$

The left sides of these two formulas are close to the two terms of $\partial(\psi * \psi)/\partial t$ in Eq. 8.2. We can make them match exactly: multiply the Schrödinger equation by $\psi*$, the conjugate equation by ψ, subtract the latter from the former, and rearrange:

$$i\hbar\psi*\frac{\partial\psi}{\partial t} = -\frac{\hbar^2}{2m}\psi*\nabla^2\psi + V(t,\mathbf{r})\psi*\psi \quad \text{and} \quad -i\hbar\frac{\partial\psi*}{\partial t}\psi = -\frac{\hbar^2}{2m}\left(\nabla^2\psi*\right)\psi$$
$$+V(t,\mathbf{r})\psi*\psi$$

$$i\hbar\psi*\frac{\partial\psi}{\partial t} + i\hbar\frac{\partial\psi*}{\partial t}\psi = -\frac{\hbar^2}{2m}\psi*\nabla^2\psi + \cancel{V(t,\mathbf{r})\psi*\psi} + \frac{\hbar^2}{2m}\left(\nabla^2\psi*\right)\psi - \cancel{V(t,\mathbf{r})\psi*\psi}$$

$$\psi*\frac{\partial\psi}{\partial t} + \frac{\partial\psi*}{\partial t}\psi = \frac{\hbar}{2im}\left[\left(\nabla^2\psi*\right)\psi - \psi*\nabla^2\psi\right].$$

We have $\partial\rho/\partial t$ on the left, so we would like to write the RHS as $\nabla\cdot\mathbf{J}$, and find an expression for $\mathbf{J}$. We can rewrite the brackets on the RHS, using $\nabla^2 \equiv \nabla\cdot\nabla$ (the first "$\nabla\cdot$" is a divergence, the second "∇" is a gradient), and a divergence product rule:

$$\nabla\cdot\left(\mathbf{A}(\mathbf{r})f(\mathbf{r})\right) = \left(\nabla\cdot\mathbf{A}\right)f + \mathbf{A}\cdot\nabla f \quad \Rightarrow$$
$$\left(\nabla^2\psi*\right)\psi - \psi*\nabla^2\psi = \left(\nabla\cdot\nabla\psi*\right)\psi - \left(\nabla\cdot\nabla\psi\right)\psi*$$
$$= \nabla\cdot\left[\left(\nabla\psi*\right)\psi - \left(\nabla\psi\right)\psi*\right] + (\text{terms that cancel}).$$

The additional product-rule terms on the RHS cancel. Substituting into Eq. 8.2, we have:

$$\frac{\partial\rho}{\partial t} = \frac{\hbar}{2im}\nabla\cdot\left[\left(\nabla\psi*\right)\psi - \left(\nabla\psi\right)\psi*\right].$$

This is in the form of the continuity equation, with (reversing the sign on the RHS):

$$\nabla\cdot\mathbf{J}(t,\mathbf{r}) = -\frac{\partial\rho}{\partial t} = \nabla\cdot\frac{\hbar}{2im}\left[\left(\nabla\psi\right)\psi* - \left(\nabla\psi*\right)\psi\right]$$
$$\textit{where} \quad \mathbf{J} \equiv \text{probability current we seek.}$$

Almost all references here declare that if the divergence of two vector fields are equal, then the vector fields are equal. This, of course, is not true, and leads to contradictions in simple examples, as shown later.

The previous equation is a differential equation for $\mathbf{J}$, with a general solution, and a particular solution determined by boundary conditions.

Certainly, *one* particular solution is the standard **probability current** (aka "number current"):

Fig. 8.1 1D tunneling: The particle approaches from the left, and interacts with the barrier. The wave-function splits into transmitted and reflected parts. Inside the barrier, $\psi(x)$ varies exponentially

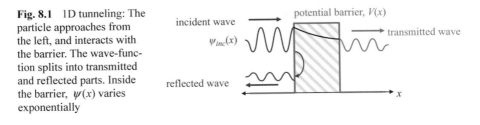

$$\mathbf{J}(t,\mathbf{r}) = \frac{\hbar}{2im}\left[(\nabla\psi)\psi^* - (\nabla\psi^*)\psi\right] \quad \text{(one particular solution)}. \quad (8.3)$$

Notice that adding any divergence-less vector field to a solution **J** yields another solution. Therefore, the general solution is:

$$\mathbf{J}(t,\mathbf{r}) = \frac{\hbar}{2im}\left[(\nabla\psi)\psi^* - (\nabla\psi^*)\psi\right] + B\mathbf{s}(t,\mathbf{r}), \quad \psi \equiv \psi(t,\mathbf{r})$$

$$\text{where} \quad B \equiv \text{arbitrary constant for fitting auxiliary conditions on } \psi(t,\mathbf{r}) \quad (8.4)$$

$$\mathbf{s}(t,\mathbf{r}) \equiv \text{purely solenoidal vector field, i.e., } \nabla\cdot\mathbf{s} = 0$$

For our purposes, the only $\mathbf{s}(t, \mathbf{r})$ we will need is a constant vector $\mathbf{s}(t, \mathbf{r}) = \mathbf{C}$.

As always with differential equations, the constants B (or $\mathbf{C}$) are determined by the boundary (or other auxiliary) conditions. Note that:

> The "standard" equation for **J** does not allow for **C**, the constant vector.

For wave functions, or parts of wave functions, that extend over an infinite space, normalization demands that the particle density must go to zero at infinity, and hence the probability current does, too. For such infinite regions, then, **C** must be 0. This is *not* required for finite regions of the wave function, since the wave function remains finite in such regions.

Note that a nonzero probability current in Eq. 8.3 requires both real and imaginary parts of ψ, since either by itself cancels (as it must, since **J** is real, and the i in the denominator requires the numerator be purely imaginary). This implies that any complex constant times a real function also has zero probability current.

8.2.2   Failure of the "Standard" Formula

We now demonstrate the need for **C** with an elementary example of 1D quantum tunneling. When an incident particle impinges on a potential barrier higher than its energy, some of the particle is reflected, and some is transmitted (tunneled), as in Fig. 8.1.

The calculation of transmission and reflection coefficients finds a stationary state, $\psi(x)$, of the time-independent Schrödinger equation that assumes the incident wave-packet is very long. This is solved in many text books ([8], p. 78), so we do not repeat it here. All we need is that we can write:

$$\psi_{inc} = \exp(ikx - i\omega t),$$

$$\psi_{reflected} = R\exp(-ikx - i\omega t), \quad \psi_{transmitted} = T\exp(ikx - i\omega t), \quad |R|^2 + |T|^2 = 1.$$

$$\text{Then} \quad J_{inc} = \frac{\hbar k}{m}, \quad J_{reflected} = |R|^2 J_{inc}, \quad J_{transmitted} = |T|^2 J_{inc}.$$

In one dimension, ρ is in particles/m, and J and C are scalars in particles/s. J is just the particle velocity.

The wave function *inside the barrier* is of the form:

$$\psi = Ae^{\kappa x} + Be^{-\kappa x}, \quad \psi^* = A^*e^{\kappa x} + B^*e^{-\kappa x}$$

In one dimension, the standard formula for the probability current inside the barrier is:

$$J(t,x) = \frac{\hbar}{2im}\left[\left(\frac{\partial}{\partial x}\psi\right)\psi^* - \left(\frac{\partial}{\partial x}\psi^*\right)\psi\right]$$

From these, we compute the relevant derivatives, and the current J:

$$\frac{\partial}{\partial x}\psi = \kappa Ae^{\kappa x} - \kappa Be^{-\kappa x} \quad \frac{\partial}{\partial x}\psi^* = \kappa A^*e^{\kappa x} - \kappa B^*e^{-\kappa x}$$

$$\left(\frac{\partial}{\partial x}\psi\right)\psi^* - \left(\frac{\partial}{\partial x}\psi^*\right)\psi = \left[\kappa AA^*e^{2\kappa x} + \kappa AB^* - \kappa A^*B - \kappa BB^*e^{-2\kappa x}\right]$$

$$- \left[\kappa AA^*e^{2\kappa x} + \kappa A^*B - \kappa AB^* - \kappa BB^*e^{-2\kappa x}\right] = 0$$

Using the "standard" formula for J, we find that $J = 0$. But this is a contradiction! At the left edge, we have a whole particle coming in, only a fraction of a particle reflected back out, and (allegedly) *nothing* going through the barrier. This violates conservation of particle density at the barrier edge.

We show later that the "standard" probability current is zero in all classically forbidden regions.

The proper solution: In 1D, the divergence $\nabla \cdot J = \partial J/\partial x$. Outside the barrier, the system extends to infinity in both directions. This requires that C outside the barrier is zero. However, inside the barrier, we have a stationary state, and $\partial \rho/\partial t = 0 = \partial J/\partial x$. The proper solution recognizes that just because the derivative of J is zero, does not mean $J = 0$. In fact, $J = C = |T|^2$, a *constant* all across the barrier (positive implies to the right). This constant is determined from the boundary condition on the left edge, which requires this constant probability current for physical consistency. Notice that this also satisfies the requirement on the right edge. Thus,

$$J = 0 + C \quad \text{where} \quad C = |T|^2 \text{ is determined from the boundary condition.}$$

8.2.3   The Meaning of Currents

Probability current, momentum, and velocity: Recall the local momentum density is given by:

$$\mathbf{p}_{local}(x) = \psi^*(x)\,\hat{\mathbf{p}}\,\psi(x) = \psi^*(x)\frac{\hbar}{i}\nabla\psi(x).$$

We notice that the first term of the standard probability current formula, Eq. 8.3, is essentially the local momentum density, $\mathbf{p}_{local}(t, \mathbf{r})$. Therefore, we rearrange the probability current formula just slightly, to make the relation explicit:

$$\mathbf{J}(t,\mathbf{r}) = \frac{\hbar}{2im}\left[(\nabla\psi)\psi^* - (\nabla\psi^*)\psi\right] = \frac{1}{2m}\left[\psi^*\left(\frac{\hbar}{i}\nabla\psi\right) + \left(-\frac{\hbar}{i}\nabla\psi^*\right)\psi\right].$$

The first term in square brackets on the RHS is exactly the local momentum density and the second term is its conjugate. Thus:

$$\mathbf{J}(t,\mathbf{r}) = \frac{1}{2m}\left[\mathbf{p}_{local}(t,\mathbf{r}) + \mathbf{p}^*_{local}(t,\mathbf{r})\right] = \frac{1}{2m}2\,\text{Re}\left[\mathbf{p}_{local}(t,\mathbf{r})\right]$$

$$= \frac{\text{Re}\left[\mathbf{p}_{local}(t,\mathbf{r})\right]}{m} = \text{Re}\left[\mathbf{v}_{local}(t,\mathbf{r})\right].$$

> The probability current is just the real part of the local velocity density.

That is quite reasonable: the local velocity is just the local momentum divided by the mass. (Recall local momentum is momentum at the point $\mathbf{r}$, averaged over all components of the wave-function.) Apparently, the real part of that corresponds to actual particle flow. Also, all local quantities are *averages*, which is why we cannot use them to compute velocity-dependent interactions, such as the magnetic force. You cannot average the velocities first, and then find the magnetic force. You must find the magnetic force separately for each velocity component, which requires using a velocity *operator*, which then defines an electric current operator.

What about the imaginary part? For an energy eigenstate, the local momentum (and therefore velocity) is either real ($E > V(x)$), or purely imaginary ($E < V(x)$). The local momentum is imaginary only in classically forbidden regions. This imaginary momentum is related to the slope of the wave-function in the forbidden region, but has no classical physical significance. Since the standard probability current is the real part of the local velocity density, it is always zero in classically forbidden regions.

Probability current is *not* a physical property of the particle. It is a mathematical construction to help us visualize the time evolution of the particle. Recall the fundamental postulate of QM: every system exists in a linear superposition of states. This demands that every physical property (i.e., observable) has a *linear* quantum operator which multiplies the wave function at each point by the local value of the physical property at that point Eq. 1.4:

$$\hat{o}\psi(x) = o_\psi(x)\psi(x) \quad where \quad o_\psi(x) \equiv \text{local value of } o \text{ at } x \text{ in the state } \psi.$$

Probability current has no linear operator. Therefore, it cannot correspond to any physical property of the system.

We show later that we cannot use this probability current to even qualitatively estimate the time evolution of the wave-function for a velocity-dependent force.

In contrast, the electric current density of a charged particle has an operator, which we find from the classical formulas. Classically, the electric current density (in C/s/m^2) is the product of charge density and velocity:

$$\mathbf{J}_e = \rho_{charge}\mathbf{v} = e\rho_{particle}\mathbf{v} \quad where \quad e \equiv \text{particle charge}, \quad \rho_{charge} = e\rho_{particle}.$$

Thus, one classical physical property of the particle leads to another: its velocity, $\mathbf{v}$, leads to the electric current density $\mathbf{J}$. A similar formula works in QM, when we replace physical properties with operators, which is equivalent to including local values of physical properties. The (nonrelativistic) quantum velocity operator is:

$$\hat{\mathbf{v}} = \frac{\hat{\mathbf{p}} - (e/c)\mathbf{A}}{m} \quad \Rightarrow$$

$$\hat{\mathbf{J}}_e = e\hat{\mathbf{v}} = \frac{e}{m}\hat{\mathbf{p}} - \frac{e^2}{mc}\mathbf{A} = \frac{e\hbar}{im}\nabla - \frac{e^2}{mc}\mathbf{A}$$

(quantum electric current density operator).

Then,

$$\mathbf{J}_e(\mathbf{r}) = \psi^*(\mathbf{r})\hat{\mathbf{J}}_e\psi(\mathbf{r}) = \frac{e\hbar}{im}\psi^*(\mathbf{r})\nabla\psi(\mathbf{r}) - \frac{e^2}{mc}\mathbf{A}\psi^*(\mathbf{r})\psi(\mathbf{r}).$$

Fig. 8.2 (*Left*) The positron wave-packet moves right; the magnetic force is down. (*Right*) The positron wave-packet moves left; the magnetic force is up

It is *this* electric current which generates, and interacts with, magnetic fields ([1], pp. 266–267; especially Eq. 13.31). Note that the particle density ($\rho = \psi^*\psi$) does not appear explicitly in the operator, because ρ is supplied by the wave function, ψ, when $\hat{\mathbf{j}}_e$ acts on it.

> Electric currents generate, and interact with, magnetic fields. Probability currents do not generate anything.

Example: Magnetic force on a positron: We now show a simple example demonstrating that particle velocity, and *not* probability current, determines the magnetic force on a moving charged particle.

Figure 8.2 (*Left*) shows a positron moving right in a magnetic field; it is deflected downward by the magnetic force. Figure 8.2 (*Right*) shows a positron moving left; it is deflected upward by the magnetic force. We know that superposition applies to the Schrödinger equation, so the solution to the superposition of both wave-functions moving in the magnetic field must be the superposition of their individual solutions. Therefore, the superposition of both left- and right-moving positrons splits the wave-packet, with one part deflected upward and the other part deflected downward. However, the superposition of the incident wave-functions is a standing wave: its probability current is zero. This proves magnetic forces cannot depend on probability current.

The magnetic interaction is determined by the particle *velocity*, as computed from linear operators on the wave-function.

> The magnetic interaction is determined by the particle *velocity*, as computed from linear operators on the wave-function.

> The magnetic interaction is *not* determined by the probability current.

The meaning of probability current: The probability current is (the real part of) the local particle velocity, which can be thought of as the weighted average of all component velocities in the wave-function. The distinction between components is lost in this local average. Therefore, for calculating quantum motion, properties, and magnetic fields, we cannot use the probability current. Instead, we must use standard quantum methods of linear operators, such as the ket $\hat{\mathbf{v}}|\psi\rangle$, and its related electric current ket $\hat{\mathbf{J}}_e|\psi\rangle$, which preserve the superposition of properties.

Furthermore, the probability current is not unique, since we can add constant vectors and solenoidal vector fields to it, and still satisfy the continuity equation. Recall that any stationary state has a constant particle density ($\psi^*\psi$) for all time, and its probability current can be taken as everywhere and everywhen zero. However, the particle may still be "moving." For example, an electron in a p-orbital around a nucleus is in a stationary state, so we can take $\mathbf{J}_{probability} = \mathbf{0}$, everywhere. However, the electron is still *orbiting* the nucleus, which is an electric current, which generates a dipole moment. This moment is a physical property of the quantum system, and is measurable, e.g., with Stern–Gerlach equipment, or in the fine structure of an atom.

[For relativistic particles, the velocity operator is no longer $\hat{\mathbf{v}} = \dfrac{\hat{\mathbf{p}}}{m} - \dfrac{e}{mc}\mathbf{A}$. For energy eigenstates, it can be taken as $\hat{\mathbf{v}} = \dfrac{c^2\gamma m\hat{\mathbf{v}}}{\gamma mc^2} = \dfrac{c^2\hat{\mathbf{p}}}{E}$. For Dirac particles, this has unexpected consequences ([17], p. 155b).]

Example: Why is there confusion about currents? We now show that for a common simple example, the velocity of a particle *does* equal its probability current; this means one can often incorrectly use probability current in a calculation, but nevertheless obtain the correct answer. This allows for confusion and misunderstanding.

Consider a 1D momentum eigenstate (with unit amplitude normalization): $\psi(t, x) = \exp[i(kx - \omega t)]$. We compute its velocity as p/m:

$$v = p/m = \hbar k/m \quad \text{(momentum eigenstate).}$$

Its probability current is (we drop the time dependence, since it does not contribute):

$$J = \frac{\hbar}{2im}\left[(\nabla\psi)\psi^* - (\nabla\psi^*)\psi\right] = \frac{\hbar}{2im}\left[(ik\psi)\psi^* - (-ik\psi^*)\psi\right]$$

$$= \frac{2ik\hbar}{2im}|\psi|^2 = \frac{\hbar k}{m} = v.$$

Thus, for simple cases, the probability current $J = v$, and the two are interchangeable. We have already noted, though, that for *stationary states*, we can *always* take $J = 0$, but we cannot (in general) take $v = 0$.

The Schrödinger equation as moderated diffusion: A particle's momentum is a measure of where the particle is going, or how it "flows" in space. Thus, we may interpret $\nabla^2 = \nabla\cdot\nabla$ as the divergence of the momentum, or the divergence of the particle flow. This must be related to $\partial\rho/\partial t$, and therefore related to $\partial\psi/\partial t$. From Schrödinger's equation, then,

$$\frac{\partial \psi}{\partial t} \propto -(const)\nabla \cdot \nabla \psi + V(x)\psi \qquad where \quad \psi \equiv \psi(t, x).$$

Near points of high magnitude momentum, $\partial \psi / \partial t < 0$, implying that particles flow *away* from regions of high momenta, with the flow moderated by the potential $V(x)$ and by $\psi(t, x)$ itself. In the second term, large positive ψ multiplying $V(x)$ tends to make $\partial \psi / \partial t < 0$. By itself, this would make ψ grow near its own already high points. However, such growth, or peaking of ψ, also increases its derivative, and thus its own outflow. When the two are in balance, we have a stationary state. This illustrates the Schrödinger equation as a kind of moderated diffusion equation.

8.2.4 Summary of Currents

The complete solution to the probability current cannot be derived from the wave function alone; it requires boundary conditions to determine the unknown vector constant **C**. That there is no unique mapping from ψ to the probability current, **J**, strongly suggests that **J** has limited physical meaning, since the wave function tells everything that can be known about a particle (or system). The nonlinear dependence of **J** on ψ further demonstrates it cannot be, nor generate, a particle property.

Probability current is primarily a mathematical construction to inform us about how the probability density of the wave-function is evolving in time.

The continuity equation becomes slightly more complicated for a charged particle in a magnetic field, which adds a vector-potential term to the momentum operator.

Magnetic fields are generated by the electric current density, which *is* a physical property of a particle and which *has* a linear operator corresponding to it. Magnetic forces are determined by the quantum velocity, which is different than the probability current.

8.3 Simple Harmonic Oscillator

8.3.1 Harmonic Oscillator Position PDF: Classical and Quantum

The PDF for a *classical* harmonic oscillator can be found by noting that the probability of being in an interval dx equals the time spent there divided by the period, where the time in dx is the distance divided by the speed:

$$\text{Pr(being in } dx) = \frac{\text{time in } dx}{T} = \frac{1}{T}\frac{dx}{|v(x)|} \qquad \text{(in each direction).} \qquad (8.5)$$

Therefore, taking a unit amplitude oscillation (without loss of generality):

$$\text{Let} \quad x(t) = \cos(2\pi t / T) \qquad \Rightarrow$$

$$v(t) = -\frac{2\pi}{T}\sin(2\pi t / T), \quad \text{and} \quad v(x) = v(t(x)), \quad t(x) = \frac{T}{2\pi}\cos^{-1}(x).$$

Each interval dx occurs twice in a cycle (once when the particle moves left and again when moving right), so Pr(being in dx) for one cycle is double Eq. 8.5. Therefore:

$$\text{pdf}_x(x) \equiv \frac{2\,\text{Pr(being in } dx)}{dx} = \frac{2\,dx}{T|v|\,dx} = \frac{1}{\pi|\sin(2\pi t / T)|}$$

$$= \frac{1}{\pi}\frac{1}{|\sin(\cos^{-1}x)|} = \frac{1}{\pi}\frac{1}{\sqrt{1-x^2}}. \qquad (8.6)$$

Notice that this PDF for the *classical* harmonic oscillator goes to infinity at the edges $x = \pm 1$. How is this possible? What about normalization of $\text{pdf}_x(x)$? The key here is that even though the probability *density* is unbounded, its integral is still finite. For example, $1/\sqrt{x}$ is unbounded toward zero, but its integral is still finite:

$$\int_0^1 x^{-1/2}\,dx = 2x^{1/2}\Big|_0^1 = 2.$$

Similarly, by construction, the PDF of the classical harmonic oscillator integrates to 1, as confirmed by straightforward integration of Eq. 8.6 (let $x = \cos\theta$).

In contrast, the quantum position PDF is just $|\psi(x)|^2$, and is everywhere finite. Nonetheless, as the energy of the oscillator increases, the quantum PDF approaches that of the classical PDF. In particular, the quantum PDF can be made arbitrarily large near the classical turning points by choosing high enough energy ([15], p. 136).

8.3.2   Quantum Harmonic Oscillator Zero Point

It can be confusing to describe the "zero-point energy" of the harmonic oscillator as nonzero. Since potential energy has an arbitrary datum (reference point), we can always define the total energy to be nonzero in any state, or to be zero in any state. What *is* significant about the ground state of the harmonic oscillator is that its average *kinetic* energy is nonzero, and its average *potential* energy is above the minimum of the potential energy $V(r)$, which we take to be zero. The kinetic part is weird because the particle cannot stay still, which in turn implies that the potential energy in the ground state is *not* the minimum PE at the equilibrium point. Because the particle is spread out in space, some of it has potential energy above the zero-point. Both these effects give greater energies to the ground state than if the particle just sat, unmoving, at the equilibrium position (like a classical particle could).

8.3.3 Raising and Lowering: How Did Dirac Do It?

The "operator method" of harmonic oscillator analysis is fundamental to all advanced QM. It foreshadows a very similar method used for angular momentum, and is the basis of a major part of QFT (including quantum EM radiation), which starts with the fields as an infinite set of quantized harmonic oscillators. The method is far from obvious, so the big question on everybody's mind is "How did Dirac do it?" How did he know how to create the creation and annihilation operators? Most references give them as "Lo! And behold!" (much like they present the Schrödinger equation out of thin air). Here is one way to do it logically, and (who knows?) perhaps is similar to how Dirac figured it out. Our goal here is not to provide the simplest derivation, but to show how such an idea might come about in the first place. In addition, this section further ties together the meaning of operators and Dirac notation, which makes this topic worth understanding. This section assumes you understand how the operator method works, but not how to motivate its development.

First, recall the stationary states of the 1D harmonic oscillator ([8], 7.18, p. 144):

$$|n\rangle \equiv u_n(x) = \frac{1}{\sqrt{2^n n!}} \left(\frac{m\omega}{\hbar\pi}\right)^{1/4} H_n\left(\sqrt{\frac{m\omega}{\hbar}}x\right) \exp\left(-\frac{m\omega}{2\hbar}x^2\right)$$

$$= C_n H_n(X) \exp\left(-\frac{1}{2}X^2\right)$$

$$where \quad X \equiv \sqrt{\frac{m\omega}{\hbar}}x \text{ is dimensionless position,} \quad C_n \equiv \text{normalization.}$$

$$H_n(X) \quad \equiv \text{Hermite polynomial of degree } n.$$

Now, consider two recursion relations between the Hermite polynomials, (which imply relations between the harmonic oscillator stationary states), which were well known long before Dirac ([7], 7.19–20, p. 144):

$$\frac{d}{dX} H_n(X) = 2n H_{n-1}(X),$$

$$\text{and} \quad 2X\, H_n(X) = H_{n+1}(X) + 2n H_{n-1}(X). \tag{8.7}$$

In the second equation, we already see the seeds of the well-known relation $\hat{x} \propto \left(\hat{a}^\dagger + \hat{a}\right)$, because it says [after multiplying through by $\exp(-X^2/2)$]:

$$\hat{x}|n\rangle \propto b|n+1\rangle + c|n-1\rangle,$$

$$\text{where} \quad b, c \text{ are as-yet unknown numbers.} \tag{8.8}$$

Also, the first recursion relation tells us something about momentum operators, since $\hat{p} \propto d/dx$:

Now Dirac is famous for "Dirac notation," which is a coordinate-free (i.e., representation-indepent) notation for quantum states, and other kets. We seek some purely conceptual relations between the state kets, which must be independent of representation. Therefore we ask: can we write the Hermite recursion relations in ket notation? We start with the first step: can we write a Hermite polynomial in ket notation? Sure:

$$u_n(x) = C_n H_n(\cdot)\exp(\cdot) \;\Rightarrow\; H_n(x) = \frac{1}{C_n}\exp\left(+X^2/2\right)\underbrace{u_n(x)}_{|n\rangle} = \frac{1}{C_n}\exp\left(+X^2/2\right)|n\rangle .$$

Since the first recursion relation in Eq. (8.7) includes a derivative, what is the derivative of a Hermite polynomial in ket notation? We simply differentiate the above equation:

$$\frac{d}{dX}H_n(X) = \frac{d}{dX}\frac{1}{C_n}\exp\left(+X^2/2\right)|n\rangle .$$

In the x-representation, $|n\rangle$ is a function of x, so we use the product rule on the RHS:

$$\frac{d}{dX}H_n(X) = \frac{1}{C_n}\left\{\left[\frac{d}{dX}\exp\left(+X^2/2\right)\right]|n\rangle + \exp\left(+X^2/2\right)\frac{d}{dX}|n\rangle\right\}$$

$$= \frac{1}{C_n}\left\{\left[X\exp\left(+X^2/2\right)\right]|n\rangle + \exp\left(+X^2/2\right)\frac{d}{dX}|n\rangle\right\}$$

$$= \frac{1}{C_n}\exp\left(+X^2/2\right)\left\{X + \frac{d}{dX}\right\}|n\rangle .$$

These results show that the first recursion relation relates $|n\rangle$ to $|n-1\rangle$; when written in ket notation, this gives us a lowering operator. After canceling the $\exp(+X^2/2)$:

$$\frac{1}{C_n}\left\{X + \frac{d}{dX}\right\}|n\rangle = 2n\frac{1}{C_{n-1}}|n-1\rangle . \tag{8.9}$$

We eliminate the normalization factors from:

$$C_n \propto 1/\sqrt{2^n n!} \;\Rightarrow\; C_n/C_{n-1} = 1/\sqrt{2n} .$$

We now return to the true position variable. Use:

$$X = \sqrt{\frac{m\omega}{\hbar}}x, \qquad \frac{d}{dX} = \sqrt{\frac{\hbar}{m\omega}}\frac{d}{dx} = \sqrt{\frac{\hbar}{m\omega}}\frac{i}{\hbar}\hat{p} .$$

To be representation-indpendent, we also take $x \to \hat{x}$. Then Eq. (8.9) becomes:

$$\left\{\sqrt{\frac{m\omega}{\hbar}}\hat{x} + i\sqrt{\frac{1}{\hbar m\omega}}\hat{p}\right\}|n\rangle = \sqrt{2n}|n-1\rangle \quad \text{or} \quad \left\{\sqrt{\frac{m\omega}{2\hbar}}\hat{x} + i\sqrt{\frac{1}{2\hbar m\omega}}\hat{p}\right\}|n\rangle = \sqrt{n}|n-1\rangle .$$

This gives the **lowering operator** in final form:

$$\hat{a}|n\rangle = \sqrt{n}|n-1\rangle \quad \text{where} \quad \hat{a} = \sqrt{\frac{m\omega}{2\hbar}}\hat{x} + i\sqrt{\frac{1}{2\hbar m\omega}}\hat{p} . \qquad (8.10)$$

As shown later, this is sufficient to also define a raising operator, which is the adjoint of $\hat{a}$. However, to continue with our wave-function derivation, we first find the raising operator directly from the recursion relations Eq. (8.7). For a raising operator, we write them with H_{n+1} on the left, and only H_n on the right:

$$H_{n+1}(X) = 2X\,H_n(X) - \frac{d}{dX}H_n(X) .$$

We already know how to convert this to ket notation. After again canceling all the $\exp(+X^2/2)$:

$$\frac{1}{C_{n+1}}|n+1\rangle = \frac{2}{C_n}\sqrt{\frac{m\omega}{\hbar}}\hat{x}|n\rangle - \frac{1}{C_n}\left\{\sqrt{\frac{m\omega}{\hbar}}\hat{x} + i\sqrt{\frac{1}{\hbar m\omega}}\hat{p}\right\}|n\rangle$$

$$= \frac{1}{C_n}\left\{\sqrt{\frac{m\omega}{\hbar}}\hat{x} - i\sqrt{\frac{1}{\hbar m\omega}}\hat{p}\right\}|n\rangle \quad \text{Use: } C_n/C_{n+1} = \sqrt{2(n+1)} \quad (8.11)$$

$$\sqrt{n+1}|n+1\rangle = \left\{\sqrt{\frac{m\omega}{2\hbar}}\hat{x} - i\sqrt{\frac{1}{2\hbar m\omega}}\hat{p}\right\}|n\rangle \quad \Rightarrow \quad \hat{a}^\dagger = \sqrt{\frac{m\omega}{2\hbar}}\hat{x} - i\sqrt{\frac{1}{2\hbar m\omega}}\hat{p} .$$

We see explicitly that the raising operator is the adjoint of the lowering operator.

Algebraic derivation of $\hat{a}^\dagger$: As noted ealier, we can derive the raising operator as the adjoint of the lowering operator more simply by using Dirac algebra. Start with Eq. (8.10), and use the definition of adjoint:

$$\hat{a}|n\rangle = \sqrt{n}|n-1\rangle \quad \Rightarrow \quad \langle n|\hat{a}^\dagger = \sqrt{n}\langle n-1| .$$

Since n is arbitrary, we can replace $n \to n+1$:

$$\langle n+1|\hat{a}^\dagger = \sqrt{n+1}\langle n| \Rightarrow \langle n+1|\hat{a}^\dagger|n\rangle = \sqrt{n+1}, \text{ and } \langle m|\hat{a}^\dagger|n\rangle = 0, m \neq n+1$$

$$\Rightarrow \quad \hat{a}^\dagger|n\rangle = \sqrt{n+1}|n+1\rangle .$$

A key property of $\hat{a}$ and $\hat{a}^\dagger$ is that $\hat{x}$ and $\hat{p}$ are not just linear combinations of them, but proportional to their sum and difference:

$$\hat{x} \propto \left(\hat{a} + \hat{a}^\dagger\right), \qquad \hat{p} \propto \left(\hat{a} - \hat{a}^\dagger\right).$$

This greatly simplifies computing inner products of $\hat{x}$ and $\hat{p}$. Also, note that $\hat{a}$ and $\hat{a}^\dagger$ are real; the appearance of i with $\hat{p}$ in Eqs. (8.10) and (8.11) cancels the i in the definition $\hat{p} \equiv (\hbar / i)\nabla$.

Of course now that we know these results, we could rederive all this much more quickly. That requires a "lo and behold" approach, where we show that a seemingly random definition of the operator $\hat{a}$ leads to a useful result. Many references supply such a derivation. Our goal here, though, was to show how one might *develop* this result from observations about well-known prior results.

8.4   Make No Mistake with a Quantum Eraser

The quantum eraser is an experimentally confirmed consequence of QM. The name derives from the fact that you can "erase" an interference pattern with a seemingly innocuous change that does not *directly* affect the particles creating the interference. This result is not obvious, and actually highly counter-intuitive. As such, understanding it provides significant insight into quantum physics, which can then be used to help understand more common phenomena. In particular, this effect provides a clear, precise description of what is sometimes called "wave-particle duality" (which is greatly misunderstood, and sometimes discussed in almost mystical terms). However, we show here that the prevention of interference obeys the already-given mathematical rules of QM, and there is no need to invoke duplicities of Nature or other exotic ideas.

Much is often made of so-called "wave-particle duality." It is claimed that a particle behaves like a wave or a particle, but not both. In the double-slit experiment, if we observe the particle going through a slit, it prevents any interference. The explanation is not Nature's defiance, but standard, multiparticle QM. It is the entanglement of the measuring device, and not our knowledge of information, which prevents the interference. This same concept applies to the quantum eraser.

In the literature, there are at least three different phenomena called a "quantum eraser." However, we believe that one of them is simply a dressed up version of the EPR (Einstein, Podolski, and Rosen) effect, and not any kind of "eraser." We describe here a recent experiment [27], which we feel well illustrate the principle. Again, there is no new physics in this experiment, just a surprising consequence of existing multiparticle QM. The result could have been predicted in the 1930s, though parametric down-converters (PDCs) did not then exist.

We proceed as follows:

- Reminder of behavior of two-particle states.
- The classical and single-photon Mach–Zehnder interferometer.
- The "PDC."

Fig. 8.3 The Mach–Zehnder interferometer: even a single photon at a time through the system produces interference

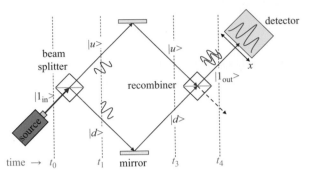
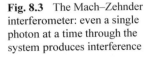

- The Hong–Ou–Mandel interferometer.
- Preventing ("erasing") the interference pattern.
- Some variations of the experiment, and how they would behave.
- Attempt to debunk some "mystical" interpretations of these QM results.
- What is not a quantum eraser.

This section assumes you are familiar with superpositions, interference, simple multiparticle states, and entanglement.

8.4.1 Reminder About Two-Particle States

Imagine a system with two distinguishable particles, A and B, each of which can be in one of two orthogonal states, say $|x\rangle$ and $|y\rangle$. Because of entanglement, a general state of the system can be described using four orthonormal basis vectors:

$$\psi = a\,|Ax, Bx\rangle + b\,|Ax, By\rangle + c\,|Ay, Bx\rangle + d\,|Ay, By\rangle$$
$$where \quad a,b,c,d \text{ are complex.}$$

The probability of measuring particle A to be in state $|Ax\rangle$ is $|a|^2 + |b|^2$. Recall that there is no possibility of interference between the a and b coefficients, because the basis vectors $|Ax, Bx\rangle$ and $|Ax, By\rangle$ are orthogonal, even though both vectors describe the state of A as $|Ax\rangle$, i.e., $\langle Ax, Bx | Ax, By\rangle = 0$.

Interference for one particle can only occur between two quantum state components with a non-zero overlap of the other particle, i.e., a non-zero inner product.

8.4.2   The classical and Single-Photon Mach–Zehnder Interferometer

A Mach–Zehnder interferometer, like all interferometers, starts by splitting a beam of light into two beams (Fig. 8.3). (We omit the label t_2 for later consistency with the time-stamps of the Hong–Ou–Mandel interferometer.)

For a classical beam, after the beam splitter, the upper and lower beams are half the intensity of the incident beam. The second "splitter" acts as a recombiner and produces two output beams. Conservation of energy (reflected in Maxwell's equations) requires that the two output beam powers add up to the two input beam powers. We can adjust the path lengths to put essentially all the power into one output, and zero in the other. The beam has a finite width of a few millimeters, and the detector records the intensity as a function of position across the beam face. Different points on the detector have slightly different path-length differences between the upper and lower paths. These differences cause the two component beams to alternately reinforce and cancel, producing the familiar interference pattern.

[The path-length difference between the upper and lower paths must be less than the coherence length of the laser. The coherence length is essentially the distance along the beam over which the phase of the light is consistent. Every laser has some limited coherence length, which can be anywhere from millimeters to kilometers, depending on the laser design. Points along the beam separated by many times the coherence length act as two separate, independent light sources. For example, if the coherence length was 1 mm, and the path length difference was 10 mm, there would be no visible interference pattern. The illumination would resemble that of shining two flashlights on a wall: a big bump in the middle, dropping off in intensity to the sides.]

As every QM physicist knows, the interference pattern still occurs even if you send in one photon at a time, and accumulate detection counts at each position (x) across the detector. The QM explanation of this is as follows:

1. Classically, a traveling EM wave can be described by its real vector A-field. In QM, we can take the *complex* A-field to be essentially the wave-function of a photon (Sect. 7.8). For a given direction, frequency, and polarization, there is only one independent component of the $\mathbf{A}(\mathbf{r})$ vector, so for simplicity, we can write the A-field as a complex *scalar* function, $A(\mathbf{r})$, much like a wave-function.
2. The photon is a wave-packet with a short length, say 3 mm. (This wave-packet length sets the laser coherence length.) The path lengths are large, say 500 mm, but the path-length difference is small, < 1 mm.
3. At time t_0 (before the beam splitter), the photon state is time evolving through a series of spatial states, and moving to the right. The photon is in a single-photon spatial state, which we represent as the ket $|1_{in}\rangle$, or $A_{in}(\mathbf{r})$, with the time dependence implied.
4. When the single photon hits the beam splitter, there is a (complex) amplitude for it to go straight through, and an amplitude for it to be reflected. At time t_1 in Fig. 8.3, the photon is in a superposition of being half on the upper path, and half on the lower path. We represent this superposition as $\sqrt{\frac{1}{2}}|u\rangle + \sqrt{\frac{1}{2}}|d\rangle$.

Fig. 8.4 A parametric down-
converter (PDC) takes a
single UV photon (~ 351 nm)
and turns it into two entangled red photons (~ 702 nm),
called "signal" and "idler."

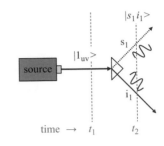

The two "basis" states are $|u\rangle$ and $|d\rangle$, and they do not overlap, i.e., the integral over space of their product is zero: $\langle u|d\rangle = 0$. At this time, the photon is time evolving through a series of superpositions of spatial states.

5. The state at t_3 is the time evolution of that at t_1, and is still a superposition of the photon being in the upper path and the lower path. Recall that $|u\rangle$ and $|d\rangle$ are two components of a single photon. If we put ideal detectors in the two paths, we would have a 50 % chance of detecting the photon on the upper path, and 50 % chance of detecting it on the lower path.

6. At the recombiner, the two quantum amplitudes for the photon add to become a single localized wave-packet again. As noted earlier, there is a dependence on the position, x: the interference wiggles are impressed across the face of the traveling wave-packet. At time t_4, we represent this spatial state as $|1_{out}\rangle$, or $A_{out}(\mathbf{r})$.

7. At the detector, the detection probability at each position x on the detector varies as the square of the wave amplitude: $\Pr \propto |A_{out}(x)|^2$. After accumulating detections from many photons, the interference pattern is visible in a histogram of detection counts vs position (x) across the face of the beam.

8.4.3   The Parametric Down-Converter

Before discussing the Hong-Ou-Mandel interferometer, we describe an essential element of it: the PDC. A beta-barium borate crystal is a nonlinear optical medium which can split a single photon into two equal-energy entangled photons, traveling in different directions. Conservation of energy requires the frequency of the two output photons be half that of the input photon (the frequency is "down-converted"). This process is called "parametric down-conversion," and the crystal is called a **PDC** (Fig. 8.4).

One of the output photons is arbitrarily called the "signal" photon, and the other is the "idler" photon. (These names have a historical origin which is not relevant to our application.)

The quantum mechanical state of the system evolves in time from a single UV photon at time t_1, to two red photons at time t_2. We denote these spatial states respectively by the kets $|1_{uv}\rangle$, and $|s_i i_i\rangle$. Note that $|1_{uv}\rangle$ is a single-particle state, and $|s_i i_i\rangle$

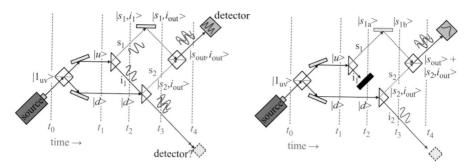

Fig. 8.5 The Hong–Ou–Mandel interferometer: (*Left*) A single photon at a time through the system produces interference. (*Right*) Blocking the idler eliminates interference between signal photons

is an entangled two-particle state. It means that if the output photon s_1 is present, then the photon i_1 must also be present. We cannot have one without the other (yet).

Before the PDC, the photon state is time evolving through a series of spatial states, and moving to the right. Similarly, after the PDC, both photons are time evolving through a series of spatial states, and moving in different directions from each other. We have chosen the states $|1_{uv}\rangle$ and $|s_1i_1\rangle$ as two representative spatial states at the representative times t_1 and t_2.

The conversion efficiency of a PDC is horrible, on the order of one converted photon out of a million, or worse. The vast majority of the incident UV photons go straight through the PDC. Those cases are of no interest to us here, so we ignore them, and consider only the occurrences where down-conversion takes place.

8.4.4   The Hong–Ou–Mandel Interferometer

The Hong–Ou–Mandel interferometer combines a Mach–Zehnder interferometer with two PDCs (Fig. 8.5, *left*). We first summarize the process in words, and then fill in the quantum details.

A single UV photon enters a beam splitter and gets an amplitude for taking the upper path and an amplitude for taking the lower path. In the upper path, any $|u\rangle$ photon is down-converted into a signal/idler pair, in state $|s_1i_1\rangle$. At the lower PDC, i_1 (the idler from the upper path) passes harmlessly through. Simultaneously, any $|d\rangle$ photon is down-converted into the state $|s_1i_1\rangle$.

As we show shortly, s_1 and s_2 are two components of the same (red) *signal* photon. At time t_4 (after the recombiner), their amplitudes add to produce a single-photon wave-packet, $|s_{out}\rangle$, just as in the simple Mach–Zehnder interferometer. Similarly, i_1 and i_2 are also two components of the same (red) *idler* photon. The amplitudes for i_1 and i_2 therefore add in the lower PDC, just as amplitudes add in a recombiner. The path lengths are set so i_1 and i_2 reinforce, and one complete (red) idler photon leaves the lower PDC, in a state we call $|i_{out}\rangle$. At time t_4, the two-photon state (signal and idler) is the combination (tensor product) of the signal photon state and the idler photon state, which we denote $|s_{out}, i_{out}\rangle$.

We summarize the spatial states of the system at various times as follows:

t_0 $|1_{uv}\rangle$ 1-UV-photon state

t_1 $\sqrt{\dfrac{1}{2}}|u\rangle + \sqrt{\dfrac{1}{2}}|d\rangle$ superposition of two (single-particle UV-photon) states

t_2 $\sqrt{\dfrac{1}{2}}|s_1, i_1\rangle + \sqrt{\dfrac{1}{2}}|d\rangle$ superposition of (entangled 2-red-photons) and 1-UV-photon

t_3 $\sqrt{\dfrac{1}{2}}|s_1, i_{out}\rangle + \sqrt{\dfrac{1}{2}}|s_2, i_{out}\rangle$ superposition of two signal states, both with the same idler state

t_4 $\left(1 + e^{i\delta(x)}\right)|s_{out}, i_{out}\rangle$ signal with interference + idler state (ignoring normalization)

where $\delta(x) \equiv$ path-length difference as a function of position (x) along the detector.

The photon states at times t_0 and t_1 are as before in the Mach–Zehnder interferometer.

At t_2, there is a superposition of a single UV photon in the lower path with a pair of red photons created from the UV component on the upper path. If we were to measure the photons along all three paths at this time, there is a 50 % chance we would find a single UV-photon on the lower path (from the $|d\rangle$ component) and a 50 % chance we would find two red photons, one on each of the upper signal and idler paths (from the $|s_1 i_1\rangle$ component).

At t_3, the two idler components have combined into a single spatial state for the idler photon, which we call $|i_{out}\rangle$. If we measure on the idler path, we are sure to detect the idler photon. If we measured along the signal paths, there is a 50 % chance we would find the signal photon on the s_1 path, and 50 % chance we would find the signal photon on the s_2 path.

At t_4, the signal components have recombined into a localized signal state, and the idler components remain in a localized idler state. As before in the simple interferometer, the signal photon has an interference pattern impressed across its face. Our (ideal) signal detector is guaranteed to detect a signal photon, and the lateral (x) probability distribution follows the interference pattern. If we measure on the idler path, we are sure to find a photon.

8.4.5   Preventing ("Erasing") the Interference Pattern

A crucial step in creating the interference pattern previously is the recombining of the idler photon components, as we now show. Figure 8.5, *right* shows a blocking screen in the i_1 path. The screen absorbs any photon (converting it to heat).

At time t_2, the photon spatial state is now $\sqrt{\frac{1}{2}}|s_1\rangle + \sqrt{\frac{1}{2}}|d\rangle$, and there is no idler. If we measure along the $|s_{1a}\rangle$ and $|d\rangle$ paths, we have a 50 % chance of measuring a red photon along s_1 and a 50 % chance of measuring a UV photon along the lower path. There is no chance of measuring i_1 anywhere.

At time t_3, the photon state is $\sqrt{\frac{1}{2}}|s_1\rangle + \sqrt{\frac{1}{2}}|s_2, i_{out}\rangle$. If we measure along all three paths, we have a 50 % chance of detecting a red photon along s_1, and a 50 % chance of detecting two photons, one each along s_2 and i_{out}.

At time t_4, the two signal paths recombine, but the quantum states *do not*. The upper path contributes a component of a signal photon to the quantum state, but no idler. The lower path contributes both a signal *and* an idler photon. The system state at this time is:

$$|\psi\rangle = \sqrt{\frac{1}{2}}|s_{out}\rangle + \sqrt{\frac{1}{2}}|s_{out}, i_{out}\rangle. \qquad (8.12)$$

If we measure at the detector, we are guaranteed to find a signal photon, since both states in the superposition include one. If we measure along the idler path, we have a 50 % chance of finding a photon. The two *components* in the superposition state are *orthogonal* because there is no overlap in their idler spatial states. The i_1 component is dead and gone from the upper path, but i_2 is present in the lower path; the idler electromagnetic state is a superposition of the vacuum state and the one-photon i_2 state: $|0\rangle + |i_2\rangle$. Orthogonal states cannot interfere: the probability of measuring an outcome (s_{out}) is the sum of the squared-magnitudes of all the orthogonal components yielding that outcome:

$$|\psi\rangle = a|s_{out}\rangle + b|s_{out}, i_{out}\rangle + \ldots \quad \Rightarrow \quad \Pr(s_{out}) = |a|^2 + |b|^2.$$

Thus, the probability of measuring a signal photon is the sum of the squared-magnitudes of the two components of the system photon state Eq. 8.12, which is 1. Each component contributes an (x) distribution that is just a central bump, with no interference, that falls off to the sides.

Note that interference is *not* "created" and then "erased." Interference either happens or it does not, just once, after the beam combiner. However, one can see, in real time, the interference pattern come and go, by simply putting your hand out of, and into, the i_1 path.

Without disturbing the signal path at all, we can change its interference pattern, by changing the relationship between components of the entangled idler photon.

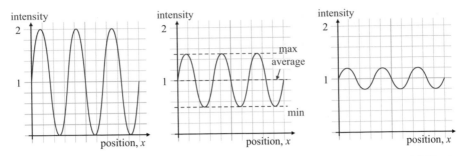

Fig. 8.6 The degree of interference is quantified by its "visibility." (*Left*) Perfect visibility is 100 %. (*Middle*) Visibility = 50 %. (*Right*) Visibility = 20 %

A second way to prevent interference is to return to the original setup (with no blocking screen), but adjust the idler path lengths so that i_1 and i_2 come out at different times, say i_1 comes out first. Then, at any given time, the i_1 wave-packet component is in a different position than the i_2 wave-packet component. In other words, there is no overlap between the two idler component spatial wave-functions (they are orthogonal). Choose time t_3 to be after i_2 comes out; the system state is

then $|\psi\rangle = \sqrt{\frac{1}{2}}|s_1, i_1\rangle + \sqrt{\frac{1}{2}}|s_2, i_2\rangle$. If we measure along all three paths (and note the

time of idler arrival), we have a 50 % chance of measuring s_1 and i_1, and a 50 %

chance of measuring s_2 and i_2. At time t_4, the system state is $\sqrt{\frac{1}{2}}|s_{\text{out}}, i_1\rangle + \sqrt{\frac{1}{2}}|s_{\text{out}}, i_2\rangle$.

These two two-particle states are orthogonal, because the idler states are orthogonal. Therefore, they cannot interfere, and the detector pattern shows a single bump, with no interference.

Note that in all cases, it does not matter whether we actually detect the idler photon or not.

8.4.6 More Variations of the Quantum Eraser

Other variations of the experiment are also instructive. Return to the setup (Fig. 8.5, *right*) with an absorbing screen. What if the absorber is not complete, but a half intensity filter? In that case, it has a 50 % chance of absorbing i_1, and a 50 % chance of passing it. That means that half the photons exhibit interference, and half do not. The detector will show some interference, but weaker than the full interferometer (Fig. 8.6, *middle*).

Physicists quantify the degree of interference with its **visibility** (or "fringe visibility"):

$$\text{visibility} \equiv \frac{max - average}{average} = \frac{max - min}{max + min}.$$

For example, if the screen absorbs 80 % of the photons, then only 20 % interfere, and the visibility is 20 %. In general, a screen that absorbs a fraction, α, of the light going through it results in a fringe visibility of $1 - \alpha$.

Return to the setup with no blocking screen (Fig. 8.5, *left*). What if we adjust the idler path lengths so that i_1 and i_2 cancel in the lower PDC, instead of reinforcing? In that case, the i_1 and i_2 components must be reflected or refracted into different directions. They no longer have any spatial overlap (they are orthogonal), and it is the same concept as adjusting the idler paths so that i_1 and i_2 come out at different times. The system state at t_3 is again $\sqrt{\frac{1}{2}}|s_1,i_1\rangle + \sqrt{\frac{1}{2}}|s_2,i_2\rangle$, and at t_4 it is again $\sqrt{\frac{1}{2}}|s_{out},i_1\rangle + \sqrt{\frac{1}{2}}|s_{out},i_2\rangle$. The two idler states are orthogonal, so there is no interference.

What if we adjust the idler path lengths so there is partial overlap between i_1 and i_2, but not complete overlap? Then there will be some fringe visibility, but less than 100 %. The more overlap, the greater the fringe visibility. If we were to continuously adjust the idler paths to vary the i_1 and i_2 spatial overlap from 100 % down to zero, the fringe visibility would smoothly decrease from 100 % to zero.

8.4.7 *"Duality" Is Misleading: Mysticism Is not Needed*

A particular class of double-slit experiment is sometimes described as a "quantum eraser." Without going into detail, it is possible to detect a particle going through one of the slits *without* changing the state of the particle; only the state of the detector is changed [Walborn et al., Phys Rev A 65 033818 (2002)]. It is found that with the detector disabled, interference occurs. With the detector enabled, so that we can "tell which slit the particle went through," interference does not occur. (This is sometimes phrased as the interference is "erased," but we have already seen that interference either occurs or it does not, but it never occurs and is then somehow later undone.)

The presence or absence of interference is sometimes said to demonstrate "duality": a particle can behave like a wave or a particle, but not both at once. If we know which way ("welter Weg" in German) the particle went, it prevents interference. The implication is that *knowledge* prevents interference. Such claims usually provide no clear reason for "duality"; it appears ad hoc. Or miraculous. Or that Nature is being obstinate. Some call it "quantum weirdness."

Such comments almost always lack a mathematical description. In fact, a proper representation of quantum states, much like the eraser described previously, explains the result. Suppose we have a photon incident on a double slit and a *microscopic* detector in front of one slit which records a passage of a photon heading toward the slit, without disturbing the photon. The detector records this event as a single quantum transition, so after the incident photon, the detector is in a superposition of two states: $\sqrt{\frac{1}{2}}|no\rangle + \sqrt{\frac{1}{2}}|yes\rangle$. The photon wave is recombined after the slit, in the usual way, and detected by a screen beyond the slits. But the two components of the photon (one from each slit) are entangled with the detector state. The quantum state of

the system, including both photon and detector is then $\sqrt{\frac{1}{2}}|S_{out}, no\rangle + \sqrt{\frac{1}{2}}|S_{out}, yes\rangle$.
This is of the same form as the previous eraser, when the two idler paths are different, or one idler is blocked: the superposition is of two distinct quantum states, and they are orthogonal. Therefore, they cannot interfere. Each component of the superposition contributes a noninterfering "bump" to the observed photon distribution on the screen. Note that we specified a microscopic photon detector, so that decoherence is not a factor in the analysis.

If the detector's two different "which-way" states, $|no\rangle$ and $|yes\rangle$, are coherently recombined into a single quantum state, then interference can again happen. This is analogous to recombining the idler components in the previous Hong–Ou–Mandel eraser. Usually, such recombination is experimentally difficult. Note that we cannot just erase the detector state by "smacking it" with some external interaction, because that entangles the external "smacking" device into the system state, which still prevents interference.

Finally, we must caution against macroscopic detectors: as noted in Sect. 1.12, all macroscopic devices acquire uncontrollable and unknowable quantum phases, due to unavoidable interactions with their environments. The resulting random phases ("decoherence") prevent interference.

8.4.8   What Is not a Quantum Eraser?

Some references' description of a quantum "eraser" starts with Young's two-slit interference experiment with light. As is well known, shining light through two closely spaced, narrow slits produces an interference pattern on a far screen. We can prevent the interference by putting polarizers in front of each slit, with perpendicular polarization directions. Either slit, with its polarizer, transmits light to the screen. Both slits together produce a pattern which is the sum of the intensities of each pattern, with *no* interference. Some claim this is a quantum eraser because the polarizers "mark" the photons with the slit they went through, by setting the photons' polarization. Since the two photons are now distinguishable, the claim is that "wave-particle duality" says they cannot behave like waves and cannot interfere.

I disagree. This is not a quantum eraser for two reasons: first, the result is completely predicted by *classical* electromagnetics. Interference only occurs between light of the same polarization, simply because only parallel components of an E (or B) field can cancel or reinforce. Any result that can be predicted with purely classical theory can hardly be called a "quantum eraser."

Secondly, we have seen that in a true quantum eraser, the prevention of interference occurs *without modifying the interfering photon components*. Putting polarizers over the slits *modifies* the components of the photon. They no longer interfere because they have been changed into noninterfering things.

Another experiment, also often called a "quantum eraser," is based on the EPR effect for entangled photons. We do not believe this qualifies as an "eraser," either, because the result comes from detecting and recording correlations. There is always interference, but we can only see it if we "count" in the right way.

Erratum to: Desultory Topics in Quantum Mechanics

E. L. Michelsen, *Quirky Quantum Concepts,* Undergraduate Lecture Notes in Physics,
DOI 10.1007/978-1-4614-9305-1_8, © Springer Science+Business Media New York 2014

**The original version of Chapter 8 was inadvertently published with errors.
Those errors have been corrected as follows:**

Page 324. "***8.3.3 Raising and Lowering: How Did Dirac Do It?***" should
read as follows.

8.3.3   Raising and Lowering: How Did Dirac Do It?

The "operator method" of harmonic oscillator analysis is fundamental to all advanced QM. It foreshadows a very similar method used for angular momentum, and
is the basis of a major part of QFT (including quantum EM radiation), which starts
with the fields as an infinite set of quantized harmonic oscillators. The method is
far from obvious, so the big question on everybody's mind is "How did Dirac do
it?" How did he know how to create the creation and annihilation operators? Most
references give them as "Lo! And behold!" (much like they present the Schrödinger
equation out of thin air). Here is one way to do it logically, and (who knows?) perhaps is similar to how Dirac figured it out. Our goal here is not to provide the simplest derivation, but to show how such an idea might come about in the first place.
In addition, this section further ties together the meaning of operators and Dirac
notation, which makes this topic worth understanding. This section assumes you understand how the operator method works, but not how to motivate its development.

First, recall the stationary states of the 1D harmonic oscillator ([8], 7.18, p. 144):

$$|n\rangle \equiv u_n(x) \quad = \frac{1}{\sqrt{2^n n!}} \left(\frac{m\omega}{\hbar\pi}\right)^{1/4} H_n\left(\sqrt{\frac{m\omega}{\hbar}}x\right) \exp\left(-\frac{m\omega}{2\hbar}x^2\right)$$

$$= C_n H_n(X) \exp\left(-\frac{1}{2}X^2\right)$$

The updated original online version of this chapter can be found at
https://doi.org/10.1007/978-1-4614-9305-1_8

E. L. Michelsen, *Quirky Quantum Concepts,* Undergraduate Lecture Notes in Physics,   E1
DOI 10.1007/978-1-4614-9305-1_9, © Springer Science+Business Media New York 2017

$$where \quad X \quad \equiv \sqrt{\frac{m\omega}{\hbar}}x \text{ is dimensionless position,} \quad C_n \equiv \text{normalization.}$$

$$H_n(X) \quad \equiv \text{Hermite polynomial of degree } n.$$

Now, consider two recursion relations between the Hermite polynomials, (which imply relations between the harmonic oscillator stationary states), which were well known long before Dirac ([7], 7.19–20, p. 144):

$$\frac{d}{dX}H_n(X) = 2nH_{n-1}(X),$$

$$\text{and} \quad 2X H_n(X) = H_{n+1}(X) + 2nH_{n-1}(X). \tag{8.7}$$

In the second equation, we already see the seeds of the well-known relation $\hat{x} \propto \left(\hat{a}^\dagger + \hat{a}\right)$, because it says [after multiplying through by $\exp(-X^2/2)$]:

$$\hat{x}|n\rangle \propto b|n+1\rangle + c|n-1\rangle,$$

$$where \quad b,c \text{ are as-yet unknown numbers.} \tag{8.8}$$

Also, the first recursion relation tells us something about momentum operators, since $\hat{p} \propto d/dx$:

Now Dirac is famous for "Dirac notation," which is a coordinate-free (i.e., representation-indepent) notation for quantum states, and other kets. We seek some purely conceptual relations between the state kets, which must be independent of representation. Therefore we ask: can we write the Hermite recursion relations in ket notation? We start with the first step: can we write a Hermite polynomial in ket notation? Sure:

$$u_n(x) = C_n H_n(\cdot)\exp(\cdot) \implies H_n(x) = \frac{1}{C_n}\exp\left(+X^2/2\right)\underbrace{u_n(x)}_{|n\rangle} = \frac{1}{C_n}\exp\left(+X^2/2\right)|n\rangle.$$

Since the first recursion relation in Eq. (8.7) includes a derivative, what is the derivative of a Hermite polynomial in ket notation? We simply differentiate the above equation:

$$\frac{d}{dX}H_n(X) = \frac{d}{dX}\frac{1}{C_n}\exp\left(+X^2/2\right)|n\rangle.$$

In the x-representation, $|n\rangle$ is a function of x, so we use the product rule on the RHS:

$$\frac{d}{dX}H_n(X) = \frac{1}{C_n}\left\{\left[\frac{d}{dX}\exp\left(+X^2/2\right)\right]|n\rangle + \exp\left(+X^2/2\right)\frac{d}{dX}|n\rangle\right\}$$

$$= \frac{1}{C_n}\left\{\left[X\exp\left(+X^2/2\right)\right]|n\rangle + \exp\left(+X^2/2\right)\frac{d}{dX}|n\rangle\right\}$$

$$= \frac{1}{C_n}\exp\left(+X^2/2\right)\left\{X + \frac{d}{dX}\right\}|n\rangle.$$

These results show that the first recursion relation relates $|n\rangle$ to $|n-1\rangle$; when written in ket notation, this gives us a lowering operator. After canceling the $\exp(+X^2/2)$:

$$\frac{1}{C_n}\left\{X+\frac{d}{dX}\right\}|n\rangle = 2n\frac{1}{C_{n-1}}|n-1\rangle. \tag{8.9}$$

We eliminate the normalization factors from:

$$C_n \propto 1/\sqrt{2^n n!} \quad \Rightarrow \quad C_n/C_{n-1} = 1/\sqrt{2n}.$$

We now return to the true position variable. Use:

$$X = \sqrt{\frac{m\omega}{\hbar}}x, \qquad \frac{d}{dX} = \sqrt{\frac{\hbar}{m\omega}}\frac{d}{dx} = \sqrt{\frac{\hbar}{m\omega}}\frac{i}{\hbar}\hat{p}.$$

To be representation-indpendent, we also take $x \to \hat{x}$. Then Eq. (8.9) becomes:

$$\left\{\sqrt{\frac{m\omega}{\hbar}}\hat{x}+i\sqrt{\frac{1}{\hbar m\omega}}\hat{p}\right\}|n\rangle = \sqrt{2n}|n-1\rangle \quad \text{or} \quad \left\{\sqrt{\frac{m\omega}{2\hbar}}\hat{x}+i\sqrt{\frac{1}{2\hbar m\omega}}\hat{p}\right\}|n\rangle = \sqrt{n}|n-1\rangle.$$

This gives the **lowering operator** in final form:

$$\hat{a}|n\rangle = \sqrt{n}|n-1\rangle \quad \text{where} \quad \hat{a} = \sqrt{\frac{m\omega}{2\hbar}}\hat{x}+i\sqrt{\frac{1}{2\hbar m\omega}}\hat{p}. \tag{8.10}$$

As shown later, this is sufficient to also define a raising operator, which is the adjoint of $\hat{a}$. However, to continue with our wave-function derivation, we first find the raising operator directly from the recursion relations Eq. (8.7). For a raising operator, we write them with H_{n+1} on the left, and only H_n on the right:

$$H_{n+1}(X) = 2X\,H_n(X) - \frac{d}{dX}H_n(X).$$

We already know how to convert this to ket notation. After again canceling all the $\exp(+X^2/2)$:

$$\frac{1}{C_{n+1}}|n+1\rangle = \frac{2}{C_n}\sqrt{\frac{m\omega}{\hbar}}\hat{x}|n\rangle - \frac{1}{C_n}\left\{\sqrt{\frac{m\omega}{\hbar}}\hat{x}+i\sqrt{\frac{1}{\hbar m\omega}}\hat{p}\right\}|n\rangle$$

$$= \frac{1}{C_n}\left\{\sqrt{\frac{m\omega}{\hbar}}\hat{x}-i\sqrt{\frac{1}{\hbar m\omega}}\hat{p}\right\}|n\rangle \quad \text{Use: } C_n/C_{n+1} = \sqrt{2(n+1)} \tag{8.11}$$

$$\sqrt{n+1}|n+1\rangle = \left\{\sqrt{\frac{m\omega}{2\hbar}}\hat{x}-i\sqrt{\frac{1}{2\hbar m\omega}}\hat{p}\right\}|n\rangle \quad \Rightarrow \quad \hat{a}^\dagger = \sqrt{\frac{m\omega}{2\hbar}}\hat{x}-i\sqrt{\frac{1}{2\hbar m\omega}}\hat{p}.$$

We see explicitly that the raising operator is the adjoint of the lowering operator.

Algebraic derivation of $\hat{a}^\dagger$: As noted ealier, we can derive the raising operator as the adjoint of the lowering operator more simply by using Dirac algebra. Start with Eq. (8.10), and use the definition of adjoint:

$$\hat{a}|n\rangle = \sqrt{n}\,|n-1\rangle \quad \Rightarrow \quad \langle n|\hat{a}^\dagger = \sqrt{n}\,\langle n-1|.$$

Since n is arbitrary, we can replace $n \to n+1$:

$$\langle n+1|\hat{a}^\dagger = \sqrt{n+1}\,\langle n| \Rightarrow \langle n+1|\hat{a}^\dagger|n\rangle = \sqrt{n+1}, \text{ and } \langle m|\hat{a}^\dagger|n\rangle = 0, m \neq n+1$$

$$\Rightarrow \quad \hat{a}^\dagger|n\rangle = \sqrt{n+1}\,|n+1\rangle.$$

A key property of $\hat{a}$ and $\hat{a}^\dagger$ is that $\hat{x}$ and $\hat{p}$ are not just linear combinations of them, but proportional to their sum and difference:

$$\hat{x} \propto \left(\hat{a} + \hat{a}^\dagger\right), \qquad \hat{p} \propto \left(\hat{a} - \hat{a}^\dagger\right).$$

This greatly simplifies computing inner products of $\hat{x}$ and $\hat{p}$. Also, note that $\hat{a}$ and $\hat{a}^\dagger$ are real; the appearance of i with $\hat{p}$ in Eqs. (8.10) and (8.11) cancels the i in the definition $\hat{p} \equiv (\hbar/i)\nabla$.

Of course now that we know these results, we could rederive all this much more quickly. That requires a "lo and behold" approach, where we show that a seemingly random definition of the operator $\hat{a}$ leads to a useful result. Many references supply such a derivation. Our goal here, though, was to show how one might *develop* this result from observations about well-known prior results.

Appendices

List of Physical Constants

Physical constants: 2006 values from NIST. For more, see http://physics.nist.gov/cuu/Constants/.

Speed of light in vacuum	$c = 299\ 792\ 458$ m s^{-1} (exact)
Gravitational constant	$G = 6.674\ 28(67) \times 10{-}11$ m^3 kg^{-1} s^{-2}
Relative standard uncertainty	$\pm 1.0 \times 10^{-4}$
Boltzmann constant	$k = 1.380\ 6504(24) \times 10^{-23}$ J K^{-1}
Stefan-Boltzmann constant	$\sigma = 5.670\ 400(40) \times 10^{-8}$ W m^{-2} K^{-4}
Relative standard uncertainty	$\pm 7.0 \times 10^{-6}$
Avogadro constant	$N_A, L = 6.022\ 141\ 79(30) \times 10^{23}$ mol^{-1}
Relative standard uncertainty	$\pm 5.0 \times 10^{-8}$
Molar gas constant	$R = 8.314\ 472(15)$ J mol^{-1} K^{-1}
Calorie	4.184 J (exact)
Electron mass	$m_e = 9.109\ 382\ 15(45) \times 10^{-31}$ kg
Proton mass	$m_p = 1.672\ 621\ 637(83) \times 10^{-27}$ kg
Proton/electron mass ratio	$m_p/m_e = 1836.152\ 672\ 47(80)$
Elementary charge	$e = 1.602\ 176\ 487(40) \times 10^{-19}$ C
Electron g-factor	$g_e = -2.002\ 319\ 304\ 3622(15)$
Proton g-factor	$g_p = 5.585\ 694\ 713(46)$
Neutron g-factor	$g_N = -3.826\ 085\ 45(90)$
Muon mass	$m_\mu = 1.883\ 531\ 30(11) \times 10^{-28}$ kg
Inverse fine structure constant	$\alpha^{-1} = 137.035\ 999\ 679(94)$
Planck constant	$h = 6.626\ 068\ 96(33) \times 10^{-34}$ J s
Planck constant over 2π	$\hbar = 1.054\ 571\ 628(53) \times 10^{-34}$ J s
Bohr radius	$a_0 = 0.529\ 177\ 208\ 59(36) \times 10^{-10}$ m
Bohr magneton	$\mu_B = 927.400\ 915(23) \times 10^{-26}$ J T^{-1}

E. L. Michelsen, *Quirky Quantum Concepts,* Undergraduate Lecture Notes in Physics, 337
DOI 10.1007/978-1-4614-9305-1, © Springer Science+Business Media New York 2014

Common Quantum Formulas

References are to Goswami's *Quantum Mechanics* [Gos].

General Commutation Relations

$[\hat{x}, \hat{p}] = i\hbar$ [3.5 p. 56] $\left[\hat{a}, \hat{a}^{\dagger}\right] = 1$ [7.26 p. 149] $\left[\hat{\phi}, \hat{L}_z\right] = i\hbar$ [9.28, p. 193]

$\left[\hat{A}\hat{B}, \hat{C}\right] = \hat{A}\left[\hat{B}, \hat{C}\right] + \left[\hat{A}, \hat{C}\right]\hat{B}$ [Q3.5 p. 72, with other commutation identities]

$\left[\hat{H}, \hat{x}\right] = \left[\dfrac{\hat{p}^2}{2m}, \hat{x}\right] = \dfrac{1}{2m}\left(\hat{p}[\hat{p}, \hat{x}] + [\hat{p}, \hat{x}]\hat{p}\right) = -\dfrac{i\hbar}{m}\hat{p}$ $\left[\hat{H}, \hat{L}_i\right] = 0$ [12.8-9 p247]

$\left[\hat{L}_x, \hat{L}_y\right] = i\hbar\hat{L}_z$ $\left[\hat{L}_y, \hat{L}_z\right] = i\hbar\hat{L}_x$ $\left[\hat{L}_z, \hat{L}_x\right] = i\hbar\hat{L}_y$ [11.6 p. 221]

$\left[\hat{L}^2, \hat{L}_i\right] = 0$ [11.10, p. 222]

$\left[\hat{J}_z, \hat{J}_+\right] = \hbar\hat{J}_+$ $\left[\hat{J}_z, \hat{J}_-\right] = -\hbar\hat{J}_-$ $\left[\hat{J}_+, \hat{J}_-\right] = 2\hbar\hat{J}_z$ [11.51, p. 235]

x-basis: free particle: $\psi(t, x) = \exp[i(kx - \omega t)]$ $\hat{p} = -i\hbar\dfrac{\partial}{\partial x}$ [3.1, p. 53]

$\hat{H}\psi = i\hbar\dfrac{\partial \psi}{\partial t} = -\dfrac{\hbar^2}{2m}\dfrac{\partial^2 \psi}{\partial x^2} + V(x)\psi$ [3.10 p58] $\langle x | \psi \rangle = \psi(x)$ [p. 121]

momentum-representation: $\hat{x} = i\hbar\dfrac{\partial}{\partial p}$ [p. 56] Energy representation: $H|E\rangle = E|E\rangle$

ψ concave toward *x*-axis when $V < E$, away when $V > E$. (correction to Q3.A4 p. 73)

Time independent Schrödinger equation:

$\left[-\dfrac{\hbar^2}{2m}\dfrac{d^2}{dx^2} + V(x)\right]u(x) = Eu(x)$ [3.11, p. 60]

in standard mathematical form: $\dfrac{d^2 u(x)}{dx^2} + \left(\dfrac{2m}{\hbar^2}\right)(E - V)u(x) = 0$ [4.2, p. 76]

General wave #: $|k| = \dfrac{\sqrt{2m(E - V)}}{\hbar}$ Bound: $\xi = k'a, \eta = \beta a, \xi^2 + \eta^2 = \dfrac{2ma^2 V_0}{\hbar^2}$

even: $k'\tan k'a = \beta$ $\eta = \xi\tan\xi$ odd: $k'\cot k'a = -\beta$ $\eta = -\xi\cot\xi$

Dirac Notation (Chap. 6)

$$\langle x|\psi\rangle = \psi(x)\ [\text{p.}121] \quad \hat{x}|x\rangle = x|x\rangle \quad [\text{p.}125] \quad \langle x|\hat{x}|x'\rangle = x'\ \delta(x-x') \quad [\text{p.}125]$$

For basis, $|\phi_i\rangle$: $\quad |\psi\rangle = \sum_{\substack{i=complete \\ set}} |\phi_i\rangle\langle\phi_i|\psi\rangle \quad\Rightarrow\quad \sum_i |\phi_i\rangle\langle\phi_i| = \mathbf{1}_{op}$

$$\text{completeness operator} \quad [6.3\text{-}4,\ \text{p.}119]$$

$$\langle j|\hat{A}|\psi\rangle = \sum_i \langle j|\hat{A}|i\rangle\langle i|\psi\rangle \qquad\qquad\qquad\qquad [6.16,\ \&\ \text{top p.}124]$$

$$\langle\phi|\hat{A}^\dagger|\psi\rangle = \langle\psi|\hat{A}|\phi\rangle^* \qquad\qquad\qquad\qquad\qquad [6.18\ \text{p.}124]$$

Harmonic Oscillator (Chap. 7)

$$H = \frac{p^2}{2m} + \frac{1}{2}kx^2 = \frac{p^2}{2m} + \frac{1}{2}m\omega^2 x^2 \quad [7.1,\ \text{p.}137], \quad x_0 = \sqrt{\frac{\hbar}{m\omega}} \qquad (\text{units of } x)$$

$$\xi = \sqrt{\frac{m\omega}{\hbar}}x = \frac{x}{x_0}, \quad \omega = \sqrt{\frac{k}{m}} \quad [\text{p.}138], \quad E = \frac{1}{2}\hbar\omega\varepsilon = \hbar\omega\left(n+\frac{1}{2}\right) \quad [7.13,\ \text{p.}142]$$

Approximate spring constant for general potential minimum: $k = \dfrac{d^2V}{dx^2}\bigg|_{x-min}$

$$[7.2,\ \text{p.}137]$$

lowering: $\hat{a} = \sqrt{\dfrac{m\omega}{2\hbar}}\hat{x} + \dfrac{i\hat{p}}{\sqrt{2m\hbar\omega}},$

raising: $\hat{a}^\dagger = \sqrt{\dfrac{m\omega}{2\hbar}}\hat{x} - \dfrac{i\hat{p}}{\sqrt{2m\hbar\omega}}$ $\qquad\qquad\qquad\qquad$ [7.25, p.149]

$$\hat{x} = \sqrt{\frac{\hbar}{2m\omega}}\left(\hat{a}^\dagger + \hat{a}\right) \qquad \hat{p} = i\sqrt{\frac{m\hbar\omega}{2}}\left(\hat{a}^\dagger - \hat{a}\right) \qquad [7.43,\ \text{p.}156]$$

$$\hat{N} = \hat{a}^\dagger\hat{a} \quad [\text{p.}152\text{t}], \qquad \hat{H} = \hbar\omega\left(\hat{N}+\frac{1}{2}\right) = \hbar\omega\left(\hat{a}^\dagger\hat{a}+\frac{1}{2}\right) \qquad [7.27,\ \text{p.}149]$$

$$\left[\hat{H},\hat{a}\right] = -\hbar\omega\hat{a} \quad [7.28,\ \text{p.}150] \qquad\qquad \left[\hat{H},\hat{a}^\dagger\right] = +\hbar\omega\hat{a}^\dagger \qquad [7.29,\ \text{p.}150]$$

$$\hat{a}^\dagger |u_n\rangle = \sqrt{n+1}\,|u_{n+1}\rangle \qquad [7.40, \text{p.}155] \qquad |u_{n+1}\rangle = \frac{1}{\sqrt{n+1}}\hat{a}^\dagger |u_n\rangle \qquad [\text{p.}154\text{b}]$$

$$\hat{a}|u_n\rangle = \sqrt{n}\,|u_{n-1}\rangle \qquad [7.42, \text{p.}155] \qquad |u_{n-1}\rangle = \frac{1}{\sqrt{n}}\hat{a}|u_n\rangle,$$

$$\hat{a}|u_0\rangle = \mathbf{0}_v \qquad\qquad\qquad\qquad\qquad\qquad\qquad\qquad\qquad [7.32, \text{p.}151]$$

Equations of Motion (Chap. 8)

$$i\hbar \frac{\partial}{\partial t} U(t,t_0) = \hat{H} U(t,t_0) \qquad\qquad\qquad\qquad\qquad [6.27, \text{p.}132]$$

$$U(t,t_0) = \exp\left[\frac{-i\hat{H}(t-t_0)}{\hbar}\right] \qquad\qquad\qquad [8.1, \text{p.}163]$$

$$|\psi\rangle(t) = \exp\left[\frac{-i\hat{H}t}{\hbar}\right]|\psi\rangle(0) \quad [8.2\ \text{p.}163] \qquad \hat{A}(t) = e^{i\hat{H}t/\hbar}\hat{A}e^{-i\hat{H}t/\hbar} \qquad [8.4, \text{p.}163]$$

Heisenberg Equation of motion: $\dfrac{d}{dt}\hat{A}(t) = \dfrac{i}{\hbar}\left[\hat{H},\hat{A}(t)\right]$ $\qquad\qquad [8.5\ \text{p.}164]$

$$\frac{d}{dt}\hat{a}(t) = -i\omega\hat{a}(t) \implies \hat{a}(t) = e^{-i\omega t}\hat{a}(0),$$

$$\frac{d}{dt}\hat{a}^\dagger(t) = i\omega\hat{a}^\dagger(t) \qquad\qquad \implies \hat{a}^\dagger(t) = e^{i\omega t}\hat{a}^\dagger(0) \qquad [\text{p.}165]$$

$$\frac{d\langle\hat{A}\rangle}{dt} = \frac{i}{\hbar}\left\langle\left[\hat{H},\hat{A}\right]\right\rangle \quad \left[2^{\text{nd}}\ 8.7,\ \text{p.}166\right] \qquad \frac{d\langle x\rangle}{dt} = \left\langle\frac{p}{m}\right\rangle \qquad [8.8, \text{p.}167]$$

$$\frac{d\langle p\rangle}{dt} = -\left\langle\frac{dV(x)}{dx}\right\rangle \qquad\qquad\qquad\qquad\qquad [8.10, \text{p.}168]$$

Particle in Two Dimensions (Chap. 9)

$$\hat{p}_x = -i\hbar\frac{\partial}{\partial x} \quad \hat{p}_y = -i\hbar\frac{\partial}{\partial y} \quad [9.3, \text{p.}184] \quad \mathbf{\hat{p}} = \hat{p}_x\mathbf{e}_x + \hat{p}_y\mathbf{e}_y = -i\hbar\nabla \quad [9.4, \text{p.}186]$$

Time independent Schrödinger eq: $\nabla^2\psi + \dfrac{2m}{\hbar^2}(E-V)\psi = 0$ $\qquad\qquad [9.8\ \text{p.}187]$

Motion on a ring: $I=m\rho^2$ [p. 191]
$$\psi(\phi)=\frac{1}{\sqrt{2\pi}}e^{im\phi}$$
[9.27, p.193]

$$\hat{L}_z = \hat{x}\hat{p}_y - \hat{y}\hat{p}_x = -i\hbar\frac{\partial}{\partial\phi}$$
[9.24-5, p.192]

Two particle system reduction to 1: $x=x_1-x_2$

$$X=\frac{m_1 x_1 + m_2 x_2}{m_1+m_2}$$
[9.32, p.194]

$$\frac{1}{\mu}=\frac{1}{m_1}+\frac{1}{m_2} \qquad M=m_1+m_2$$
[9.33, 9.34+ p.194]

Time dependent Schrödinger eq:

$$i\hbar\frac{\partial\psi(t,\mathbf{r})}{\partial t}=-\frac{\hbar^2}{2m}\nabla^2\psi(t,\mathbf{r})+V(\mathbf{r})\psi(t,\mathbf{r})$$
[9.45, p.198]

Angular Momentum (Chap. 11)

$$\hat{L}_z = \hat{x}\hat{p}_y - \hat{y}\hat{p}_x = -i\hbar\frac{\partial}{\partial\phi}$$
[11.4, p. 220, 11.19, p. 225]

$$L^2 = r^2 p^2 \quad [11.2, \text{p. }219] \quad H = L^2/2I$$
[11.3, p. 219]

$$\hat{L}^2 = -\hbar\left[\frac{1}{\sin\theta}\frac{\partial}{\partial\theta}\left(\sin\theta\frac{\partial}{\partial\theta}\right)+\frac{1}{\sin^2\theta}\frac{\partial^2}{\partial\phi^2}\right]$$
[11.20, p. 226]

$$\hat{L}_z|l\,m\rangle=m\hbar|l\,m\rangle, \qquad \hat{L}^2|l\,m\rangle=l(l+1)\hbar^2|l\,m\rangle, \quad l=0,1,2,...$$
[p. 229]

$$\langle\theta,\phi|l\,m\rangle=Y_{lm}(\theta,\phi), \quad l=0,1,2,...; \quad m=-l,....,+l$$
[p. 230]

Legendre polynomials:

$$\zeta=\cos\theta \,(11.27,\text{p. }226)\, P_0(\zeta)=1, P_1(\zeta)=\zeta, P_2(\zeta)=\tfrac{1}{2}(3\zeta^2-1)$$
[11.34, p. 228]

$$(l+1)P_{l+1}(\zeta)=(2l+1)\zeta P_l(\zeta)-lP_{l-1}(\zeta) \qquad \frac{d}{d\zeta}(1-\zeta^2)\frac{dP_l}{d\zeta}=-l\zeta P_l+lP_{l-1}(\zeta)$$

Recursion :

$$P_{lm}(\zeta)=\left(1-\zeta^2\right)^{m/2}\frac{d^m}{d\zeta^m}P_l(\zeta)=\frac{1}{2^l\cdot l!}\left(1-\zeta^2\right)^{m/2}\frac{d^{l+m}}{d\zeta^{l+m}}\left(\zeta^2-1\right)^l$$
[11.35 p. 228]

$$m > 0: \quad Y_{lm}(\theta, \phi) = (-1)^m \sqrt{\frac{2l+1}{4\pi} \cdot \frac{(l-m)!}{(l+m)!}} e^{im\phi} P_{lm}(\cos\theta)$$
[11.41 p. 231]

$$Y_{l,-m} = (-1)^m Y_{lm}^*$$
[11.40 p. 231]

$$\hat{J}_+ \equiv \hat{J}_x + i\hat{J}_y \qquad \hat{J}_+ \equiv \hat{J}_x + i\hat{J}_y$$
$\big[$11.50, p. 234$\big]$

$$\hat{J}_+\hat{J}_- = \hat{J}^2 - J_z^2 + \hbar\hat{J}_z \qquad \hat{J}_-\hat{J}_+ = \hat{J}^2 - J_z^2 - \hbar\hat{J}_z$$
[11.52-3, p. 235]

$$\hat{J}_+ | j, m \rangle = \hbar\sqrt{j(j+1) - m(m+1)} | j, m+1 \rangle$$
$$\hat{J}_- | j, m \rangle = \hbar\sqrt{j(j+1) - m(m-1)} | j, m-1 \rangle$$
[11.62, p. 238]

Motion in a Central Potential (Chap. 12)

$$H = \frac{P^2}{2M} + \frac{p^2}{2\mu} + V(|r|)$$
[12.2, p. 245]

$$\hat{H} = \frac{p^2}{2\mu} + V(|r|) = -\frac{\hbar^2}{2\mu}\nabla^2 + V(r)$$
[12.3, p. 245]

$$\hat{p}^2 = -\hbar^2\nabla^2 \quad [12.12] \qquad \underbrace{\hat{p}_r = -i\hbar\frac{\partial}{\partial r}}_{\textit{not hermitian}}$$
[12.15, p. 249]

$$V_{eff}(r) = V(r) + \hbar^2\frac{l(l+1)}{2\mu r^2}$$
[12.25, p. 252]

Hydrogen Atom (Chap. 13)

$$V(r) = -\frac{Ze^2}{r} \quad [\text{p. 268}] \qquad E_n = \frac{-\mu Z^2 e^4}{2\hbar^2 n^2}$$
[13.12, p. 271]

$$n = n_r + l + 1, \quad l < n$$
[p. 271]

Spin and Matrices (Chap. 15)

$$|z+\rangle = [1,\ 0]^T \qquad |z-\rangle = [0,\ 1]^T$$
[15.12, p. 309]

$$(\text{z-basis}) \; |x+\rangle = \frac{1}{\sqrt{2}}\begin{bmatrix}1\\1\end{bmatrix} \qquad |x-\rangle = \frac{1}{\sqrt{2}}\begin{bmatrix}1\\-1\end{bmatrix} \qquad [15.16, \text{p.}\,312, \&\,\text{p.}\,316]$$

$$|y+\rangle = \frac{1}{\sqrt{2}}\begin{bmatrix}1\\i\end{bmatrix} \qquad |y-\rangle = \frac{1}{\sqrt{2}}\begin{bmatrix}1\\-i\end{bmatrix} \qquad [\text{p.}\,314]$$

Pauli matrices: $\quad \sigma_x = \begin{bmatrix}0&1\\1&0\end{bmatrix}, \quad \sigma_y = \begin{bmatrix}0&-i\\i&0\end{bmatrix}, \quad \sigma_z = \begin{bmatrix}1&0\\0&-1\end{bmatrix}$ $\quad[15.27\ \text{p}316]$

$$\sigma_i^2 = 1_2 \qquad \{\sigma_i,\sigma_j\} \equiv \sigma_i\sigma_j + \sigma_j\sigma_i = 0 \;(i \ne j), \qquad [\sigma_x,\sigma_y] = 2i\sigma_z$$
$$(x,y,z \text{ cyclic})$$
$$[15.29\ \text{p}317]$$

Addition of Angular Momentum (Chap. 17)

$$\mathbf{J} = \mathbf{L} + \mathbf{S} \qquad \mathbf{L}\cdot\mathbf{S} = L_z S_z + \tfrac{1}{2}\left(L_+ S_- + L_- S_+\right) \qquad\qquad [17.13, \text{p.}\,355]$$

General Clebsh-Gordon coefficient:

$$\langle j_1\, m_1; j_2\, m_2 | J\, M\rangle$$

$$= \sqrt{\frac{(J+j_1-j_2)!(J-j_1+j_2)!(j_1+j_2-J)!(J+M)!(J-M)!(2J+1)}{(J+j_1+j_2+1)!(j_1-m_1)!(j_1+m_1)!(j_2-m_2)!(j_2+m_2)!}}$$

$$\times \delta(m_1+m_2, M)\sum_k \frac{(-1)^{k+j_2+m_2}(J+j_2+m_1-k)!(j_1-m_1+k)!}{(J-j_1+j_2-k)!(J+M-k)!(k!)(k+j_1+j_2-M)!}$$

$$[\text{Con \& Sho } 14^3 5, \text{p.}\,75]$$

Perturbation Theory (Chapter 18)

$$E_n^{(1)} = \langle \varphi_n | \hat{H}_1 | \varphi_n\rangle \qquad\qquad [18.10, \text{p.}\,380]$$

$$E_n^{(2)} = \sum_{k \ne n} \frac{\left|\langle \varphi_k | \hat{H}_1 | \varphi_n\rangle\right|^2}{E_n^{(0)} - E_k^{(0)}} \qquad\qquad [18.14, \text{p.}\,381]$$

$$\left|\psi_n^{(1)}\right\rangle = \sum_{k \neq n} \left[\frac{\left\langle \varphi_k|\hat{H}_1|\varphi_n\right\rangle}{E_n^{(0)} - E_k^{(0)}}\right]\left|\varphi_k\right\rangle \qquad\qquad \text{[18.12 p380]}$$

$$\Gamma = \frac{2\pi}{\hbar}\left|\langle f|V|i\rangle\right|^2 \rho(E_f) \qquad\qquad \text{[Bay 12-20, p. 251]}$$

Spherical Harmonics and Their Friends

All spherical harmonics are functions of θ and ϕ. They can be considered functions of all space by ignoring the radial coordinate: $Y_{lm}(r, \theta, \phi) \equiv Y_{lm}(\theta, \phi)$, and can thus be written in rectangular coordinates as $Y_{lm}(x, y, z)$. The transformations require only the conversions in the right column, here:

$$x = r\sin\theta\cos\phi \qquad\qquad \cos\theta = z/r$$

$$y = r\sin\theta\sin\phi \qquad\qquad e^{i\phi}\sin\theta = \frac{x+iy}{r}$$

$$z = r\cos\theta \qquad\qquad e^{-i\phi}\sin\theta = \frac{x-iy}{r}$$

$$Y_{00} = \sqrt{\frac{1}{4\pi}} \qquad\qquad Y_{l0}(\theta) = \sqrt{\frac{2l+1}{4\pi}}P_l\cos\theta$$

Plane wave expansion:

$$e^{ikz} = \sum_{l=0}^{\infty} i^l\left(2l+1\right)j_l(kr)P_l(\cos\theta) = \sum_{l=0}^{\infty} i^l\sqrt{4\pi\left(2l+1\right)}j_l(kr)Y_{l0}(\theta)$$

The functions below use the Condon-Shortley phase (the $(-1)^m$ factor) [Con & Sho 4³17 p. 52]:

$$Y_{lm}(\theta,\phi) \equiv \begin{cases} (-1)^m\sqrt{\dfrac{(2l+1)}{2}\dfrac{(l-m)!}{(l+m)!}}P_{lm}(\cos\theta)\dfrac{e^{im\phi}}{\sqrt{2\pi}}, & m \geq 0, \\[1em] \sqrt{\dfrac{(2l+1)}{2}\dfrac{(l-|m|)!}{(l+|m|)!}}P_{l|m|}(\cos\theta)\dfrac{e^{im\phi}}{\sqrt{2\pi}}, & m < 0, \end{cases} \qquad \text{[Wyl 3.6.5, p. 96]}$$

$P_{lm}(x)$ is the associated Legendre function, $\quad l = 0,1,2...,\quad m = -l, -l+1, ...l-1,l$.

$$Y_{lm}(\theta,\phi) = (-1)^m Y_{l,-m}^*(\theta,\phi), \qquad \forall\, m \text{ (positive and negative)}.$$

$$Y_{11} = -\sqrt{\frac{3}{8\pi}}\sin\theta e^{i\phi} = \sqrt{\frac{3}{8\pi}}\frac{x+iy}{r} \qquad Y_{10} = \sqrt{\frac{3}{4\pi}}\cos\theta = \sqrt{\frac{3}{4\pi}}\frac{z}{r}$$

$$Y_{1,-1} = \sqrt{\frac{3}{8\pi}}\sin\theta e^{-i\phi} = \sqrt{\frac{3}{8\pi}}\frac{x-iy}{r}$$

$$Y_{22} = \sqrt{\frac{15}{32\pi}} \sin^2 \theta e^{2i\phi} = \sqrt{\frac{15}{32\pi}} \frac{(x+iy)^2}{r^2}$$

$$Y_{21} = -\sqrt{\frac{15}{8\pi}} \sin \theta \cos \theta e^{i\phi} = -\sqrt{\frac{15}{8\pi}} \frac{(x+iy)z}{r^2}$$

$$Y_{20} = \sqrt{\frac{5}{16\pi}} \left(3\cos^2 \theta - 1 \right) = \sqrt{\frac{5}{16\pi}} \frac{-x^2 - y^2 + 2z^2}{r^2} = \sqrt{\frac{5}{16\pi}} \left(3\frac{z^2}{r^2} - 1 \right)$$

$$Y_{2,-1} = \sqrt{\frac{15}{8\pi}} \sin \theta \cos \theta e^{-i\phi} = \sqrt{\frac{15}{8\pi}} \frac{(x-iy)z}{r^2}$$

$$Y_{2,-2} = \sqrt{\frac{15}{32\pi}} \sin^2 \theta e^{-2i\phi} = \sqrt{\frac{15}{32\pi}} \frac{(x-iy)^2}{r^2}$$

Spherical Hankel Functions

[h_l and h_l^* are sometimes written as $h_l^{(1)}$ and $h_l^{(2)}$.]

$$h_l(\rho) = j_l(\rho) + i\eta_l(\rho) \quad \text{(outgoing)} \qquad h_l^*(\rho) = j_l(\rho) - i\eta_l(\rho) \quad \text{(ingoing)}$$

$$h_l(\rho \rightarrow \infty) \rightarrow -i\frac{e^{i(\rho - l\pi/2)}}{\rho} \qquad h_l^*(\rho \rightarrow \infty) \rightarrow +i\frac{e^{-i(\rho - l\pi/2)}}{\rho} \qquad \text{[Wyl 6.2.23-24 p183]}$$

$$j_l(\rho \rightarrow 0) = \frac{2^l l!}{(2l+1)!}\rho^l = \frac{\rho^l}{(2l+1)!!} \qquad \text{[Wyl 6.2.21 p183, Bay 6-82 p166]}$$

$$\eta_l(\rho \rightarrow 0) = -\frac{(2l)!}{2^l l!}\frac{1}{\rho^{l+1}} = -\frac{(2l-1)!!}{\rho^{l+1}} \qquad \text{[Wyl 6.2.22 p183, Bay 6-82 p166]}$$

Integrals (See also en.wikipedia.org/wiki/Lists_of_integrals)

$$\int_{-\infty}^{\infty} dx\, e^{-ax^2} = \sqrt{\frac{\pi}{a}} \qquad\qquad \int_{-\infty}^{\infty} dx\, x^2 e^{-ax^2} = \frac{1}{2}\sqrt{\frac{\pi}{a^3}}$$

$$\int_0^{\infty} dr\, r^3 e^{-ar^2} = \frac{1}{2a^2} \qquad \text{More generally:} \qquad \int_0^{\infty} dr\, r^n e^{-ar^p} = \frac{\Gamma\big((n+1)/p\big)}{pa^{(n+1)/p}}$$

[Tal 370, p. 235]

General Mathematical Formulas

completing the square: $ax^2 + bx = \left(\sqrt{a}x + \dfrac{b}{2\sqrt{a}}\right)^2 - \dfrac{b^2}{4a}$ $\qquad$ (x-shift $= -b/2a$)

$$\nabla^2 = \frac{\partial^2}{\partial x^2} + \frac{\partial^2}{\partial y^2} + \frac{\partial^2}{\partial z^2} \qquad\qquad \nabla^2 = \frac{1}{r}\frac{\partial}{\partial r}\left(r\frac{\partial}{\partial r}\right) + \frac{1}{r^2}\frac{\partial^2}{\partial \phi^2} + \frac{\partial^2}{\partial z^2}$$

$$\nabla^2 = \frac{1}{r^2}\frac{\partial}{\partial r}\left(r^2\frac{\partial}{\partial r}\right) + \frac{1}{r^2 \sin\theta}\frac{\partial}{\partial \theta}\left(\sin\theta\frac{\partial}{\partial \theta}\right) + \frac{1}{r^2 \sin^2\theta}\frac{\partial^2}{\partial \phi^2},$$

$$\frac{1}{r^2}\frac{\partial}{\partial r}\left(r^2\frac{\partial}{\partial r}\right) = \frac{1}{r}\frac{\partial^2}{\partial r^2}(r\cdot)$$

$$\cos 2a = \cos^2 a - \sin^2 a = 2\cos^2 a - 1 = 1 - 2\sin^2 a \qquad \sin 2a = 2\sin a\, \cos a$$

$$\cos\frac{a}{2} = \sqrt{\frac{1 + \cos a}{2}} \qquad\qquad\qquad \sin\frac{a}{2} = \sqrt{\frac{1 - \cos a}{2}}$$

$\cos(a \pm b) = \cos a \cos b \pm \sin a \sin b \qquad\qquad \sin(a \pm b) = \sin a \cos b \pm \cos a \sin b$

$\cos^2 a = \tfrac{1}{2}\left(1 + \cos 2a\right) \qquad\qquad\qquad\quad \sin^2 a = \tfrac{1}{2}\left(1 - \cos 2a\right)$

$\cos a \cos b = \tfrac{1}{2}\left[\cos(a + b) + \cos(a - b)\right] \qquad \sin a \sin b = \tfrac{1}{2}\left[\cos(a + b) - \cos(a - b)\right]$

$\cosh^2 x - \sinh^2 x = 1$

$\cosh 2x = \cosh^2 x + \sinh^2 x = 2\cosh^2 x - 1 = 1 + 2\sinh^2 x \qquad \sinh 2x = 2\sinh x \cosh x$

$e^{ikz} = \displaystyle\sum_{l=0}^{\infty} i^l (2l + 1)\, j_l(kr) P_l(\cos\theta) \qquad\qquad\qquad$ [Gos 23.34 p500b]

Glossary

Definitions of common Quantum Mechanics terms

<x> The average (sometimes called "expectation") value of 'x'

Abelian group A commutative group: $a+b=b+a$.

Adjoint The adjoint of an operator produces a bra from a bra in the same way the original operator produces a ket from a ket: $\hat{O}|\psi\rangle = |\phi\rangle \Rightarrow \langle\psi|\hat{O}^\dagger = \langle\phi|, \ \forall|\psi\rangle$. Some mathematics references use "adjoint" differently: the "adjoint matrix" $\equiv$ transpose of the cofactor matrix.

aka Also known as.

Amplitude A complex number specifying the magnitude and phase of a quantum value. This is quite different from most applications, where "amplitude" is a *real* number giving the maximum value of a wave. In QM, an "amplitude" can be considered a phasor.

arg(z) The angle of the complex number, z, in polar form, measured counter-clockwise from the positive real axis.

Azimuthal quantum number This is an anachronism for l, the *orbital* angular momentum quantum number. However, [Bay p156m] defines it as m, the z-component of angular momentum, which is consistent with the term "azimuthal."

Baryons 3-quark particles, including protons, neutrons, and others. All baryons are hadrons.

Basis (plural: bases) A set of vectors used to construct arbitrary vectors in the vector space.

Bra A vector in the Hilbert space dual to kets: this means we can take an inner product of a bra with a ket to get a complex number (a scalar).

By definition In the very nature of the definition itself, without requiring any logical steps. To be distinguished from "by implication."

By implication Combining the definition, and other true statements, a conclusion can be shown by implication.

E. L. Michelsen, *Quirky Quantum Concepts,* Undergraduate Lecture Notes in Physics, 347
DOI 10.1007/978-1-4614-9305-1, © Springer Science+Business Media New York 2014

C-number "commuting number," or more realistically, "complex number". In general, it would be a "scalar" in the mathematical vector-space of kets.

c'est la vie French for "that's life."

canonical momentum In Lagrangian mechanics, canonical momentum is a generalize momentum conjugate to a generalized coordinate, defined as the derivative of the lagrangian with respect to the coordinate: $p_{canonical} \equiv \partial L / \partial q$. (Conjugate here is *not* complex conjugate.)

cf "compare to." Abbreviation of Latin "confer."

CGS Centimeter-gram-second: a system of measuring distance, mass, and time. There are two different CGS systems, with different units of charge.

Closed interval Between c and d is written "$[c, d]$", and means the range of numbers from c to d *including* c and d.

Complex Having a real an imaginary component: the sum of a real and imaginary number.

Complex plane A 2-D graph of complex numbers, with the real part on the abscissa (horizontal axis), and imaginary part on the ordinate (vertical axis).

Complicated Involved or intricate.

Component One vector, usually a basis vector, in a superposition of vectors. Each vector in the superposition has a complex coefficient, so the superposition is called "coherent."

Comprise To include. e.g., "An insect comprises three parts: a head, thorax, and abdomen." We could say "An insect is composed of three parts," but there is no "comprised of".

Conjugate bilinear A function of two mathematical objects which scales as the conjugate of the amplitude of the first, and linearly with the amplitude of the second. For example, see "inner product."

Constituent As used here: one of the states, $|\psi_k\rangle$, of a mixture (distinct from "component").

Continuous Having the property that between any two elements there are an infinite number of other elements, e.g., real numbers are continuous.

Contrapositive The contrapositive of the statement "If A then B" is "If not B then not A." The contrapositive is equivalent to the statement: if the statement is true (or false), the contrapositive is true (or false). If the contrapositive is true (or false), the statement is true (or false).

Converse The converse of the statement "If A then B" is "If B then A". In general, if a statement is true, its converse may be either true or false. The converse is the contrapositive of the inverse, and hence the converse and inverse are equivalent statements.

Correlated In QM, correlated is used "colloquially" (i.e., incorrectly) to mean "dependent." A better word is "entangled." See "entangled." Properly, two sequences of numbers are correlated if there is any component of a linear relationship between the sequences, i.e., if the covariance is nonzero. See *Funky Mathematical Physics Concepts* for more.

Density Some quantity per unit volume, e.g., probability density. Outside QM, density can also be per unit mass.

Dependent Two quantum systems are dependent (aka entangled) if information about one of them provides information about the other.

Dimension (1) The number of bases in a complete set; (2) units of measure, e.g., the dimensions of voltage in the MKSA system are J/C, or $kg\text{-}m^2/(C\text{-}s^2)$.

Discrete Having the property that between two adjacent elements, there are no other elements, e.g., integers are discrete.

Dot product Inner product.

Dual vector In this context, a "dual vector" is one which can be combined with a "vector" to produce a scalar (via the inner product). (Mathematics defines at least two other meanings for "dual.")

Ensemble A hypothetical set of identically prepared systems.

Entangled The quantum states of two systems are entangled when neither system has a definite value of some property, but the values of the two systems' properties are interdependent. This means a measurement of one system provides information about the other.

Esu Electrostatic unit: a unit of charge which repels a like unit with a force of 1 dyne when spaced by 1 m. AKA stat-coulomb.

Expectation value Not used in this book. Usually, the average value.

Fact A small piece of information backed by solid evidence (in hard science, usually repeatable evidence). If someone disputes a fact, it is still a fact. "If a thousand people say a foolish thing, it is still a foolish thing." See also "speculation," and "theory."

Field (1) In mathematics: a set of elements and two operators in which simultaneous linear equations can be solved. Infinite fields have an infinite number of elements. (2) In physics, a (scalar or vector) function of space.

Flux Particles per second (distinguish from "flux-density"). Some references use "flux" to mean "flux density."

Flux density Flux per unit area, i.e., particles per second per area (distinguish from "flux"). Some references use "flux" to mean "flux density."

Forbidden Forbidden *in first order perturbation theory*. See "strictly forbidden."

Hadrons Particles that interact via the strong force e.g., pions and baryons, but not leptons. 3-quark particles are both hadrons and baryons. 2-quark particles are both hadrons and mesons.

Hermitian Her-mish'-un: an operator whose eigenvalues are real; equivalently, a self-adjoint operator.

Hilbert space Physics: a vector space, of finite or infinite dimension, with a metric (dot product). Mathematicians require that a Hilbert space be infinite dimensional.

Idempotent An operator is idempotent if $\hat{O}^2 = \hat{O}$, such as a projection operator, $|\psi\rangle\langle\psi|$. By extension, $\hat{O}^n = \hat{O}$ for all positive integers n.

Idler photon One of the two photons produced by a parametric down-converter. See also signal photon.

Iff If and only if; both necessary and sufficient. Used in definitions.

Implies Guarantees. In conversation, "implies" means "suggests." But in math and science, "implies" is stronger.

Inflection A change in curvature (from up to down, or down to up).

Inner product A conjugate bilinear function of two vectors producing a scalar, i.e., the inner product scales as the conjugate of the amplitude of the first vector, and linearly with the amplitude of the second vector.

Instantaneous amplitude The magnitude of a wave at a given point in space at a given point in time.

Inverse The inverse of the statement "If A then B" is "If not A then not B." In general, if a statement is true, its inverse may be either true or false. The inverse is the contrapositive of the converse, and hence the converse and inverse are equivalent statements.

Ket A vector in the Hilbert space of quantum states and of the space of vectors resulting from operators acting on states. In wave mechanics, a ket is a function of space. In matrix mechanics, a ket is a vector with discrete components. The ket space is dual to the bra space, which means we can take an inner product of a bra and a ket to get a complex number (a scalar).

Logarithmic derivative The ratio of a derivative to the function, i.e., $f'(x)/f(x) = d/dx$ ($\ln f(x)$). This is the *fractional* rate of change with x, with units of $[x]^{-1}$.

MKSA Meter-kilogram-second-ampere: a subset of the SI system for measuring mechanical and electromagnetic phenomena.

Momentum representation A wave function expressed as a function of momentum, often written $a(p)$.

NIST National Institute of Standards and Technology: the US government body which establishes US standards and units of measurement. Works closely with ISO.

Number current Particle number current density, synonym: probability current.

Observation A measurement.

Occupation number The number of particles in a single-particle quantum state, e.g., photons in a given mode.

Old quantum mechanics aka "Wilson-Sommerfeld quantization": the incorrect notion that the classical action of a periodic or quasi-periodic system is quantized to multiples of $\hbar$. The hydrogen atom disproves this, but for higher energies, W-S quantization corresponds to the WKB approximation.

Open interval Between c and d is written as (c, d), and means the range of numbers from c to d *exclusive* of c and d.

Orthogonal Having a dot product of zero.

PDF Probability distribution function (or probability density function): e.g., pdf(x)=probability per unit interval of x, for differentially small intervals, 'dx'. Mathematically, pdf$(x)\, dx$=Pr(value being in the region $[x, x+dx]$).

Phasor A complex number that represents the amplitude and phase of a sinusoid. The sinusoid frequency is *not* part of the phasor, and must be known from other sources.

Polar form A complex number expressed as a magnitude and angle, $z=(r, \theta)=re^{i\theta}$.

Position representation A wave function expressed as a function of positions in space.

pr(event) The probability of 'event'.

Probability amplitude A complex number whose magnitude-squared is a probability. Probability amplitudes generally add as complex numbers, then the magnitude-squared of the sum is a probability.

QED quod erat demonstrandum. Latin for "which was to be demonstrated."

Quantum Field Theory The physics of relativistic particles and fields (such as EM fields). Massive particles are not conserved, since they may be created or destroyed in interactions.

Quantum mechanics The physics of nonrelativistic particles and systems at microscopic scales. Massive particles are conserved, i.e., they are not created or destroyed.

Quantum state A complex valued function of all space, combined with a spin state, that defines all the properties of a particle; in particular, the state defines the probability amplitudes, and thus density functions, for all values of all measurements of all observables of the particle.

Probability current Particle number current density, synonym: number current.

Positive definite A matrix or operator which is >0 for all *nonzero* operands. It may be 0 when acting on a "zero" operand, such as the zero vector. This implies that all eigenvalues>0. e.g., L^2-hat is a positive definite operator when acting on eigenstates $|l\ m\rangle$, when $l \geq 1$.

Positive semidefinite A matrix or operator which is ≥ 0 for all *nonzero* operands. It may be 0 when acting on a nonzero operands. This implies that all eigenvalues ≥ 0. For example, L^2-hat is a positive semidefinite operator when acting on all eigenstates $|l\ m\rangle$, for all l.

Probability amplitude A quantum amplitude (a complex number) whose squared-magnitude is the probability of something [1, p. 8t].

Rectangular form A complex number expressed as the sum of a real and imaginary component, $z = x + iy$

RHS Right hand side.

Signal photon One of the two photons produced by a parametric down-converter. See also idler photon.

Spatial state The wave-function of a particle or system, which represents all its properties except spin-related ones.

Speculation A guess, possibly hinted at by evidence, but not well supported. Every scientific fact and theory started as a speculation. See also "fact," and "theory."

Spherical wave A wave which can be written as $(e^{ikr}/r)f(\theta,\ \phi)$. A spherical wave is *not*, in general, spherically symmetric, as its amplitude and phase may vary with $(\theta,\ \phi)$.

Spin The intrinsic angular momentum of a particle. Most particles cannot avoid having spin (electrons, protons, neutrons). It is a property of the particle.

Spin-state The quantum state of a particle's spin. It can be represented as a spinor.

Spinor A vector in a 2-dimensional Hilbert space representing a particle's spin. It can be written as a column of two complex numbers. In relativistic QM, a spinor has four components.

Square-integrable A function whose squared magnitude integrates over all space to a finite value: $\int_{-\infty}^{\infty} dx\,[f(x)]^2 = C$.

Stat-coulomb An esu, or electrostatic unit: a unit of charge in the CGS (aka Gaussian) electromagnetic unit system.

Static Not moving, compare to "stationary." A uniformly rotating sphere is stationary, but not static.

Stationary Properties constant in time. compare to "static." A uniformly rotating sphere is stationary, but not static.

Strictly forbidden Can never happen at any order, such as one which violates conservation of angular momentum. See "forbidden."

Superposition A linear combination of vectors, each with a complex coefficient: $a_1|b_1\rangle + a_2|b_2\rangle + \dots$.

Symmetry In general, an invariant property under a given transformation.

Theory The highest level of scientific achievement: a quantitative, predictive, testable model which unifies and relates a body of facts. A theory becomes accepted science only after being supported by overwhelming evidence. A theory is not a speculation, e.g., Maxwell's electromagnetic theory. See also "fact," and "speculation."

Thought experiment A way to test the logical consistency of a theory against itself, and other trusted theories, by following the theory to some end. For example, a thought experiment might compare the predictions of QM against those of Special Relativity. If there is a contradiction, then one of the theories must have an error. (No such contradiction is known.)

Trace The trace of a square matrix is the sum of its diagonal elements.

TwisterTM The game that ties you up in knots (a stocking feet game).

Unit The base measure of a measurement, e.g., the unit of distance is the meter.

Vectors Mathematically, abstract entities that meet the requirements of vector elements of a vector space. In QM, vectors are usually wave-functions or discrete vectors, aka "kets."

Vector space A mathematical set (often infinite) of a "field" of scalars and a "group" of vectors, with algebraic rules that allow solving linear equations. See text for complete definition.

Wave-function The spatial part of a state: a complex-valued function of space that defines everything there is to know about a particle, except spin, including the probabilities of measuring every value of every observable property of the particle. The complex value at each point is the *probability amplitude*.

Wave-number The spatial frequency of a sinusoid, typically in radians per meter.

Wave-vector A vector describing a propagating sinusoid, whose magnitude is the wave-number and direction is the direction of propagation.

welter Weg "which way" in German.

Wilson-Sommerfeld Quantization see "old quantum mechanics".

WKB Wentzel-Kramers-Brillouin: an approximation method for solving the time-independent Schrödinger equation.

WLOG Without loss of generality.

x-representation Sometimes used as a synonym for "position representation"

References

Books

1. [Bay] G. Baym, in *Lectures on Quantum Mechanics* (Westview Press, 1990). ISBN 0-805-30667-6
2. [Blo] D. I. Blokhintsev, in *Quantum Mechanics,* 1st edn. (Springer, 31 July 1964). English language, ISBN-10: 9027701040, ISBN-13: 978-9027701046
3. [Con & Sho] E. U. Condon, G. H. Shortley, in *The Theory of Atomic Spectra* (Cambridge, Cambridge University Press, 1957)
4. [Dir] P. A. M. Dirac, in *Quantum Mechanics*, 4th edn. (London, Oxford University Press, 1958)
5. [Eis & Res] R. Eisberg, R. Resnick, in *Quantum Physics of Atoms, Molecules, Solids, Nuclei, and Particles*, 2nd edn. (Wiley, 1985). ISBN 0-471-87373-X
6. [Gas] S. Gasiorowicz, in *Quantum Physics* (Wiley, 1974). ISBN: 047129280X (ISBN13: 9780471292807)
7. [Gla] R. J. Glauber, *The quantum theory of optical coherence.* Phy. Rev. 130, 2529–2539 (1963)
8. [Gos] A. Goswami, in *Quantum Mechanics*, 2nd edn. (Wm. C. Brown Publishers, 1997). ISBN 0-697-15797-0
9. [Gri] D. J. Griffiths, in *Introduction to Quantum Mechanics,* 2nd edn (Benjamin Cummings, 10 Apr 2004). ISBN-13 978-0131118928
10. [Gry, Asp & Fab] G. Grynberg, A. Aspect, C. Fabre, in *Introduction to Quantum Optics: From the Semi-classical Approach to Quantized Light* (Cambridge University Press, 18 Oct 2010). ISBN-13: 978-0521551120
11. [Lou] R. Loudon, in *The Quantum Theory of Light*, 3rd edn. (USA, Oxford University Press, 23 Nov 2000). ISBN-13: 978-0198501763
12. [Nis] National Institute of Science and Technology (2006), http://physics.nist.gov/cuu/Constants/
13. [Paa] H. P. Paar, in *An Introduction to Advanced Quantum Physics,* 1st edn. (Wiley, 17 May 2010). ISBN-13: 978-0470686751
14. [Pes & Sch] M. E. Peskin, D. V. Schroeder, in *An Introduction to Quantum Field Theory* (Westview Press, October 2, 1995). ISBN: 13: 978-0201503975
15. [Pow & Cra] J. I. Powell, B. Craseman, in *Quantum Mechanics* (Addison-Wesley Publishing Company, Inc, 1961)
16. [Sak] J. J. Sakurai, in *Advanced Quantum Mechanics* (Reading, Addison-Wesley Publishing Co., 1967). ISBN 0201067102
17. [Sak2] J. J. Sakurai, in *Modern Quantum Mechanics*, revised edition (Addison-Wesley Publishing Co., 1994). ISBN 0-201-53929-2

E. L. Michelsen, *Quirky Quantum Concepts,* Undergraduate Lecture Notes in Physics,   355
DOI 10.1007/978-1-4614-9305-1, © Springer Science+Business Media New York 2014

18. [Sch] L. I. Schiff, in *Quantum Mechanics,* 3 edn. (Mcgraw-Hill College, June 1968). ISBN 13: 9780070856431
19. [Sch & Wes] B. Schumacher, M. Westmoreland, in *Quantum Processes, Systems, and Information* (Cambridge University Press, April 26, 2010). ISBN: 13: 978-0521875349
20. [Sha] R. Shankar, in *Principles of Quantum Mechanics*, 2nd edn. (New York, Plenum Press, 1994). ISBN 0-306-44790-8
21. [Tow] J. S. Townsend, in *A Modern Approach to Quantum Mechanics* (McGraw-Hill, Inc., 1992). ISBN 0-07-065119-1
22. [Wal] S. P. Walborn et al., *Double-slit quantum eraser*. Phy. Rev. A65, 033818 (2002)
23. [Wei] S. Weinberg, in *Lectures on Quantum Mechanics* (Cambridge University Press, November 30, 2012). ISBN: 13: 978-1107028722
24. [Wyl] Wyld, H. W., in *Mathematical Methods for Physics* (Perseus Books Publishing, LLC, 1999). ISBN 0-7382-0125-1
25. [Zee] A. Zee, in *Quantum Field Theory in a Nutshell* (Princeton University Press, 2003). ISBN 9780691010199

Papers

26. [Jen] Jensen, William B., The Origin of the s, p, d, f Orbital Labels. J. Chem. Educ. **84**, 757--758 (2007)
27. [Man] L. Mandel, Quantum effects in one-photon and two-photon interference, Rev. Mod. Phys. **71**(2) (Centenary 1999)
28. [Pfl] R. L. Pfleeger, L. Mandel, Phys. Rev. **159**, p1084 (1967)

Index

E. L. Michelsen, *Quirky Quantum Concepts,* Undergraduate Lecture Notes in Physics,   357
DOI 10.1007/978-1-4614-9305-1, © Springer Science+Business Media New York 2014